Minerals Bioprocessing II

Minerals Bioprocessing II

Proceedings of the
Engineering Foundation Conference
"Minerals Processing II"
held in Snowbird, Utah
from July 10 - 15, 1995

Edited by

David S. Holmes

Ross W. Smith

a Publication of

A Publication of The Minerals, Metals & Materials Society
420 Commonwealth Drive
Warrendale, Pennsylvania 15086
(412) 776-9000

Printed in the United States of America
Library of Congress Catalog Number 95-78349
ISBN Number 0-87339-301-5

If you are interested in purchasing a copy of this book, or if you would like to receive the latest TMS publications catalog, please telephone 1-800-759-4867.

CONTENTS

III. Bioleaching

IV. Biosorption

V. Fossil Fuels and Byproducts: Bioconversion, Processing and Bioremediation

VI. Microorganisms in Mineral Processing Operations and Remediation

PREFACE

Advances in Minerals Bioprocessing

This book represents the published proceedings of the Engineering Foundation Conference "Minerals Bioprocessing II" held in Snowbird, Utah from July 10 -15, 1995. Forty three participants from fourteen different countries attended. The countries represented, in addition to the United States, were Australia, Belgium, Bulgaria, Canada, Chile, France, Germany, Italy, Netherlands, Norway, Poland, South Africa, and Spain. Eighteen of the participants, nearly 1/2, were from outside the USA emphasizing the strong international interest in the subject of the conference.

The papers presented at the conference roughly fell into six different categories: I. GENETIC STUDIES OF MICROORGANISMS OF INTEREST IN MINERALS BIOPROCESSING; II. DESIGN AND KINETICS OF BIOOXIDATION OPERATIONS; III. BIOLEACHING; IV. BIOSORPTION; V. FOSSIL FUELS AND BYPRODUCTS: BIOCONVERSION, PROCESSING AND BIOREMEDIATION; VI. MICROORGANISMS IN MINERAL PROCESSING OPERATIONS AND REMEDIATION.

Papers in category I included the introduction of quite sophisticated kinetic models of the biooxidation processes in terms of the sub-processes involved such as in "Recent Developments in Modelling Bio-Oxidation Kinetics, Part II Kinetic Modelling of the Bio-Oxidation of Sulphide Minerals in Terms of the Critical Sub-Processes Involved" by Boon et al.

Among the papers in category II is the paper "Genetic and Phenotypic Instability in *Thiobacillus ferrooxidans* and Why these Phenomena are Important to Understand" by D. S. Holmes on the importance of insertion sequences in *T. ferrooxidans*. The presence of such sequences implies that the microorganism genetically modifies rather easily and often spontaneously and, thus, may spontaneously adapt to a normally hostile environment. However, their presence also suggests that a strain improved in the laboratory may spontaneously revert in the field to a less desirable strain. Other papers in the category further elucidated the genetics of *Thiobacillus* sp. including the effect of phosphate starvation on gene expression by C. Jerez et al. of the University of Chile and a paper describing progress in sequencing and understanding the gene for rusticyanin by N. Guiliani et al. from the CNRS, Marseilles, France.

Included in the sessions on bioleaching (category III) was the paper "Pilot Scale Microbial Leaching of Gold and Silver from an Oxide Ore in Elshitza Mine, Bulgaria" by Groudev et al. in which results were reported on the use of thiosulfate plus heterotrophic bacteria as leaching agents. The apparent role of the bacteria was to produce certain amino acids which aided the dissolution process. In the paper "Previous Studies of Characterization and Bioleaching of a Novel Thermophilic Microorganism" by Gomez et al. a particularly promising organism for the dissolution of metallic sulfide minerals was studied. Unique about the microorganism is its tolerance toward metallic cations.

Work on sorption of Mn onto both prokaryotic and eukaryotic immobilized microorganisms was investigated in the paper "Immobilization of Microbial cells for Metal Adsorption and Desorption" (Ercole et al.). It was found that *Arthrobacter* sp. were better biosorbents than were green algae. Other papers in category IV investigated the sorption of metal ions onto to yeast cells and chitosan polymers.

In category V, a paper "The Flood/Drain Bioreactor for Coal and Mineral Processing" (Andrews and Noah) explores the use of a reactor that can be thought of as intermediate in nature and cost between heap bioleaching with low costs but slow leaching rates and conventional bioreactors with higher rates but, also, higher costs. Encouraging results were obtained in laboratory studies of pyrite degradation and removal using the reactor. Other papers in the category included papers on bioconversion and bioremediation of fossil fuel contaminated soil.

In category VI the use of sulfate reducing bacteria for the treatment of acid mine water was discussed in the paper "Evaluation of Sulphate Reducing Bacteria and Related Process Parameters for Developing a Passive Treatment Method" by Kuyucak and St-Germain. In particular, both the advantages of the process over conventional hydroxide precipitation and the limits of the process were noted. Other papers considered microorganisms and their direct use in mineral processing including the paper "Microorganisms in Mineral Processing" by Smith et al. which discussed, among several items, a scheme whereby coal can be specifically flocculated from coal pyrite using a hydrophobic bacterium as flocculant.

In addition to scheduled conference presentations, two simultaneous *ad hoc* workshops were convened entitled "What Can Be Done to Advance Minerals Bioprocessing from the Perspective of Engineers?" chaired by Geoff Hansford of the University of Cape Town, South Africa and Frank Crundwell, University of the Witwatersrand, South Africa and "What can be Done to Advance Minerals Bioprocessing from the Perspective of Biologists?", chaired by Carlos Jerez and Eugenia Jedlicki, both of the University of Chile, Santiago, Chile. Summaries of the respective workshops were presented to the full audience by the workshop chairs and a lively discussion between biologists and engineers ensued. Since the contents of the workshop do not appear as part of the written proceedings we append here a brief description of the issues addressed and the discussion that followed.

Can a model of biological solubilization of metals be devised that incorporates all the important parameters for engineering design and optimization? Such a model would include phenomena such as temperature, pulp density, particle size, particle mineralogy, and O_2 and CO_2 diffusion rates, etc. It would also have to include a rate constant(s) for the biological reaction. This rate constant, in turn, depends on a number of biological and biochemical parameters including, among others, the number and composition of the microorganisms involved and the relative contributions to the process of direct bioleaching versus indirect bioleaching.

No one knows the number or composition of the microorganisms involved in a leaching operation nor, obviously, their relative contributions to metal solubilization in heaps or in dumps. Almost certainly, the relevant microorganisms exist in biofilms over the rocks, as well as free in the solution, and one of the problems is that biofilms are very difficult to model in the laboratory. The biofilms form because as bacteria come close to and then adhere to solid surfaces they not only colonize, but also produce copious amounts of polysaccharides. The binding provided by the polysaccharides aids both bacterial adhesion to the surface and the structural integrity of the biofilm. Bill Costerton of the Biofilm Engineering Center of Montana

State University, Bozeman, Montana, summarized the current thinking on the structure of biofilms. Microorganisms are not necessarily spread as a more or less uniform film over the surface of rocks or mineral particles. Rather, they are thought to occur as "pods" which may or may not connect depending on the circumstances as is illustrated in figure 1.

One of the striking features of this type of organization is its complexity. An individual pod could have an exterior that has an oxygen content in equilibrium with the surrounding solution and an interior that is depleted in oxygen or lacking oxygen entirely. The organisms that compose the pods may vary from aerobes on the outside to extreme anaerobes on the inside according to the availability of oxygen. They could also vary depending on their nutritional requirements and on the type of symbiotic and mutualistic interactions established with other members of the community. Neighboring pods could also interact with each other establishing pH and Eh and/or O_2 gradients. A consequence of this is that electrons could flow between pods down an Eh gradient via the solution or through minerals in the ore with concomittant oxidation/reduction of metals. Pods could also harbor *Thiobacillus* species and other microorganisms that could contribute to the solubilization. In addition, the pods could affect the activity, either positively or negatively, of microorganisms that are situated away from the pod in solution or attached to the ore minerals in a non-biofilm manner. It is obviously important to understand the structure, composition, and the activity of such biofilms and to determine their relative contributions to the bioleaching of dumps and heaps (biofilms probably don't have the right conditions or sufficient time to develop sufficiently during tank leaching). The correct "ecological management" of such biofilms is also important to promote bioleaching.

The concept of the biofilm also further blurs the distinction between direct and indirect mechanisms of metal solubilization: a distinction that was already beginning to be fuzzy from a practical point of view. One can define the direct mechanism as the direct oxidation of a metal by an enzyme(s) situated on or in a microorganism, requiring direct contact between the microorganism and the mineral, and the indirect mechanism as the oxidation of the metal by a byproduct of a microorganism (usually considered to be Fe(III) but could theoretically be an enzyme in solution). The rate constants for indirect versus direct oxidation would be very different, but if a microorganism is attached directly to the substrate or indirectly via a biofilm, then the local concentration of the biologically produced Fe(III) could be very high altering the rate constant for Fe(III) induced oxidation. In fact, a complete spectrum of rate constants might be present in a leaching operation ranging from essentially zero order to pseudo second order, as is illustrated in figure 2. Not only will there be spatial variations of the rate constant, but there will also be temporal variations depending on the growth of the biofilm and on the changes that occur to the substrate as it is leached. This makes life very complicated for the engineer trying to model a dump or heap leaching operation.

Life is also more difficult for the biologist, who must now think in terms of ecological consortia, some of whose members may be very difficult to grow and/or understand because they are strictly anaerobes or because they behave differently as consortia than they do as individuals. Even under well defined laboratory conditions, where everything, including microorganisms and substrate, is nicely in solution and using well studied microorganisms, it is very difficult to model the activities and interactions of two or three microorganisms. So, it is not surprising that it is even more difficult to model a leaching operation containing not only many different microorganisms, but that also of a solid and variable substrate, and with striking variations of temperature, oxygen and carbon dioxide concentrations, pH and Eh.

What is also important is that many of these issues are also of concern to environmental biotechnologists who are using indigenous or laboratory strains of microorganisms to clean the environment and also to scientists studying biofilms that corrode materials or that can be used for biotechnological applications such as in the fossil fuel industry. Hopefully, advances made in these areas regarding the nature and uses of biofilms will be applicable to the mineral industry. Meanwhile, the engineers will continue to use largely empirically derived rate constants in their models and the biologists will struggle to understand the nature of the catalysts (microorganisms) with a view to improving their performance. With all the best engineering optimization in the world the reaction rate will never exceed a certain theoretical value dictated by intrinsic properties of the catalyst. Ultimately, it will be up to the biologist to alter the intrinsic properties of the catalyst to break through this ceiling.

The conference was supported, in part, from funds from the Engineering Foundation, the National Science Foundation and the Noranda Mining Company.

David S. Holmes, University of Chile, Santiago, Chile
Ross W. Smith, University of Nevada, Reno

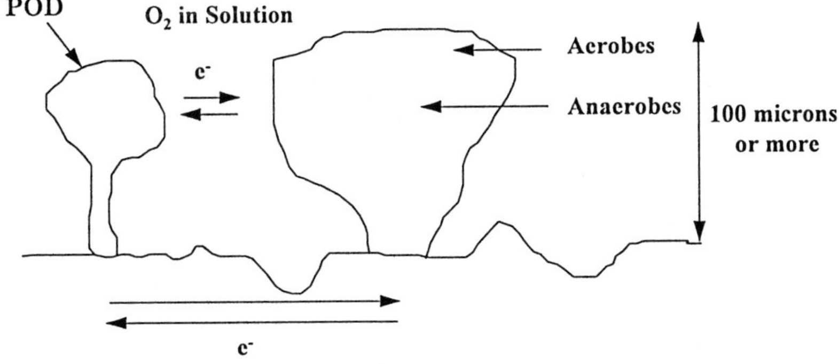

Figure 1 Biofilm in Pod Form on the Surface of a Sulfide Mineral

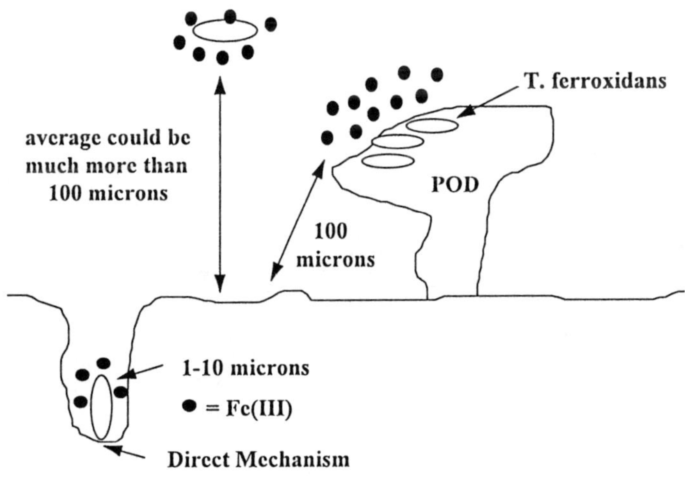

Figure 2 Possible Combination of Mechanisms in Biooxidation Leading to Combination Rate Constants

I.

DESIGN AND KINETICS OF BIOOXIDATION OPERATIONS

GOLD BIOHYDROMETALLURGY:
CURRENT DESIGN and OPERATION of BIO-OXIDATION PLANTS, NEW RESEARCH TOOLS and CHALLENGES

G.S.Hansford[1]
Department of Chemical Engineering
University of Cape Town
Rondebosch 7700
South Africa

SUMMARY

Bio-oxidation processes and plants for the pretreatment of refractory gold ores and concentrates are described. The empirical logistic equation and its application in describing steady state batch and continuous kinetics is presented. The need for mechanistically based steady state and dynamic kinetic models is discussed. Theoretical and experimental tools for the study of mechanisms and kinetics of the important sub-processes are introduced.

INTRODUCTION

The bacterial oxidation of sulphide concentrates has become established as a pretreatment method for refractory gold ores particularly those containing arsenopyrite. These refractory gold ores occur in many parts of the world; Australia, Brazil, Ghana, South Africa, the United States and counties of Central Asia and the Pacific Rim. There are several full-scale plants in operation, and more under construction and commissioning.

Bio-oxidation is one of several methods used for the pretreatment of refractory arsenical gold ores. The most common alternative methods of pretreatment are roasting, and pressure leaching (Thomas, 1994; Perkins *et al.*, 1994), with these processes still being implemented. Each process has its proponents and advantages and disadvantages. One of these is the question of scale, where it has been shown that bio-oxidation is profitable for medium to small-scale plants (Nicholson *et al.*, 1993).

---

[1]. Until January 1995 on sabbatical leave at:
Department of Biochemical Engineering, Kluyver Laboratory of Biotechnology, Technische Universiteit Delft, Julianalaan 67, 2628BC Delft, The Netherlands

Mineral Bioprocessing II
Edited by David S. Holmes and Ross W. Smith
The Minerals, Metals & Materials Society, 1995

These plants have been designed based on information from laboratory and pilot-plant tests. The reactors have been sized using the logistic equation (Pinches *et al.*,1988) to describe the kinetics and the correlation of Mills *et al.*(1987) to predict the gas-liquid mass transfer coefficients.
This paper describes the current plants, demonstrating that bio-oxidation is an established technology for the pretreatment of refractory of gold ores and concentrates. It then goes on to show the need for a mechanistically based model and describes the theoretical and experimental tools being used to develop such models. The detailed background and application of these tools are described in subsequent papers (Boon *et al.*,1994a and 1994b)

BIO-OXIDATION PLANTS

The major mining company to develop bio-oxidation has been GENCOR in South Africa. This led to the construction of the first Biox plant at Fairview Gold Mine treating 20 tons per day of arsenopyrite-pyrite concentrate amounting to half of the mines production. In time the Fairview Biox plant has been enlarged to 40 tons per day treating the entire production of the mine and replacing the Edwards roasters used previously. The second production-scale bio-oxidation facility was at Sao Bento in Brazil also using the Biox process to pretreat a refractory arsenopyrite-pyrite-pyrrhotite concentrate prior to pressure oxidation. Subsequently bio-oxidation plants have been installed at Harbour Lights and Wiluna in Australia (van Aswegen, 1993; Brown, 1994)and the very large plant for Ashanti in Ghana started operation earlier this year (Smith *et al.*, 1994) with a plant for Youanmi in Australia having reached the design stage (Budden and Bunyard, 1994; Brierley and Brans, 1994).

In addition to these full-scale operations, Mintek in South Africa has been active in developing processes for the bio-oxidation of refractory gold ores and have operated a pilot plant on site at their research facilities outside Johannesburg and in cooperation with the AngloAmerican Corporation operated a large demonstration plant at the Vaal Reefs Gold Mine in the Western Transvaal (Dempsey *et al.*,1990; Neale *et al.*,1991; Pinches *et al.*, 1991). These two plants were highly instrumented and enabled the process to be monitored using off-gas analysis to measure oxygen and carbon dioxide utilization rates and to study power requirements and solids suspension. Similar pilot plant plant studies have been carried out by CRA in Australia (Hoffman *et al.*, 1992). With the exception of the Sao Bento plant all the full-scale plants are multi stage.

Typical operating conditions are a concentration of solids of about 200 $kg.m^{-3}$ and a residence time of 2 days per stage. The pH is controlled at 1.8 using lime or limestone which also can provide carbon dioxide. The reactors operate at temperatures of about 40°C. The baffled tanks are agitated and aerated to maintain a minimum dissolved concentration of about 2 $mg.l^{-1}$. In the earlier plants, six-bladed Rushton turbines were used for agitation, but most recently, there is considerable interest in Lightning A315 impellers (van Aswegen, 1993).

Bio-oxidation is also to used for the treatment of whole ores in heaps. Extensive trials have been run on pilot-scale heaps of up to 30 000 tons (Brierley and Luinstra, 1993) and a heap of one million tons treating 10 000 tons of ores per day is planned to be operational by the end of this year (Brierley, 1994).

MECHANISM and KINETICS

In industrial situations, the bio-oxidation of sulphide minerals is thought to be carried out by a mixed population of thiobacilli, of which *Thiobacillus ferrooxidans, Thiobacillus thiooxidans,* and *Leptospirillum ferrooxidans* are believed to the be principal species. The role of each species is not certain but they are thought to oxidize arsenopyrite and pyrite mainly by the direct mechanism whereby close association of the bacteria with the mineral surface is necessary for bio-oxidation to proceed. There appears to be virtually no oxidation of pyrite by ferric iron using the indirect mechanism (Boon and Heijnen, 1993). In the direct mechanism, recent evidence suggests that there is not irreversible attachment of the bacteria to the sulphide mineral surface but rather that they are closely associated with the surface and that all the bacteria in the system have an equal chance of participating in the bio-oxidation reactions. Several presentations at this conference will address the question of direct and indirect mechanisms for the attack of thiobacilli on sulphide minerals, but this is still an open question. Although rate expressions based on the direct mechanism have been proposed by several authors (Hoffman *et al.*, 1981; Panin *et al.*,1985; Pinches *et al.*,1988; Gormely *et al.*, 1975; Chang and Myerson, 1983; Sanmugasunderam *et al.*,1985) none of these models has found widespread use in practice, and research on the development of such a mechanistic model of practical value is currently underway (Boon *et al.,* 1994a,1994b).

The rate expression which has found the most use in modelling both batch and continuous bio-oxidation kinetics in practice is the logistic equation (Pinches *et al.*,1988). This has been shown to fit both batch and continuous data for several pyrite and arsenopyrite-pyrite concentrates (Miller and Hansford, 1992a,1992b; Hansford and Chapman, 1992 and Hansford and Bailey, 1992, Katsikaros *et al.*,1990, Dew 1993). The logistic equation is not mechanistically based, but has proved very useful in modelling data from laboratory-, pilot- and full-scale plants. The basic rate equation is written in terms of sulphide conversion as:

$$\frac{dX}{dt} = kX\left[1 - \frac{X}{X_m}\right] \tag{1}$$

where conversion, X is given by:

$$X = \frac{c_0 - c}{c_0} \tag{2}$$

so that in terms of sulphide concentration, the logistic rate expression becomes;

$$\frac{dc}{dt} = -k\,(c_0 - c)\left[\frac{c - c_m}{c_0 - c_m}\right] \tag{3}$$

For a continuous bio-oxidation reactor the conversion is given by;

$$X = X_m\left[1 - \frac{1}{k\tau}\right] \tag{4}$$

Hansford and Miller (1992) have shown that the logistic equation can be used to model the bio-oxidation of both pyrite and arsenopyrite in a series of bioreactors, with higher values of the rate constant for arsenopyrite which is attacked more rapidly than pyrite. The logistic equation has been used for the prediction of performance of industrial scale bio-oxidation reactors (Dew, 1993).

Another important aspect of bio-oxidation as a method for pretreatment of refractory gold concentrates is the finding that attack of the sulphide mineral is selective in the gold-rich regions resulting in high levels of gold liberation for only moderate extents of bio-oxidation. The degree of bio-oxidation necessary to give satisfactory gold liberation is however dependent on the type of concentrate and quite variable. It is one of the important pieces of information necessary for a feasiblity study.

This selective attack is due to two factors; the occurence of the gold in the sulphide and the manner of attack of the thiobacilli on the mineral. Swash (1980) and Claassen *et al.*(1991) have shown that gold is not usually uniformly distributed through the sulphide, but accumulated at the grain boundaries or within the arsenic-rich regions. Hansford and Drossou (1988) have shown that bio-oxidation does not proceed by a shrinking particle mechanism but by the formation of hexagonal holes which develop in the sulphide particles. They have postulated that these holes develop in the gold-rich regions due to the fact that the sub-microscopic gold particles strain the crystal lattice. It is of interest to note that the bacteria appear to attach to the gangue material rather than the sulphide mineral being oxidized.

GAS-LIQUID MASS TRANSFER

It has been established that the supply of oxygen to the bio-oxidation process is a major cost factor (van Weert, 1989). Optimal reactor selection and design to maximize oxygen transfer efficiency is not well understood. The oxygen and carbon dioxide requirements can be calculated from the rate of sulphide mineral oxidation as shown later.

In order to predict the rate of gas-liquid mass transfer into mineral slurries, use has been made of two correlations (Mills *et al.*,1987; Oguz *et al.*,1987). This last contains a term involving the viscosity of the slurry which from the data of Rao (1966) is seen to rise steeply at a volume fraction corresponding to about 200 $kg.m^{-3}$ for -75μm pyrite. This causes a sharp decrease in the mass transfer coefficient. Hansford and Bailey (1994) have shown that this means that at solids concentrations below about 200 $kg.m^{-3}$ the rate of pyrite oxidation is reaction rate limited. while above 200 $kg.m^{-3}$ the rate is mass transfer limited.

The literature does not contain a detailed investigation of gas-liquid mass transfer into mineral slurries in bioreactors and this subject needs furhter investigation, in particular the effect of reactor length to diameter ratio and the comparison of impeller types such as six-bladed Rushton turbines and Lightning A315 impellers, and the use of pachuca-type air-agitated reactors for bio-oxidation.

Only a few reports exist of different systems for aerating bio-oxidation reactors (Hardwick *et al.*, 1988, van Weert and Snoek, 1993 and Hoffman *et al., 1993*) The latter CRA bioreactor is claimed to have a significant improvement in oxygenation efficiency.

CHALLENGES

As the use of bio-oxidation processes expands, there is a need for mechanistically based models to describe the reaction mechanisms and kinetics and the dynamic behaviour of the system.

It is important to be able to describe how solution conditions, in particular the ferrous to ferric iron ratio and pH affect the kinetics. At present there is no clear understanding of the mechanisms by which thiobacilli oxidize sulphide minerals, whether a direct or indirect mechanism applies, what intermediates are formed and how process conditions affect the kinetics. Neither are the rate limiting sub-processes understood.

There is also the need for a dynamic model in order to describe how a bio-oxidation reactor responds during start-up or following perturbations of the process conditions.

Other areas where a greater understanding is required is the role of the different species of thiobacilli, galvanic interactions between mixed minerals and the nature of the association between the thiobacilli and the sulphide mineral surfaces.

Some of the theoretical tools available for studying the mechanisms will be introduced here. Their detailed description and use are given elsewhere (Boon *et al.*, 1994a and 1994b) together with the description of the use of experimental tools such as off-gas analysis and off-line oxygen utilization rate measurements.

THEORETICAL TOOLS

Theoretical tools currently being used to describe heterotrophic and autotrophic processes of industrial importance can also be applied to the bio-oxidation of sulphide minerals by thiobacilli. These tools include the description of the stoichiometry and thermodynamics of microbial growth and metabolism, the degree of reduction balances, bioenergetics and biokinetics and the use of Gibbs free energy dissipation in predicting theoretical yields and maintenance requirements.

In designing a bio-oxidation process it is useful to be able to predict the oxygen, carbon and nitrogen and other nutrient requirements, and to relate the rate of pyrite oxidation to the rates of oxygen and carbon dioxide utilization. This can be done using the stoichiometric equation for bacterial growth on pyrite. Starting with the stoichiometric equation for the purely chemical oxidation of pyrite:

$$15O_2 + 4FeS_2 + 2H_2O = 4Fe^{3+} + 4H^+ + 8SO_4^{2-}$$

which provides energy for both growth and maintenance of the bacteria. Assuming a stoichiometric formula for biomass as $CH_{1.8}O_{0.5}N_{0.2}$ which is close to the formula of $CH_{1.808}O_{0.55}N_{0.1757}$ reported by Jones and Kelly (1983) and that the sources of carbon and nitrogen are carbon dioxide and ammonium ions respectively, the rate of pyrite oxidation can be related to the rates of oxygen and carbon dioxide utilization by means of the degree of reduction balance (Roels,1987; Heijnen, 1991):

$$-15r_{FeS2} = -4rO_2 - 4.2r_{CO2}$$

This relationship is of value in checking data for mass balance consistency, as well as for predicting oxygen and carbon requirements for a given pyrite oxidation rate, or for determining the pyrite oxidation rate from the oxygen and carbon dioxide utilization rates measured by off-gas analysis.

Including the production of biomass into the stoichiometric equation:

$$CO_2 + 0.2NH_4^+ + aO_2 + (1/Y_{XS})FeS_2 + bH_2O$$
$$=$$
$$CH_{1.8}O_{0.5}N_{0.2} + (1/Y_{XS})Fe^{3+} + cH^+ + (2/Y_{XS})SO_4^{2-}$$

written in terms of a yield, $Y_{XS,}$ of biomass on pyrite and solving the elemental and charge balances, gives:

$$CO_2 + 0.2NH_4^+ + \{(15 - 4.2Y_{XS})/4Y_{XS}\}O_2 + 1/Y_{XS}FeS_2 - (0.6 + 0.5/Y_{XS})H_2O$$
$$=$$
$$CH_{1.8}O_{0.5}N_{0.2} + (1/Y_{XS})Fe^{3+} + (0.2 + 1/Y_{XS})H^+ + (2/Y_{XS})SO_4^{2+}$$

Note that biomass is reported as moles of carbon. This means that only the yield, $Y_{XS,}$ as moles of carbon per mole of pyrite needs to be known in order to define the stoichiometry of the reaction. The yield, $Y_{XS,}$ can be determined from the rate of carbon dioxide uptake using off-gas analysis and the rate of pyrite oxidation, then the mass balance for the process is known. This gives the oxygen, carbon and nitrogen requirements for the process. Using a value of $Y_{XS} = 0.25$ (Boon, 1994)the stoichiometric equation for bacterial growth on pyrite is:

$$0.25CO_2 + 0.05NH_4^+ + 3.225O_2 + FeS_2 + 0.65H_2O$$
$$=$$
$$0.25CH_{1.80}O_{0.5}N_{0.2} + Fe^{3+} + 1.05H^+ + 2SO_4^=$$

The following mass balance can be calculated:

	gmole	kg
FeS_2	1	1000
CO_2	0.25	92
NH_4^+	0.05	7.5
O_2	3.5	940
H_2SO_4	2	1650
biomass	0.25	50

Similarly if heats of formation are known then the heat load of the process can be calculated as -1438 $kJ(gmoleFeS_2)^{-1}$ or 12 $MJ(tonFeS_2)^{-1}$.

From the Gibbs free energies of the reactants and products, it is possible to calculate the theoretical yield and the maintenance requirements of bacteria. This has been developed by Heijnen and coworkers(Heijnen *et al.*,1992; Tijhuis *et al.*,1993) and shown to be successful for aerobic and anaerobic heterotrophs and some autotrophs. Current research is testing the applicability of this technique to iron , sulphide and sulphur oxidation by thiobacilli.

EXPERIMENTAL TOOLS

An experimental tool which provides valuable information on the rate of growth and metabolic activity of the thiobacilli in bio-oxidation is the use of off-gas analysis to measure the rates of oxygen and carbon dioxide utilization. Its use has been reported by

Pinches and coworkers (Dempsey *et al.*, 1990; Neale *et al.*,1991; Pinches *et al.*,1991) and Nagpal *et al.*,(1993). By combining off-gas analysis with carbon as a measure of biomass, Boon *et al.* (1994a and 1994b) have been able to determine biomass concentration and specific oxygen and sulphide utilization rates and growth rate.

Off-line respirometry for monitoring bio-oxidation has been reported (Lizama and Suzuki,1989). However, Boon *et al.* (1994a and 1994b) have refined its use to determine not only specific oxygen consumption rates but also to determine the maximum specific rates as a measure of bacterial activity under different conditions.

CONCLUSIONS

Bio-oxidation for the pretreatment of refractory gold ores and concentrates is an established technology with several large scale plants in operation.

Current modelling of the kinetics of bio-oxidation is inadequate. there is a need for mechanistically based kinetic models to describe both steady state and dynamic behaviour. Theoretical and experimental tools are available for research in elucidating the mechanisms and kinetics of the rate limiting sub-processes.

ACKNOWLEDGEMENTS

The author wishes to thank Dr D.Dew, GENCOR; Dr C.Brierley, Vistatech,; Dr J.Brierley, Newmont and Prof J.Heijnen of TUDelft for their assistance in providing material for this paper.
Financial assistance from the Foundation for Research Development, South Africa and the Engineering Foundation, United States of America is gratefully acknowledged.

REFERENCES

Bailey,A.D. and G.S.Hansford (1993), "Factors affecting bio-oxidation of sulphide minerals at high concentrations of solids: A review", Biotechnol. Bioeng., **42**, 1164-1174.

Bailey,A.D. and G.S.Hansford (1994), "Oxygen mass transfer limitation of batch bio-oxidation at high solids concentration",Minerals Engineering, **7**, 293-304.

Boon,M. (1994), personal communication, Kluyver Laboratory for Biotechnology, Technische Universiteit Delft.

Boon,M. and J.J.Heijnen (1993),"Mechanisms and rate limiting steps in bioleaching of sphalerite, chalcopyrite and pyrite with *Thiobacillus ferrooxidans* ". Biohydrometallurgical Technologies Vol I, A.E.Torma, J.E.Wey and V.L.Lakshmanan (Eds.) The Minerals, Metals and Materials Society, Warrendale, PA, 469-478.

Boon,M., J.J.Heijnen and G.S.Hansford (1994a),"Recent developments in the modelling of bio-oxidation kinetics and their implications in practice: Part I, Measurement methods in bio-oxidation kinetics", paper presented at the Engineering Foundation Conference, Mineral Bioprocessing, Snowbird Utah (July 1994).

Boon,M., G.S.Hansford and J.J.Heijnen (1994b),"Recent developments in the modelling of bio-oxidation kinetics and their implications in practice: Part II, A mechanistic approach to the modelling of the bio-oxidation of sulphide minerals in terms of the critical sub-processes involved", paper presented at the Engineering Foundation Conference, Mineral Bioprocessing, Snowbird Utah (July 1994).

Brierley,C.L. and R.Brans (1994),"Selection of BACTECH's thermophilic bio-oxidation process for Youanmi Mine", paper presented at BIOMINE'94, Applications of Biotechnology to the Minerals Industry, Perth, Australia (September 1994).

Brierley,J. and L.Luinstra (1993),"Bio-oxidation-heap concept for pretreatment of refractory gold ore", Biohydrometallurgical Technologies Vol I, A.E.Torma, J.E.Wey and V.L.Lakshmanan (Eds.) The Minerals, Metals and Materials Society, Warrendale, PA, 437-448.

Brierley,J.A. (1994),"Bio-oxidation heap technology for pretreatment of refractory sulphidic gold ore", paper presented at BIOMINE'94, Applications of Biotechnology to the Minerals Industry, Perth, Australia (September 1994).

Brown,R.G. (1994),"Bioleaching, Wiluna operating experience", paper presented at BIOMINE'94, Applications of Biotechnology to the Minerals Industry, Perth, Australia (September 1994).

Budden,J.R. and M.J.Bunyard (1994),"Pilot plant testwork for the BACTECH bacterial oxidation plant at Youanmi Gold Mine", paper presented at BIOMINE'94, Applications of Biotechnology to the Minerals Industry, Perth, Australia (September 1994).

Chang,Y.C. and A.S.Myerson (1983),"Growth models for the continuous bacterial leaching of iron pyrite by *Thiobacillus ferrooxidans*", Biotechnol.Bioeng., **25**, 2981-2990.

Claassen,R., C.T.Logan and C.P.Snyman (1993),"Bio-oxidation of refractory gold-bearing arsenopyritic ores", Biohydrometallurgical Technologies Vol I, 469-478, A.E.Torma, J.E.Wey and V.L.Lakshmanan (Eds.) The Minerals, Metals and Materials Society, Warrendale, PA, 479-498.

Dempsey,P., P.Human, A.Pinches and J.W.Neale (1990),"Bacterial oxidation at Vaal Reefs",paper presented at International Deep Mining Conference: Innovations in Metallurgical Plant, South African Institute of Mining and Metallurgy, Johannesburg.

Dew,D. (1993), private communication, GENCOR Process Research, Johannesburg, South Africa.

Gormely,L.S., D.W.Duncan, R.M.R.Branion and K.L.Pinder (1975),"Continuous culture of *Thiobacillus ferrooxidans* on a zinc sulphide concentrate", Biotechnol. Bioeng., **17**, 31-49.

Hansford,G.S. and M.Drossou (1988), "A propagating-pore model for the batch bioleach kinetics of refractory gold-bearing pyrite". Biohydrometallurgy, Proceedings of the International Symposium, Warwick (July 1987) Norris,P.R. and D.P.Kelly (Eds), STL, Kew, Surrey, UK, 345-358.

Hansford,G.S. and J.T.Chapman (1992), "Batch and continuous bio-oxidation kinetics of a refractory gold-bearing pyrite concentrate". Minerals Engineering, **5**, 597-612.

Hansford,G.S. and A.D.Bailey (1992), "The logistic equation for modelling bio-oxidation kinetics", Minerals Engineering, **5**, 1355-1364.

Hansford,G.S. and A.D.Bailey (1993), "Oxygen transfer limitation of bio-oxidation at high solids concentration", Biohydrometallurgical Technologies Vol I, 469-478, A.E.Torma, J.E.Wey and V.L.Lakshmanan (Eds.) The Minerals, Metals and Materials Society, Warrendale, PA, 469-478.

Hardwick,W.E., M.T.Errington and P.C.Miller (1988),"A brief techno-economic assessment of the provision of oxygen to bio-oxidation reaction systems for refractory sulphide ores", paper presented at Randol Gold Forum, Scottsdale AZ, (January 1988).

Heijnen,J.J. (1991),"Bioenergetics and Biokinetics of Microbial Growth and Product formation", Lecture Notes, Department of Biochemical Engineering, Technische Universiteit delft, The Netherlands.

Heijnen,J.J., M.C.M.van Loosdrecht and L.Tijhuis (1992),"A black box mathematical model to calculate auto- and heterotrophic biomass yields based on Gibbs energy dissipation", Biotechnol. Bioeng., **40**, 1139-1154.

Hoffman,M.R., B.C.Faust, F.A.Panda, H.H.Koo and H.Tsuchiya (1981), "Kinetics of the removal of iron pyrite from coal by microbial catalysis", Appl.Environ.Microbiol., **42**, 259-271.

Hoffman,W., R.Batterham and D.Conochie (1993), "A novel low energy Bioreactor for refractory gold ores", paper presented at Randol Gold Forum, Beaver Creek CO, (August 1993).

Jones,C.A. and D.P.Kelly (1983),"Growth of *Thiobacillus ferrooxidans* on ferrous iron in chemostat culture: Influence of product and substrate inhibition", J.Chem.Technol.Biotechnol., **33B**, 241-261.

Katsikaros,N., G.Davis, K.Chouzadjian, B.Kelley and M.Mavatoi (1990),"Bioleaching process for Bougainville Copper Limited", paper presented at Gold'90, A.I.M.E..

Lizama,H.M. and I.Suzuki (1989),"Rate equations and kinetic parameters of the reactions involved in pyrite oxidation by *Thiobacillus ferrooxidans*", Appl.Environ.Microbiol. **55**, 2918-2923.

Miller,D.M. and G.S.Hansford (1992a), "Batch bio-oxidation of a gold-bearing arsenopyrite-pyrite concentrate". Minerals Engineering, **5**, 613-630.

Miller,D.M. and G.S.Hansford (1992b), "The use of the logistic equation for modelling the performance of a bio-oxidation pilot plant treating a gold-bearing arsenopyrite-pyrite concentrate". Minerals Engineering, **5**, 737-750.

Mills,D.B., R.Bar and D.J.Kirwan (1987), "Effect of solids on oxygen transfer in agitated three-phase systems", A.I.Ch.E.Jour., **33**, 1542-1549.

Nagpal,S., D.A.Dahlstrom,M.L.Free and T.Oolman (1993),"Effect of sodium xanthate on the bioleaching of a pyrite-arsenopyrite ore concentrate", Biohydrometallurgical Technologies Vol I, A.E.Torma, J.E.Wey and V.L.Lakshmanan (Eds.) The Minerals, Metals and Materials Society, Warrendale, PA, 449-458.

Neale,J.W., A.Pinches, H.H.Muller, N.H.Hannweg and P.Dempsey (1991),"Long term bacterial oxidation pilot plant operation at MINTEK and Vaal Reefs", paper presented at Bacterial Oxidation Colloquium, South African Institute of Mining and Metallurgy, Johannesburg (June 1991).

Nicholson,H., S.Oti-Atakorah, D.J.Lunt and I.C.Ritchie (1993),"Selection of a refractory gold treatment process for the Sansu project", Applications of Biotechnology to the Minerals Industry, Proceedings of BIOMINE '93, , Adelaide , Australia (March 1993), Australian Mineral Foundation, Glenside South Australia.

Oguz,H., A.Brehm and W-D.Deckwer (1987), "Gas-liquid mass transfer in sparged agitated slurries", Chem.Eng.Sci., **42**, 1815-1822.

Panin,V.V., G.I.Karavaiko and S.I.Pol'kin (1985),"Mechanism and kinetics of bacterial oxidation of sulphide minerals", Biogeotechnology of Metals, Karavaiko,G.I. and S.N.Groudev (Eds.), Centre for International Projects, GKNT, Moscow, 197-215.

Perkins,J., I.C.Ritchie, H.Marais and T.P.Weston (1994),"Concentrate bio-oxidation (and roasting) in a hypersaline enviroment for Kanowna Belle", paper presented at BIOMINE'94, Applications of Biotechnology to the Minerals Industry, Perth, Australia (September 1994).

Pinches.A., J.T.Chapman, W.A.M.te Riele and M.van Staden (1988),"The performance of bacterial leach reactors for the pre-oxidation of refractory gold-bearing concentrates", Biohydrometallurgy, Proceedings of the International Symposium, Warwick (July 1987), Norris,P.R. and D.P.Kelly (Eds), STL, Kew, Surrey, UK, 329-344.

Pinches,A., R.Huberts, M.van Staden and R.M.Muhlbauer (1991),"Process options and parameters in the development and optimization of bacterial oxidation processes for the pre-oxidation of refractory sulphide gold ores", paper presented at Bacterial Oxidation Colloquium, South African Institute of Mining and metallurgy, Johannesburg (June 1991).

Rao,T.C. (1966),"Mineral Crushing and Grinding Circuits, Lynch,A.J.,(Ed.). Elsevier, Amsterdam, 88.

Sanmugasunderam,V., R.M.R.Branion and D.W.Duncan (1985),"A growth model for the continuous microbiological leaching of a zinc sulphide concentrate by *Thiobacillus ferrooxidans*", Biotechnol. Bioeng., **27**, 1173-1184.

Smith,G.R., R.J.Stewart and F.W.Kock (1994),"Designing and Commissioning Ashanti's Sansu plant", paper presented at BIOMINE'94, Applications of Biotechnology to the Minerals Industry, Perth, Australia (September 1994).

Swash,P.M. (1980),"A mineralogical investigation of refractory gold ores and their beneficiation, with special reference to arsenical ores", J.South African Inst.Mining Metall. **88**,173-180.

Tijhuis,L., M.C.M.van Loosdrecht and J.J.Heijnen (1993),"A thermodynamically based correlation for maintenance Gibbs energy requirements in aerobic and anaerobic chemotrophic growth", Biotechnol. Bioeng., **42**, 509-519.

Thomas,K.G. (1994),"Research, Engineering Design and Operation of a Pressure Hydrometallurgy Facility for Gold Extraction", PhD Thesis, Technische Universiteit Delft.

van Aswegen,P.C. (1993),"Bio-oxidation of refractory gold ores, the GENMIN experience", Applications of Biotechnology to the Minerals Industry, Proceedings of BIOMINE'93, Adelaide, Australia (March 1993), Australian Mineral Foundation, Glenside South Australia.

van Weert,G. (1989),"The cost of oxygen in hydrometallurgical processing of refractory gold sulphides" Proc.International Gold Expo., Reno NV, Eng.Mining Jour., 206-215.

van Weert,G. and J.A.Snoek (1993),"Oxygen transfer from air in the Delft inclined plate (DIP) bioreactor", Biohydrometallurgical Technologies Vol I, A.E.Torma, J.E.Wey and V.L.Lakshmanan (Eds.) The Minerals, Metals and Materials Society, Warrendale, PA, 237-248.

LIST of SYMBOLS

c	concentration of pyrite, $kg.m^{-3}$
c_o	initial concentration of pyrite, $kg.m^{-3}$
c_m	maximum concentration of pyrite, $kg.m^{-3}$
k	logistic rate constant, h^{-1}
r_{FeS2}	pyrite oxidation rate, $mole.l^{-1}.s^{-1}$
r_{O2}	oxygen consumption rate, $mole.l^{-1}.s^{-1}$
r_{CO2}	carbon dioxide consumption rate, $mole.l^{-1}.s^{-1}$
t	time, day
τ	residence time, day
X	pyrite conversion
X_m	maximum pyrite conversion
Y_{XS}	yield of biomass on pyrite, moleC.(mole FeS_2)$^{-1}$

THE PREDICTION OF THE BIOLEACHING OF REFRACTORY GOLD ORES IN A CONTINUOUS PLANT FROM THE BATCH DATA

Frank K. Crundwell
Department of Chemical Engineering
University of the Witwatersrand
Private Bag 3
WITS 2050
South Africa

email: fkc@metchem.chmt.wits.ac.za

ABSTRACT

Bacterial oxidation of sulphidic minerals is a well-known phenomenon and has been commercially exploited in dump-leaching processes and in the pretreatment of refractory gold ores. A mathematical model of the bacterial leaching operation has been developed to assist in the design and piloting of new operations, and the improvement of the efficiency of an existing plant.

It has been assumed in this work that bacterial leaching of the sulphidic minerals present in refractory gold ores occurs by direct leaching, and that the growth of bacteria occurs mainly on the surface of the sulphide mineral. By accounting for this growth process, and the shrinkage of the mineral particle due to reaction, a model for the batch leaching of the refractory gold ores has been derived. This model describes the batch leaching data.

The leaching of minerals in a continuous operation is more complex, since the particles spend different times in the reactor depending on the particle residence-time distribution. The population balance is used to describe the leaching of particles in a continuous reactor. The effects of particle-size distribution, leaching kinetics, and residence-time distribution are accounted for in the population balance. A bacterial cell balance accounts for the growth of bacteria on the mineral surface.

For two sets of experiments, the values of parameters which are obtained from the batch data are used in the model of the continuous operation. The continuous model is therefore an *a priori* prediction of continuous leaching operation. This model prediction compares favourably with data from continuous experiments in the laboratory and with data from a continuous pilot plant. This work indicates that this model is an accurate model of continuous bioleaching of pyrite.

Mineral Bioprocessing II
Edited by David S. Holmes and Ross W. Smith
The Minerals, Metals & Materials Society, 1995

1. Introduction

The gold in sulphidic ores is often not amenable to the cyanidation process. This gold is finely disseminated within the pyritic and arsenopyritic minerals of these ores. In order to liberate the gold in these ores, the sulphidic minerals need to be oxidized.

Various process options are available for the treatment of refractory gold ores prior to their treatment by the conventional cyanidation process. These options include bacterial leaching, roasting and pressure leaching. Bacterial leaching offers the advantage of significantly reduced capital costs while producing environmentally acceptable effluents.

Although bacterial leaching of sulphidic minerals is a well-known phenomenon, it is only in the last ten years that full-scale bacterial leaching plants have been commissioned. Dew *et al.* (1) report that presently there are full-scale plants at the Fairview Gold Mine (South Africa), the Sao Bento Mine (Brazil), the Harbour Lights Mine (Australia) and the Wiluna Mine (Australia). A plant at the Ashanti Mine (Ghana) is scheduled to be commissioned during the second half of 1994. There are large-scale pilot plants at Mintek (South Africa) and Vaal Reefs Gold Mine (South Africa) (2).

In order for bacterial leaching to compete successfully with other treatment processes for refractory gold ores, it needs to be efficient. To meet this requirement, the design and operation of the plant must be based on a thorough understanding of the important phenomena occurring in bacterial leaching. In spite of the large amount of research effort that has been directed towards the study of bacterial leaching, kinetic and process models for the bacterial leaching of sulphidic ores are underdeveloped.

The purpose of this paper is to propose a model for the kinetics of bacterial growth and of the consumption of mineral particles. These kinetic expressions are used to derive a model for batch and continuous bacterial leaching of pyritic minerals. The model accounts for the phenomena of bacterial growth, particle shrinkage by reaction and, in the case of continuous leaching, the flow of particles of different sizes into and out of the reactor.

2. Review of previous models of batch and continuous leaching

Previous models of bacterial leaching have been concerned mainly with the batch leaching of sulphidic minerals. Little attempt has been made to describe the continuous leaching of sulphidic minerals, or to apply the model parameters obtained from the batch data to predict the performance of continuous reactors.

In batch leaching the particle-size distribution changes because of the leaching reaction. The nature of continuous leaching is more complex than that of batch leaching. In continuous leaching, the particle-size distribution changes not only as a result of the kinetics of the leaching reaction, but also as a result of the flow of fresh particles into and partially-reacted particles out of the reactor.

Gormely *et al.* (3) assumed that bacterial attachment to the surface of the mineral is a prerequisite for bacterial growth. Their model described the growth of the bacteria, and no attempt was made to describe the leaching conversion. They observed that both the bacterial growth rate and the leaching rate were linearly dependent on the available surface area.

Chang and Myerson (4) and Myerson and Kline (5) proposed a model for the continuous leaching of pyrite. Their model is an extension of the model of Gormely *et al.* (3) but they allow for the growth of bacteria both on the surface and in solution. Their model focusses on the growth of bacteria and ignors the shrinkage of particles, in particular the aspect that fresh particles flow into the reactor and partially-reacted particles flow from the reactor.

Bacterial growth oftens displays three phases of growth. Initially, the growth rate increases from zero to the maximum growth rate; this is referred to as the lag phase. Subsequent to this, the growth rate remains at the maximum rate; this is referred to as the exponential phase. Lastly, in the stationary phase the growth rate decreases and finally reaches zero. This results in a sigmoidal curve, which can be described by the logistic equation (6), as well as by other models (7).

Pinches *et al* (8) proposed a modification of the logistic equation. The logistic equation is an empirical equation that describes the exponential growth phase and the stationary phase. Pinches *et al* (8) proposed that the bacterial number (or biomass) in the logistic equation is proportional to the conversion of pyrite. Thus, they obtained the logistic equation for bacterial leaching:

$$\frac{dX}{dt} = k_m X\left(1 - \frac{X}{X_m}\right) \qquad 1.$$

where X is the conversion of pyrite, X_m is the maximum conversion achieved in the batch experiment and k_m is the rate constant.

Pinches *et al* (8) presented a continuous form of this equation. The conversion achieved in a single tank reactor, X, is given by:

$$X = X_m - \frac{1}{k_m \tau} \qquad 2.$$

where τ is the mean residence time in the tank reactor.

As in the case of the models of Gormeley *et al.* (3) and Chang and Myerson (4), this model focusses on the bacterial population, simply assuming that the size of the bacterial population is proportional to the conversion. This model also ignors the complex phenomena involved in continuous leaching.

The application of the logistic equation to bacterial leaching has been studied extensively (9 - 13). The parameter of importance is the rate constant, k_m. Values for the batch and continuous data of Chapman (9) and Miller (10) are given in Tables 1 and 2. While the model appears to describe the shape of the batch and continuous curves, Hansford and Chapman (11) concluded that there is no correspondence between the parameters for the batch and continuous operation.

Table 1 Logistic equation parameters for the data of Chapman (9)

	k_m (days^{-1})	
Size fraction	Batch	Continuous
-75 +53 μm	0.259	0.276
-53 +38 μm	0.277	0.290
-38 μm	0.265	0.348

Table 2 Logistic equation parameters for the data of Miller (10)

Batch		Continuous	
Size fraction	k_m (days^{-1})	Reactor stage	k_m (days^{-1})
Bulk	0.683	Primary stages	1.35
+75 μm	0.373	Stage 2	0.97
-75 +53 μm	0.428	Stage 3	1.30
-53 +38 μm	0.462	Stage 4	0.31
-38 +25 μm	0.556		
-25 μm	0.907		

The logistic equation fails to account for the effect of different sizes of particles on the leaching rate. The rate constant for the continuous operation should be the same throughout the cascade of reactors, and, in addition, the rate constant should correspond to that obtained from the batch data.

Models of continuous bacterial leaching at steady-state may be improved by accounting for the change in size distribution as a result of reaction and flow phenomena in the reactor. Crundwell (14) presented a general model for the batch and continuous bacterial leaching of sulphide minerals. The model incorporates the population balance approach to particulate leaching, and the bacterial cell balance. The model accounts for bacterial populations on the mineral surface and in the solution. The population balance model of leaching accounts for the effects of particle-size distribution, particle kinetics, and the residence-time distribution of the reactor. Bacterial balances account for the growth of bacteria, both on the mineral surface and in the solution, and for the bacterial attachment to and detachment from the mineral surface.

The application of this model (14) to batch and continuous bacterial leaching of refractory gold ores is discussed in the following sections.

3. Mathematical modelling of the bacterial leaching of pyritic ores

The leaching of pyrite by bacteria is thought to occur mainly by direct bacterial oxidation. This is a mechanism in which bacteria attach to the surface of the mineral and oxidise the pyrite at the point of attachment. The bacteria thrive, the population increases in size, and as a result the rate of consumption of pyrite increases. The factors that directly influence the rate of bacterial leaching of pyrite are the growth rate of the bacterial population, the availability of pyrite as a substrate, and the availability of mineral surface area.

3.1 Bacterial leaching of pyrite in a batch reactor

It is proposed that the rate of growth of the bacterial population is proportional to the size of the bacterial population present on the surface of the bacteria and proportional to the amount of surface area that is available for bacterial attachment and growth. Thus, the rate of growth of bacteria on the mineral surface, r_b, is given by:

$$r_b = k_b M(M_{max} - M) \qquad 3.$$

where M is the number of bacteria on the mineral surface per unit volume of mineral slurry, M_{max} is the maximum bacterial population that can be supported, and k_b is the rate constant for bacterial growth.

The balance on the bacterial number in a batch reactor in which the volume is constant gives:

$$\frac{dM}{dt} = k_b M(M_{max} - M) \qquad 4.$$

The rate of consumption of pyrite is proportional to the rate of bacterial growth:

$$F_o \frac{dX}{dt} = k_{dl} M(M_{max} - M) \qquad 5.$$

where F_o is the initial molar concentration of pyrite per unit volume, X is the conversion of pyrite, and k_{dl} is a rate constant for the direct leaching of pyrite.

For particles initially of a narrow size range, F_o is given by $n_p \rho L^3$, where n_p is the number of particles of pyrite initially of size L per unit volume of slurry, and ρ is the molar density of the mineral. The conversion is related to the size of the particles, ℓ, and for particles initially of a narrow size range in a batch reactor it is given by:

$$X = 1 - \frac{\ell^3}{L^3} \qquad 6.$$

Equating equations [4] and [5] and rearranging we obtain the equation:

$$\frac{1}{k_b}\frac{dM}{dt} = \frac{F_o}{k_{d1}}\frac{dX}{dt} \qquad 7.$$

Integration of this equation using the initial condition that $M = M_o$, where M_o is the initial cell number, gives:

$$\frac{M}{M_o} = 1 + \frac{k_b F_o X}{k_{d1} M_o} \qquad 8.$$

Subtituting equation [8] into equation [5] and rearranging gives:

$$F_o \frac{dX}{dt} = k_{d1} M_o M_{max} \left(1 + \frac{k_b F_o X}{k_{d1} M_o}\right)\left\{1 - \frac{M_o}{M_{max}}\left(1 + \frac{k_b F_o X}{k_{d1} M_o}\right)\right\} \qquad 9.$$

The maximum number of bacteria that can be supported on the surface of the pyrite is proportional to the maximum number of bacteria per unit surface area, N_{max}, and the total surface area per unit volume, A:

$$M_{max} = N_{max} A \qquad 10.$$

Substituting equation [10] into equation [9], we obtain the equation:

$$F_o \frac{dX}{dt} = k_{d1} M_o N_{max} A \left(1 + \frac{k_b F_o X}{k_{d1} M_o}\right)\left\{1 - \frac{M_o}{N_{max} A}\left(1 + \frac{k_b F_o X}{k_{d1} M_o}\right)\right\} \qquad 11.$$

The surface area per unit volume of the pyrite particles in the batch reactor, A, is dependent on the extent of the reaction, and, for particles of a narrow size range, is given by:

$$A = A_o (1 - X)^{2/3} \qquad 12.$$

where A_o is the initial mineral surface area per unit volume, and is given by $n_p \phi L^2$. ϕ is a shape factor, which is assumed to be constant for all the particles.

Substitution of equation [12] and $A_o/F_o = \phi/\rho L$, into equation [11] gives the equation:

$$\frac{dX}{dt} = \frac{k_D}{L}(1 - X)^{2/3}(1 + k_B X)\left(1 - \frac{k_{MAX}(1 + k_B X)}{(1 - X)^{2/3}}\right) \qquad 13.$$

where:

$$k_D = \frac{k_{d1}\phi N_{max} M_o}{\rho} \qquad k_B = \frac{k_b F_o}{k_{d1} M_o} \qquad k_{MAX} = \frac{M_o}{N_{max} A_o}$$

Equation [13] describes the bacterial leaching of a pyrite sample of initial size L in a batch reactor. At low conversions, $X \ll 1$ and equation [13] indicates that conversion increases exponentially with time. At higher conversions, the rate of conversion decreases since the terms

$$(1 - X)^{2/3} \quad \text{and} \quad \left(1 - \frac{k_{MAX}(1 + k_B X)}{(1 - X)^{2/3}}\right)$$

both approach zero as X approaches 1.

The solution to equation [13] is obtained by numerical integration. A program, using the Runge-Kutta-Fehlberg technique, was written in C (15) and executed on a personal computer. The parameters were obtained by minimizing the sum of squared errors between the data and the model solution. The minimization routine used was the downhill simplex method of Nelder and Mead (16) written in C (17).

3.2 Bacterial leaching of pyrite in a continuous reactor

The analysis of leaching in a continuous reactor is fundamentally different from the analysis of a continuous homogeneous reactor. In a homogeneous reactor the molecular reactants of the reactor can have one of two states: they are either unreacted or they are reacted. However, in the leaching reactor, the feed particles change their size on reaction. As a consequence, a particle could be partially reacted if its size has not been reduced to zero. In addition, particles may spend different amounts of time in the reactor, so that even if all the particles entering the reactor were of the same size, the size of the particles leaving the reactor would be distributed. This exit distribution arises as a result of the kinetics of the reaction and the residence-time distribution of the reactor.

3.2.1 The population balance model of leaching reactors

Leaching reactors have been described by the method of the population balance (also referred to as the number balance) (18,20,21). This method is a statistical-mechanical description of the change in the distribution of particle size on reaction (19).

The population balance recognizes that the particles in a leaching reactor are countable entities, and that their numbers are conserved. Consider the processes which influence the number of particles in a particular size class in the reactor. Particles of this particular size may enter or leave the reactor, therefore influencing the number in that size class; or particles that are in the reactor may shrink as a consequence of the leaching reaction into or out of that particular size class. For leaching in a continuous reactor, the population balance is therefore:

$$\left\{\begin{array}{l}\text{Number into size class}\\ \text{and reactor by flow}\\ \text{into reactor}\end{array}\right\} + \left\{\begin{array}{l}\text{Number into size class}\\ \text{by reaction of particles}\\ \text{in reactor of larger class}\end{array}\right\}$$

$$= \left\{\begin{array}{c}\text{Number out of size class}\\ \text{and reactor by flow from}\\ \text{reactor}\end{array}\right\} + \left\{\begin{array}{l}\text{Number out of size class}\\ \text{by reaction of particles in}\\ \text{reactor into smaller class}\end{array}\right\}$$

If $n(\ell)d\ell$ represents the number of particles per unit volume of slurry in the size range ℓ to $\ell + d\ell$, then the population balance for a steady-state, continuous-stirred-tank leaching reactor may be derived:

$$Qn_f(\ell) = Qn(\ell) + V\frac{d\{R(\ell)n(\ell)\}}{d\ell} \qquad 14.$$

with the boundary condition that $R(\ell)n(\ell) \rightarrow 0$ as $\ell \rightarrow \infty$. Q is the volumetric flowrate of slurry, V is the volume of the reactor, $R(\ell)$ is the rate of change of size of particles, and $n_f(\ell)$ is the number distribution of the particle size in the feed to the reactor.

Equation [14] is a first-order differential equation, the solution of which is found using the method of integrating factors. The solution is given by:

$$n(\ell) = \frac{1}{R(\ell)\tau} \exp(-T/\tau) \int_{\infty}^{\ell} \exp(T/\tau) n_f(\ell) d\ell \qquad 15.$$

where $\tau = V/Q$ is the mean residence time, and

$$T = \int_{o}^{\ell} 1/R(\ell) d\ell$$

The solution of the population balance gives $n(\ell)$, from which the conversion, X, can be obtained from the equation:

$$1 - X = \frac{\int_{o}^{\infty} \ell^3 n(\ell) d\ell}{\int_{o}^{\infty} \ell^3 n_f(\ell) d\ell} \qquad 16.$$

$R(\ell)$ describes the kinetics of the leaching reaction, and is dependent on the concentration of bacteria on the surface of the pyrite particles. The kinetics of the leaching reaction are described by equation [5]. Since the conversion for particles of a narrow size distribution in a batch reactor is given by equation [6], equation [5] may be written as:

$$n_p \rho \, 3 \, \ell^2 \frac{d\ell}{dt} = k_{dl} M_o \, N_{max} \, n_p \phi \, \ell^2 \left(\frac{M}{M_o} \right) \left(1 - \frac{M_o}{N_{max} A} \frac{M}{M_o} \right) \qquad 17.$$

A is the total area of particles per unit volume in the reactor. For continuous reactor, A is proportional to the second moment of the distribution, μ_2, and is given by:

$$A = \phi \mu_2 = \phi \int_{\infty}^{o} \ell^2 n(\ell) d\ell \qquad 18.$$

Therefore, equation [17] may be rearranged to give:

$$R(\ell) = \frac{d\ell}{dt} = \frac{k_D}{3} \mathbf{M} \left(1 - \frac{k_{MAX} A_o \mathbf{M}}{\phi \mu_2} \right) \qquad 19.$$

where $\mathbf{M} = M/M_o$.

3.2.2 Balance for the bacterial population

The model for the bacterial leaching of pyrite is completed by considering the growth of the bacterial population on the surface of the pyrite. The basis of the model of the bacterial growth is that the rate of growth is proportional to the rate of pyrite consumption. This establishes a pseudo-stoichiometry between the size of the bacterial population and the amount of pyrite oxidized. Since no bacteria are fed to the primary reactor of a reactor cascade, the number of bacteria on the mineral surface in a continuous reactor is given by:

$$M = M_o + \frac{k_b F_f}{k_{dl}} X \qquad 20.$$

where F_f is the molar concentration of pyrite in the feed to the first reactor.

Dividing equation [20] by M_o, we obtain:

$$M = 1 + \frac{F_f}{F_o} k_B X \qquad 21.$$

3.2.3 Size distribution for the mineral feed

The mass distribution of the particle size of the feed may be described by the Rosin-Rammler distribution. The corresponding density function is given by:

$$m_f(\ell) = \frac{p \, \ell^{p-1}}{q^p} \exp\left[-\left(\frac{\ell}{q} \right)^p \right] \qquad 22.$$

where p and q are constants.

The mean size (on a mass basis) for this density function is given by:

$$\bar{\ell} = q \, \Gamma \left(\frac{p+1}{p} \right) \qquad 23.$$

The density function on a number basis, $n_f(\ell)$, may be calculated from the density function on a mass basis using the equation:

$$n_f(\ell) = \frac{\ell^{-3} m_f(\ell)}{\int_{\infty}^{o} \ell^{-3} m_f(\ell) d\ell} \qquad 24.$$

3.2.4 Procedure for the solution of the continuous model

The solution to the equations describing the continuous model is as follows:

(i) Obtain the size distribution of the feed particles to the first reactor. Calculate the size distribution on a number basis using equation [24].
(ii) Estimate **M** for the reactor.
(iii) Calculate $n(\ell)$ from equation [15] using $R(\ell)$ given by equation [19] with parameters k_D and k_{MAX} from batch analysis.
(iv) Calculate X from equation [16].
(v) Calculate **M** from equation [21] using parameter k_B from batch anlysis.
(vi) Compare **M** from (v) with that from (ii). Calculate new estimate of **M** using secant or regula falsi methods and return to (iii).
(vii) Repeat until (ii) and (v) agree with a selected tolerance.
(viii) This gives the conversion, X, and the bacterial population, **M**, for the continuous leaching tank.

This procedure was programmed in C and run on an IBM-compatible personal computer.

4. Application of the model to batch and continuous bacterial leaching

Experimental data were presented by Chapman (9) and Miller (10) for the batch and continuous bacterial leaching of pyrite and arsenopyrite ores.

Chapman (9,11,12) performed tests using a pyrite concentrate from Crown Mines (South Africa). The pyrite content of the concentrate was 95%. The leaching tests were performed in 5 litre aerated vessels at 30°C and pH 1.8. The solids concentration in each batch test was adjusted to give the same initial surface area. A laboratory culture of *Thiobacillus ferrooxidans* was used as the inoculum.

Miller (10,13) performed tests using a pyrite-arsenopyrite concentrate obtained from the Fairview Mine (South Africa), and reported the results from a series of pilot-plant trials. The batch leach tests were performed in 300 litre vessels using a mixed *Thiobacillus* culture. The pH was adjusted to 1.6. The pilot plant consisted of four leaching stages. The primary stage had a volume of 10 m^3 and the following three stages had volumes of 5 m^3. Miller (10) reported the size distribution on a mass basis for the feed to the pilot plant.

4.1 Analysis of the data of Chapman (9,11,12)

The batch leaching tests of Chapman (9,11,12) were analysed by fitting the parameters of equation [13] to the data. Figure 1 illustrates the results of the solution to equation [13] using the parameters given in Table 3. The figure indicates that the model is a good description of the bacterial leaching reaction in a batch reactor.

Chapman (9) used different amounts of pyrite in each batch test. This procedure was followed so that the initial surface area available for bacterial attachment and leaching was constant for each test. The parameter k_B in the equation [13] is dependent on the initial amount of solids in the batch test. As a result, the parameter k_B/F_o should be the same for each of Chapman's batch experiments. The parameters given in Table 3 indicate that k_B/F_o is the same for the batch experiments using different size fractions.

The parameter k_{MAX} determines the maximum conversion that is achieved in the batch reactor. Equation [13] indicates that this parameter is dependent on the surface area that is initially available, given by A_o. Therefore, the theoretical development of equation [13] indicates that $k_{MAX}A_o$ should be constant for the batch experiments using different size fractions. The values of $k_{MAX}A_o$ are given in Table 3. These values are not the same as each other for this set of experiments. The values for -75 +53 μm fraction and the -38 μm fraction are similar, but the value for the -53 +38 μm fraction is much larger than the others. Chapman (9) measured the BET surface areas for these fractions. The values for A_o based on the BET surface areas, given in Table 3, indicate that the available area for the -53 +38 μm fraction was lower than that for the other two size fractions. The anomalous BET surface area for this size fraction could account for the anomaly in the $k_{MAX}A_o$ values.

Chapman (9) performed continuous tests using the same material as the batch tests. The results of these continuous experiments were analysed using equations [15], [16] and [21]. The parameters for these equations are given in Table 3. The results are illustrated in Figure 2, which shows that the continuous model is a good description of bacterial leaching.

The parameters k_B and k_D used in the modelling of the continuous results were those obtained from the analysis of the batch experiments. The correspondence between the parameters for models of the batch and the continuous experiments suggests that these models have accounted correctly for the most important phenomena that occur during bacterial leaching of pyrite.

ls36

Table 3 Values of parameters fitted to data of Chapman (9)			
Analysis of data for the batch experiments			
	Size class		
Parameter	-75 +53 μm	-53 +38 μm	-38 μm
k_B (-)	12.77	9.075	4.994
k_D (μm/day)	0.862	0.862	0.862
k_{MAX} (-)	0.059	0.060	0.030
lag time (day)	2.49	1.93	2.08
k_B/F_o (L/kg)	90.75	90.75	90.75
$k_{MAX}A_o$ (m^2/L)	0.153	0.235	0.122
Conditions for batch experiments			
L (μm)	64.0	45.5	25.0
F_o (kg/L)	0.1407	0.100	0.055
A_o (m^2/L)	2.59	3.90	4.10
BET area (m^2/L)	33.8	20.0	30.8
Analysis of the data for the continuous experiments			
Parameter			
k_B (-)	12.77	9.075	4.994
k_D (μm/day)	0.862	0.862	0.862
k_{MAX} (-)	0.03	0.02	0.018
Conditions for continuous experiments			
$\bar{\ell}$ (μm)	64.0	45.5	25.0
p (equ 22)	4.4	4.4	4.4
F_f (kg/L)	0.1	0.1	0.1

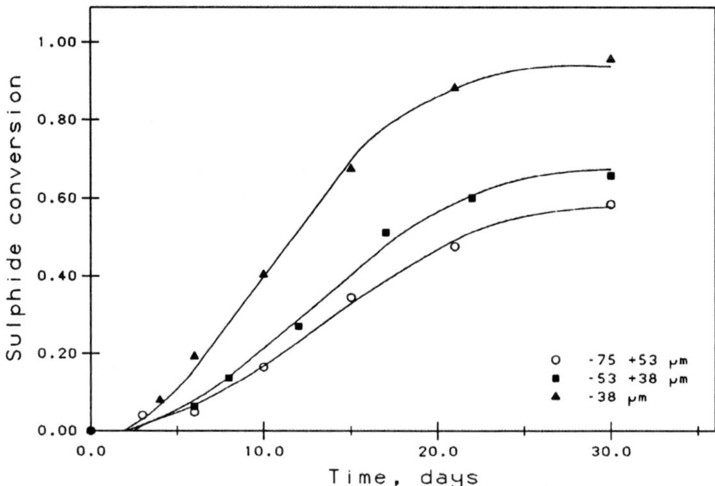

Figure 1. The bacterial leaching of a pyrite concentrate in a batch reactor. The data are from Chapman (9) and the lines represent the solution to the model, given by equation [13], using the parameters given in Table 3.

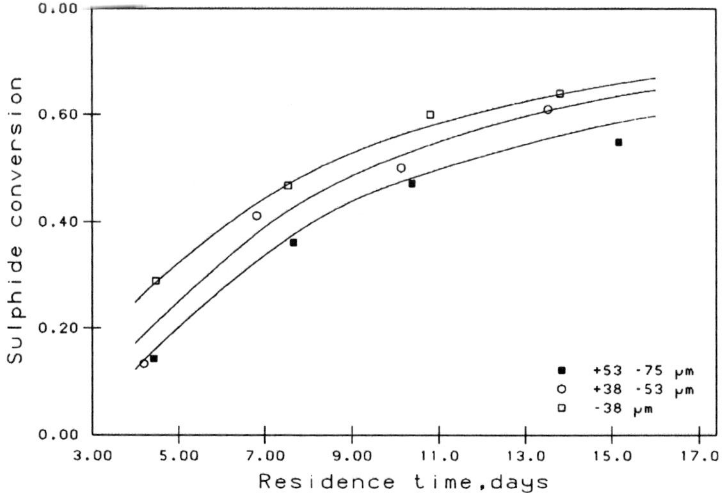

Figure 2. The bacterial leaching of a pyrite concentrate in a continuous reactor. The data are from Chapman (9) and the lines represent the solution to the model given by equations [15], [16] and [21], using the parameters given in Table 3.

The value of the parameter k_{MAX} used to model the continuous experiments is smaller than that used to model the batch experiments. This suggests that the effect that this parameter has on the process is diminished in the case of continuous leaching. Further experimental data is required to explain this phenomenon fully.

4.2 Analysis of the data of Miller (10,13)

The batch leaching experiments of Miller (10) were analysed using equation [13]. The solution to equation [13] using the parameters given in Table 4 is shown in Figure 3. This figure indicates that the batch model is a good description of Miller's data. In addition, the rate parameters, k_B and k_D, are the same for the various size classes, as predicted by the theory.

Table 4 Values of parameters fitted to data of Miller (10)			
Analysis of data for the batch experiments			
	Size class		
Parameter	-75 +53 μm	-53 +38 μm	-38 +25 μm
k_B (-)	14.79	14.79	14.79
k_D (μm/day)	1.074	1.074	1.074
k_{MAX} (-)	0.0052	0.0063	0.0044
lag time (day)	28.5	4.2	11.4
$k_{MAX}A_o$ (m^2/L)	0.0329	0.0329	0.0329
Conditions for batch experiments			
L (μm)	64.0	45.5	25.0
F_o (kg/L)	0.12	0.12	0.12
A_o (m^2/L)	6.35	5.22	7.43
Analysis of the data for the continuous experiments			
Parameter			
k_B (-)	14.79		
k_D (μm/day)	1.074		
k_{MAX} (-)	0.001		
Conditions for continuous experiments			
$\bar{\ell}$ (μm)	25.0		
p (equ 22)	1.84		
F_f (kg/L)	0.12		

The parameters k_B and k_D obtained from the batch experiments were used to predict the operation of the continuous pilot plant. The results are given in Figure 4, which shows good agreement between the data and the model prediction.

The parameter k_{MAX} was found to be smaller in the case of continuous leaching than for that of leaching in a batch reactor. This was also found to be true for the data of Chapman (9). Further work is required to fully understand this phenomenon.

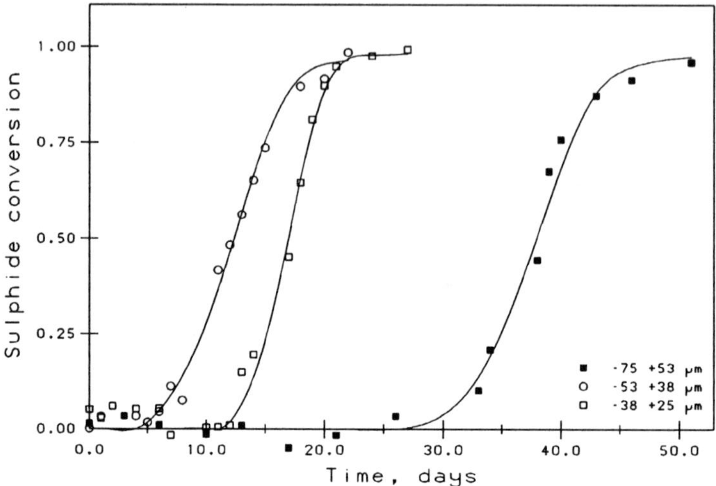

Figure 3. The bacterial leaching of a pyrite concentrate in a batch reactor. The data are from Miller (10) and the lines represent the solution to the model, given by equation [13], using the parameters given in Table 4.

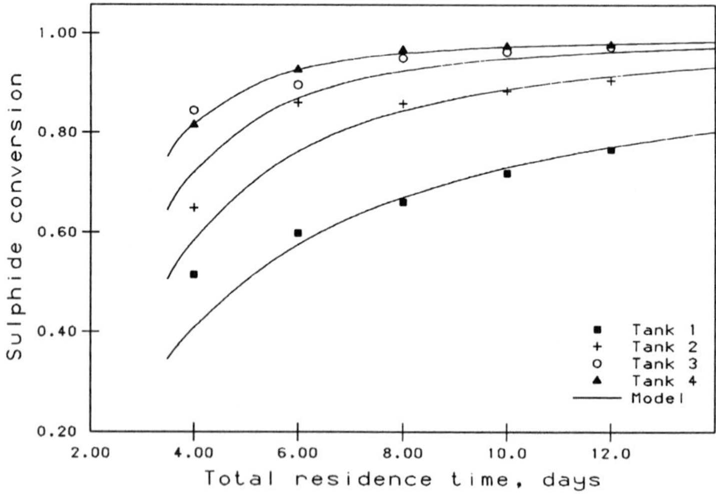

Figure 4. The bacterial leaching of a pyrite concentrate in a series of continuous reactors. The data are from Miller (10) and the lines represent the solution to the model given by equations [15], [16] and [21] using the paramters of Table 4.

5. Discussion

Figures 2 and 4 illustrate the solution to the models for continuous operation based on the parameters obtained from the batch data. These figures indicate that there is good agreement between the model and the experimental data. This agreement is particularly close for the data and model shown in Figure 4. The only difference between the parameters obtained for the analysis of the batch and the continuous data is that the value of k_{MAX} is smaller in the case of continuous leaching for the data of Miller (10).

Since the value of k_{MAX} for the continuous experiments of Chapman (9) is lower than that obtained for the corresponding batch experiments, the maximum concentration of bacteria on the mineral surface may be higher in continuous experiments than that in batch experiments.

6. Conclusions

A model has been developed here that accounts for the complex phenomena occurring during the batch and continuous leaching of minerals by bacteria. This model is based on the population balance model of leaching (18,20,21). The model has been shown to be a good description of both the batch and continuous leaching of pyritic minerals. The parameters obtained from the batch curves were used to describe the continuous results. This model could provide benefits in the design and optimization of bacterial leaching plants. The usefulness of the pilot-plant stage in the development of a new process plant could be significantly increased by the use of a model such as this.

References

1. Dew, D.W., Miller, D.M. and van Aswegen, P.C. GENMIN's commercialization of the bacterial oxidation process for the treatment of refractory gold concentrates. Proceedings of the International Gold Conference, Beaver Creek 1993. Randol, Golden, Colorado, pp 229-237.

2. Pinches, A., Neale, J., Huberts, R, and Dempsey, P. Development of the Mintek bacterial oxidation process (MINBAC). Proceedings of the International Gold Conference, Beaver Creek 1993. Randol, Golden, Colorado, pp 221-228A.

3. Gormely, L.S., Duncan, D.W., Branion, R.M.R., and Pindler, K.L. Continuous culture of *Thiobacillis ferrooxidans* on a zinc sulphide concentrate. Biotech. Bioeng. 17 (1975) 34-49.

4. Chang, Y.C. and Myerson, A.S. Growth models of the continuous bacterial leaching of iron pyrite by *Thiobacillus ferrooxidans*. Biotech. Bioeng. 24 (1982) 889-902.

5. Myerson, A.S. and Kline, P.C. Continuous bacterial coal desulphurization employing *Tiobacillus ferrooxidans*. Biotech. Bioeng. 26 (1984) 92-99.

6. La Motta, E.J. Kinetics of continuous growth cultures using the logistic curve. Biotech. Bioeng. 18 (1976) 1029-1032.

7. Zwietering, M.H., Jongenburger, I., Rombouts, F.M., and van 't Riet, K. Modelling of the bacterial growth curve. Appl. Environ. Microbiol. 56 (1990) 1875-1881.

8. Pinches, A., Chapman, J.T., te Riele, W.A.M., and van Staden, M. The performance of bacterial leach reactors for the preoxidation of refractory gold-bearing sulphide concentrates. Proc. Int. Symp. Biohydrometallurgy, Warwick (1987). Eds. Norris, P.R. and Kelly, D.P., Science and Technology Letters, Kew, U.K., pp 329-344.

9. Chapman, J.T. The batch and continuous bacterial leaching kinetics of a refractory gold-bearing pyrite concentrate. M.Sc. dissertation (1989), University of Cape Town, South Africa.

10. Miller, D.M. Biooxidation of a gold-bearing pyrite/arsenopyrite concentrate. M.Sc. dissertation (1990), University of Cape Town, South Africa.

11. Hansford, G.S. and Chapman, J.T. Batch and continuous biooxidation kinetics of a refractory gold-bearing pyrite concentrate, Miner. Engng. 5 (1992) 597-612.

12. Hansford, G.S. and Chapman, J.T. Batch and continuous biooxidation of a gold-bearing pyrite concentrate. S.A.J.Chem.Eng.

6 (1994) 1-10.

13. Miller, D.M. and Hansford, G.S. Batch oxidation of a gold-bearing pyrite-arsenopyrite concentrate. Miner. Engng. 6 (1992) 613-629.

14. Crundwell, F.K. Mathematical modelling of batch and continuous bacterial leaching, Accepted for publication, The Chemical Engineering Journal (1994).

15. Baker, L. C Tools for scientists and engineers. McGraw-Hill, N.Y. (1989) pp 271-286.

16. Nelder, J.A. and Mead, R. A simplex method for function minimization. Computer Journal 7 (1965) 308-313.

17. Press, W.H., Teukolsky, S.A., Vetterling, W.T. and Flannery, B.P. Numerical Recipes in C, Cambridge University Press, 2nd ed. (1992) pp 408-412.

18. Sepulveda, J.E. and Herbst, J. A population balance approach to the modelling of multistage continuous leaching systems. AIChE Symp. Ser. 57 (1979) 41-55.

19. Hulbert, H.M and Katz, S. Some problems in particle technology: a statistical formulation. Chem. Eng. Sci. 19 (1964) 555-574.

20. Crundwell, F.K. and Bryson, A.W. The modelling of particulate leaching reactors - the popualtion balance approach. Hydrometall. 29 (1992) 275-295.

21. Crundwell, F.K. Micromixing in continuous particulate reactors. Accepted by Chem. Eng. Sci (1994).

Nomenclature

A	Total surface area of mineral particles per unit volume of slurry, m^2 L^{-1}
A_o	Total surface area of mineral particles per unit volume of slurry at time t = 0, m^2 L^{-1}
F	Concentration of mineral to the reactor, mol L^{-1}
F_o	Concentration of mineral in the reactor at time t = 0, mol L^{-1}
F_f	Concentration of mineral in the feed to the reactor, mol L^{-1}
k	Rate constant
k_b	Rate constant for the growth of bacteria, L $cells^{-1}$ $days^{-1}$
k_B	Rate parameter in equation [13], unitless
k_{d1}	Rate constant for the rate of pyrite consumption, $L^2 cells^{-2} days^{-1}$
k_D	Rate parameter in equation [13], μm $days^{-1}$
k_m	Rate constant in the logistic equation, $days^{-1}$
k_{MAX}	Rate parameter in equation [13], unitless
ℓ	Particle size, m
L	Particle size at time t = 0, m
M	Total number of bacteria per unit volume on surface of pyrite, cells L^{-1}
M_{max}	Maximum number of bacteria per unit volime on the surface of pyrite, cells L^{-1}
M_o	Total number of bacteria per volume on surface of pyrite at time t = 0, cells L^{-1}
$\mathbf{M}$	M/M_o
$m_f(\ell)$	Mass density of particles in feed of size ℓ, m^{-1} L^{-1}
$n(\ell)$	Number density of particles of size ℓ, m^{-1} L^{-1}
$n_f(\ell)$	Number density of particles in feed of size ℓ, m^{-1} L^{-1}
n_p	Number of pyrite particles per unit volume, L^{-1}
N	Number of bacteria on mineral surface per unit area of mineral surface, cells m^{-2}
N_{max}	Maximum concentration of bacteria that can exist on the mineral surface, cells m^{-2}
p	Parameter in Rosin-Rammler distribution, unitless
q	Parameter in Rosin-Rammler distribution, unitless

Q	Flowrate of slurry to the reactor, L day^{-1}
Q_s	Flowrate of solution to the reactor, L day^{-1}
r_b	Rate of growth of bacteria attached to the mineral, cells L^{-1} day^{-1}
$R(\ell)$	Rate of shrinkage of the mineral particle, m day^{-1}
t	time, days
T	Time for particle of initial size L to be leached to size ℓ, days
V	Volume of the reactor, L
V_s	Volume of solution in the reactor, L
X	Conversion of mineral, unitless
X_m	Maximum conversion of mineral obtained according the logistic equation, unitless

Greek

Γ	Gamma function
μ_2	Second moment of the number density, m^2L^{-1}
ρ	Molar density, mol m^{-3}
τ	Mean residence time, days
ϕ	Particle shape factor, unitless

RECENT DEVELOPMENTS IN MODELLING BIO-OXIDATION KINETICS PART I: MEASUREMENT METHODS

M. Boon, J.J. Heijnen and G.S.Hansford[1]
Department of Biochemical Engineering
Kluyver Laboratory of Biotechnology
Julianalaan 67, 2628 BC Delft, The Netherlands

SUMMARY

A theoretical and experimental approach to studying the growth of thiobacilli on ferrous iron and pyrite is presented. The balanced stoichiometric equation for bacterial growth is given for the production of one C-mole of biomass. Biomass is measured in terms of its carbon content either as Total Organic Carbon or the amount of carbon dioxide taken up. The rate of substrate utilization is related to oxygen and carbon dioxide utilization by means of a degree of reduction balance, and yields of biomass production based on substrate and oxygen are defined. Specific rates are defined as rates per unit of biomass and give measures of the growth rate and metabolic activity of the bacteria under different conditions. The concept of maintenance is introduced, so that substrate consumption provides energy for both growth and maintenance. It is the possible to define kinetic parameters, maximum yield and maintenance coefficient based either on substrate or oxygen and to derive relationships between these quantities.
Off-gas analysis in order to measure oxygen and carbon dioxide utilization rates and a biological oxygen monitor for oxygen utilization rate measurement off-line are the two main experimental tools used. The equipment is described and its use illustrated with data from batch and continuous culture of thiobacilli on ferrous iron and batch culture on pyrite.

INTRODUCTION

In recent years there has been a considerable growth in the application of bio-oxidation for the pretreatment of refractory sulphide gold concentrates, particularly those containing arsenopyrite (van Aswegen, 1993; Hansford, 1994). In addition to its use in the pretreatment of refractory gold ores and concentrates, bio-oxidation is widely used in the bioleaching of copper from low-grade ores. There is also considerable interest being shown in its application to the bioleaching of cobalt, nickel and zinc from sulphidic ores. Bio-oxidation is also being considered as a means of coal desulphurization and has potential in the decontamination of soils and sludges containing heavy metals.

Up until now, the basis for the stirred bioreactor design has been the use of the logistic equation (Pinches *et al.*,1988). This has been shown to fit both batch and continuous data and to predict multistage pilot plant performance (Hansford and

---

[1]on sabbatical leave from Department of Chemical Engineering, University of Cape Town, Rondebosch, 7700, South Africa

Mineral Bioprocessing II
Edited by David S. Holmes and Ross W. Smith
The Minerals, Metals & Materials Society, 1995

Bailey, 1992, and Hansford and Miller, 1993). While successful in fitting continuous steady state data, the empirical logistic equation does not contain terms reflecting the effect of particle size, concentration nor pH and ferrous or ferric iron concentrations. In other words, the major limitation of the logistic equation is that it is not mechanistically-based and therefore does not provide information of the fundamental mechanisms, kinetics and rate limiting factors.

With the successful commercialization of bio-oxidation for refractory gold and for low-grade copper, and the potential for its used in a wide number of minerals processing and environmental control operations, there is an considerable incentive to gain further understanding of the sub-processes involved, their mechanisms and kinetics in order to identify the rate limiting sub-processes.

It is the aim of this paper to show the application of general theory and experimental methods used in other fields of biotechnology, to kinetic experiments in biohydrometallurgy. The paper presents an approach to studying bio-oxidation kinetics which makes use of the concept of a balanced stoichiometric equation for bacterial growth (Roels, 1983), that the primary process is the derivation of energy from substrate oxidation which is used for growth and maintenance (Pirt, 1982). From the stoichiometric equation a relationship between substrate consumption, oxygen utilization, carbon dioxide utilization and biomass production is derived. Experimental tools which have been used are; the measurement of oxygen utilization and carbon dioxide utilization by means of off-gas analysis, and respirometric measurements to determine the oxygen consumption under various conditions and also the maximum oxygen utilization rate per unit of biomass. Biomass is defined in terms of moles of carbon and is determined by a total organic carbon measurement or by the summation of the carbon dioxide absorbed from the air. Knowing the biomass concentration it is possible to calculate specific rate oxygen consumption rate (moles O_2/C-mole/h) and the maximum specific oxygen consumption rate of the bacteria. Biomass specific rates can be used as a measure of their activity. The use of these theoretical and experimental tools is illustrated by examples from a study of the growth kinetics of thiobacilli on ferrous iron and pyrite. The details of the background theory and experimental methods and further discussion of the results will be published in the PhD thesis of M.Boon in 1995 and in papers submitted to the journal *Biotechnology and Bioengineering*.

THEORY

Balances

A stoichiometric equation for bacterial growth on ferrous iron can be derived from the elemental balances on C, H, O, N, Fe, and the charge balance. Introduction of $Y_{Fe,x}$ as the biomass yield on ferrous iron, and assuming that biomass composition is represented by $CH_{1.8}O_{0.5}N_{0.2}$, results in the following stoichiometric equation:

$$CO_2 + 0.2NH_4^+ + \frac{(1-4.2Y_{Fe,x})}{4Y_{Fe,x}}O_2 + \frac{1}{Y_{Fe,x}}Fe^{2+} + (\frac{1}{Y_{Fe,x}} - 0.2)H^+ = CH_{1.8}O_{0.5}N_{0.2} + \frac{1}{Y_{Fe,x}}Fe^{3+} + (\frac{1}{2Y_{Fe,x}} - 0.6)H_2O \quad (1)$$

According to the definition, $1/Y_{Fe,x}$ is the amount of ferrous iron to be oxidized to produce one C-mole of biomass. Similarly $1/Y_{ox}$ is defined as the amount of oxygen that is to be consumed to produce one C-mole of biomass. So, the production rate of biomass equals the oxygen consumption rate times the yield of biomass on oxygen, Y_{ox}. The production rate of bacteria, r_x, equals the consumption rate of carbon dioxide, r_{CO2}:

$$-r_{CO_2} = r_x \quad (2)$$

This equation shows that the biomass production rate can be determined directly from the carbon dioxide consumption rate, and therefore the total amount of biomass can be derived from the total amount of carbon dioxide consumed. The relation between the oxidation rate of ferrous iron and the oxygen and carbon dioxide consumption rates is called the degree of reduction balance and follows from the stoichiometry:

$$-r_{Fe^{2+}} = -4r_{O_2} - 4.2r_{CO_2} \quad (3)$$

The degree of reduction balance on the bio-oxidation of pyrite, FeS_2, is :

$$-15r_{FeS_2} = -4r_{O_2} - 4.2r_{CO_2} \quad (4)$$

It will be shown that the degree of reduction balance is useful in predicting ferrous or pyrite oxidation rates and the degree of conversion of these substrates from measured oxygen and carbon dioxide consumption rates. Using the degree of reduction balance a relation between $Y_{Fe,x}$ and Y_{ox} can be derived for the bio-oxidation of ferrous iron:

$$\frac{1}{Y_{ox}} = \frac{1-4.2Y_{Fe,x}}{4Y_{Fe,x}} \quad (5)$$

Specific Rates

For the description of the bacterial oxidation kinetics biomass specific rates (mol/C-mol/h) are commonly used. For example, the biomass specific oxygen consumption rate, q_{O2}, being the oxygen consumption rate per C-mole of biomass, is defined by: The biomass specific substrate consumption rate, q_{Fe2+} and the biomass specific growth rate, μ, can be similarly defined. These specific rates are measures of the activity of the bacteria (mole/C-mole/h). They are defined in the kinetic model in terms of the rate equations which describe their dependence on the process conditions such as ferrous and ferric iron concentration, pH, temperature etc..

Yield and Maintenance

The primary energy generating process is considered to be the oxidation of substrate (ferrous iron) defined by q_{O2} and q_{Fe2+}. This generates energy which is used by the bacteria for growth and maintenance. Using the maximum yield of biomass on oxygen, Y_{ox}^{max}, and the maintenance coefficient of biomass on oxygen, m_o, the oxygen consumption rate required for growth and maintenance of the bacteria according to the Pirt equation is:

$$-r_{O_2} = \frac{r_x}{Y_{ox}^{max}} + m_o * C_x \qquad (6)$$

which can be rewritten as:

$$\frac{1}{Y_{ox}} = \frac{1}{Y_{ox}^{max}} + \frac{m_o}{\mu} \qquad (7)$$

Therefore, the actual yield of biomass on substrate, Y_{sx}, and oxygen, Y_{ox}, depends on the specific growth rate, μ. The actual yield increases at increasing growth rate.

EXPERIMENTAL METHODS

Batch and Continuous Cultures

Continuous and batch culture experiments were performed in baffled stirred fermenters (Applikon, ADI Holland) with a working volume of 2 litres and a geometry of H/D=1, thermostated at 30°C. In batch experiments pH control was applied by regular addition of acid or base. The pH in the continuous culture experiments with ferrous iron was controlled between 1.8-1.9 by setting the pH of the influent ferrous iron solution. In continuous cultures influent medium was transported with a peristaltic pump, and the flow rate was calculated from the weight decrease of the medium vessel. Figure 1 shows the fermenter and off-gas analysis equipment.

Off-Gas Analysis

The oxygen and carbon dioxide consumption rates in batch and continuous cultures can be determined using mass flow controllers on the air supply and by analyzing the reference air and the off-gas from the reactors by means of an infra-red carbon dioxide analyzer and a paramagnetic oxygen analyzer. The carbon dioxide consumption data are used to determine the biomass concentration and growth rate. The oxygen consumption data are used to determine the specific oxygen consumption rate, which gives a measure of the bacterial activity. These are quantities of major importance and which have proved difficult to determine by other means. It will be shown that biomass concentrations calculated from CO_2 analysis agree with total organic carbon measurements (TOC). Using the degree of reduction balance the oxidation rate and the degree of oxidation of ferrous iron or pyrite or other mineral sulfides can be determined from the gas analysis data as shown in the examples below.

BOM Measurements

A biological oxygen monitor (BOM) was used to perform respiration measurements. The BOM consists of a thermostated vessel, a Clark cell oxygen electrode (Orbisphere Model 2715) and a monitor. The vessel that contains the sample is closed air tight by removing all the air in the head space with a stopper that holds the electrode and a mechanical stirrer (Figure 2). A measurement only takes 10-15 minutes and the measurements are reproducible. The oxygen consumption rate is found to be independent on dissolved oxygen concentrations down to 2 mg O_2 per liter (Figure 3). Several authors in bio-oxidation research have used respiration experiments to measure the initial kinetics (Jones and Kelly ,1983 and Lizama and Suzuki, 1989). This experimental tool gives the opportunity to measure the specific oxygen consumption rates of the bacteria as a function of different process conditions such as ferrous and ferric iron concentrations, pH, inhibitiors, etc..

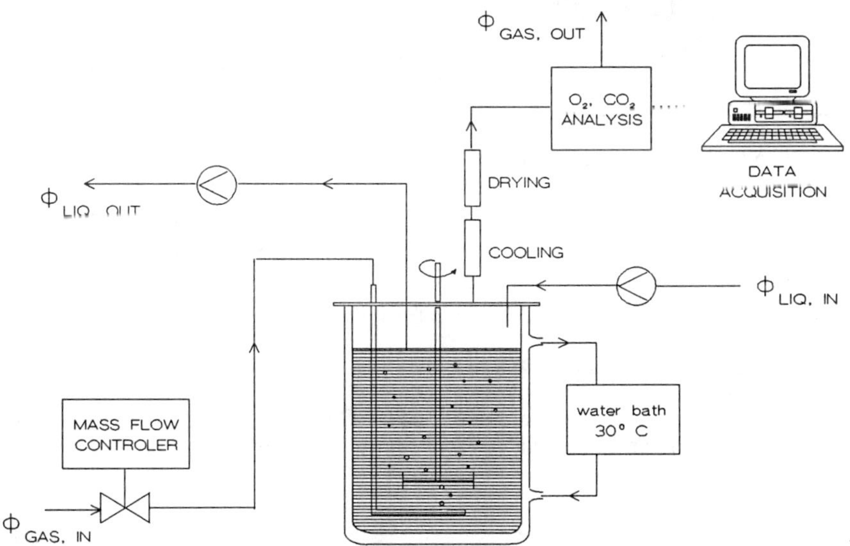

Figure 1: Fermenter equipment.

In this work BOM measurements were performed with cells from batch and continuous cultures at known oxygen and carbon dioxide consumption rates, which enables to measure the effects of changing concentrations or inhibitory effects on the cells. Respiration measurements were also used to determine the maximum specific oxygen consumption rate, q_{O2}^{max}, of bacteria in batch and continuous cultures. In these measurements samples were taken from batch or continuous cultures and diluted with an excess of ferrous iron solution to produce a high ratio of ferrous to ferric iron. The maximum oxygen utilization rate, q_{O2}^{max} is measured if no further

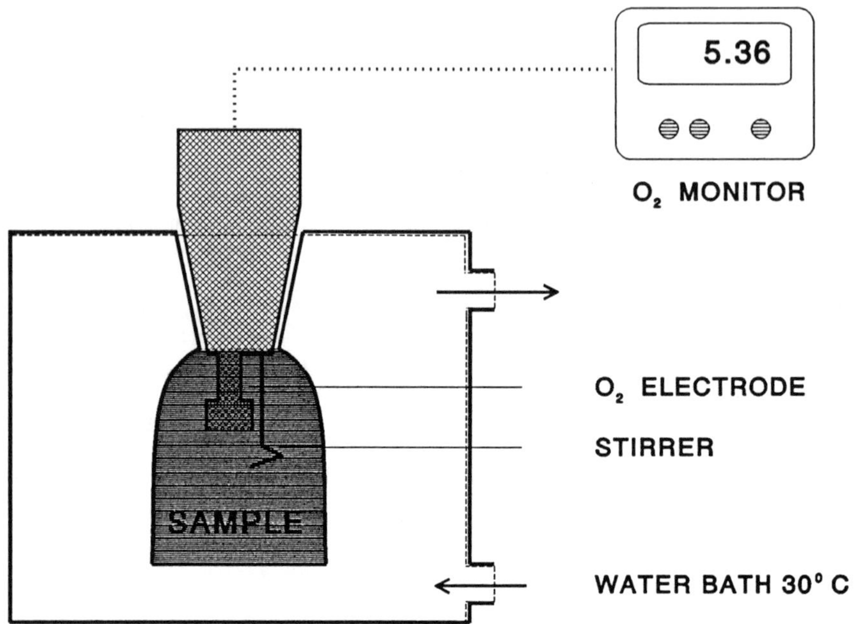

Figure 2: The biological oxygen monitor, BOM.

increase of q_{O2} occurs at increasing ferrous and decreasing ferric iron concentrations. Because off-gas analysis is applied the cell suspension contains a known concentration of biomass, which enables the determination of q_{O2} and q_{O2}^{max} from the oxygen consumption rate, r_{O2}, measured in the biological oxygen monitor.

C_x in Continuous and Batch Cultures

Because $r_x = -r_{CO2}$, the biomass concentration, C_x (in C-mole/l), in a continuous culture at steady state ($\mu = D$) can be calculated from the carbon dioxide consumption rate, r_{CO2}:

$$C_{x,CO_2} = \frac{-r_{CO_2}}{D} \tag{8}$$

The biomass concentration in a batch culture at time t can be derived from integration of the measured carbon dioxide consumption rate:

$$C_{x,CO_2}(t) = C_x(0) - \sum_{t_i=0}^{t_{i+1}=t} \frac{r_{CO_2}(t_i) + r_{CO_2}*(t_{i+1})}{2} * (t_{i+1} - t_i) \tag{9}$$

In this equation t_{i+1}-t_i is the time interval between two succesive carbon dioxide analyses, yielding the carbon dioxide consumption rates $r_{CO2}(t_i)$ and $r_{CO2}(t_{i+1})$.

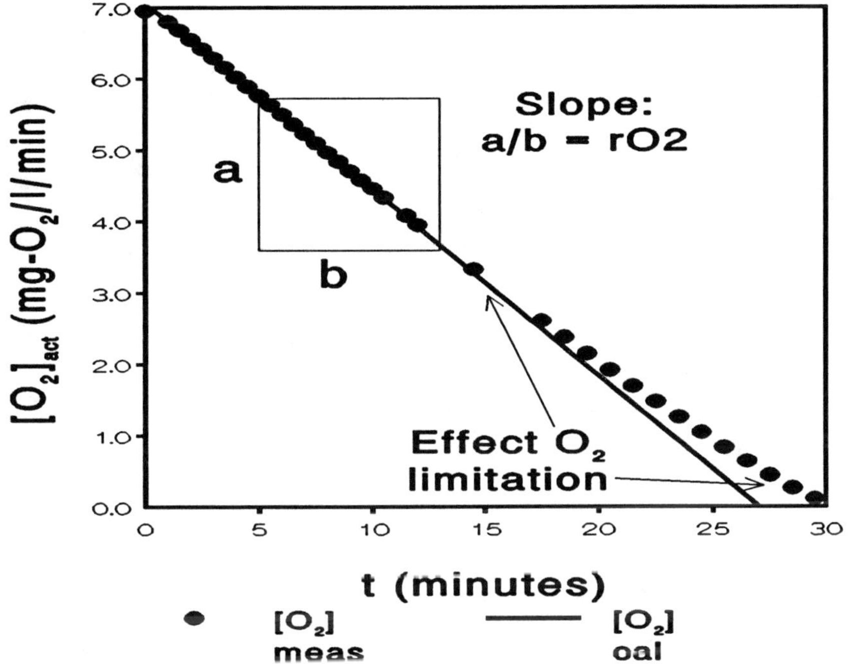

Figure 3: Typical BOM measurement.

Because the cells only divide 2 or 3 times during the course of a batch culture experiment, the specific rates can only be determined accurately if the initial biomass concentration is accurately known. For that reason cells from a continuous culture on ferrous iron in steady state were used to inoculate batch cultures on ferrous iron.

EXAMPLES

Examples on Bacterial Oxidation of Ferrous Iron in Batch Cultures

Figures 4 to 8 give examples of measurements and calculations on one batch culture experiment at a total iron concentration of 0.21 mole/l ([Fe] = 12 g/l), that is performed with 10% inoculum from a continuous ferrous iron culture in steady state at a dilution rate of 0.033 h^{-1} and a total iron concentration also of 0.21 mole/l .
In Figure 4 the measured oxygen and carbon dioxide consumption rates as calculated from the off-gas analyses are plotted against time. Over the course of a batch about 100 data points are measured by means of the off-gas analyses. The plot shows the typical behaviour of a batch culture experiment: exponential increase of the oxygen and carbon dioxide consumption rates in the initial phase and a steep drop in the final phase. Note that the surface area under the r_{CO2} curve between t=0 and t=t, is the total amount of CO_2 consumed (per litre) between t=0 and t=t, and therefore the

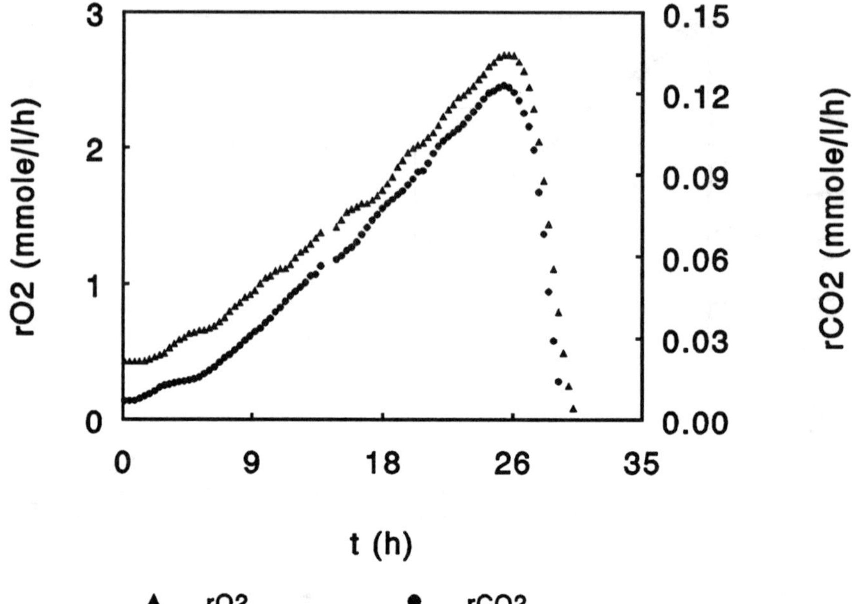

Figure 4: Oxygen and carbon dioxide consumption rates in a batch culture on ferrous iron.

total amount of biomass, $C_x(t)$-$C_x(0)$, that is produced up to time, t. The surface area under the r_{O2} curve between t=0 and t=t, is the total amount of oxygen consumed at time t (per litre), SUM $[O_2(t)]$. Using the degree of reduction balance, Equation 3, the total amount of ferrous iron consumed per litre at time t can be calculated:

$$[Fe^{2+}(0)]-[Fe^{2+}(t)] = 4SUM\ [O_2(t)] + 4.2(C_x(t)-C_x(0))$$

Note also that in the very last phase of the experiment the carbon dioxide consumption rate drops to zero faster than the oxygen consumption rate. This effect is caused by the maintenance requirement of the bacteria: in the last phase the concentration of bacteria is high, therefore relatively more substrate and oxygen are required for maintenance of the bacteria ($m_o.C_x$) and less energy is available for growth of the bacteria. Figure 5 shows the biomass concentration in the batch culture calculated from the carbon dioxide consumption rate against time, illustrating typical batch behaviour.

The ferrous iron concentration as analyzed with the *o*-phenantroline analysis, shown as data points, and the ferrous iron concentration as calculated from the integrated degree of reduction balance, shown as a line are plotted in Figure 6. This shows that measured and calculated data on the ferrous iron concentration coincide, and therefore that the degree of reduction balance applies. This enables the prediction of

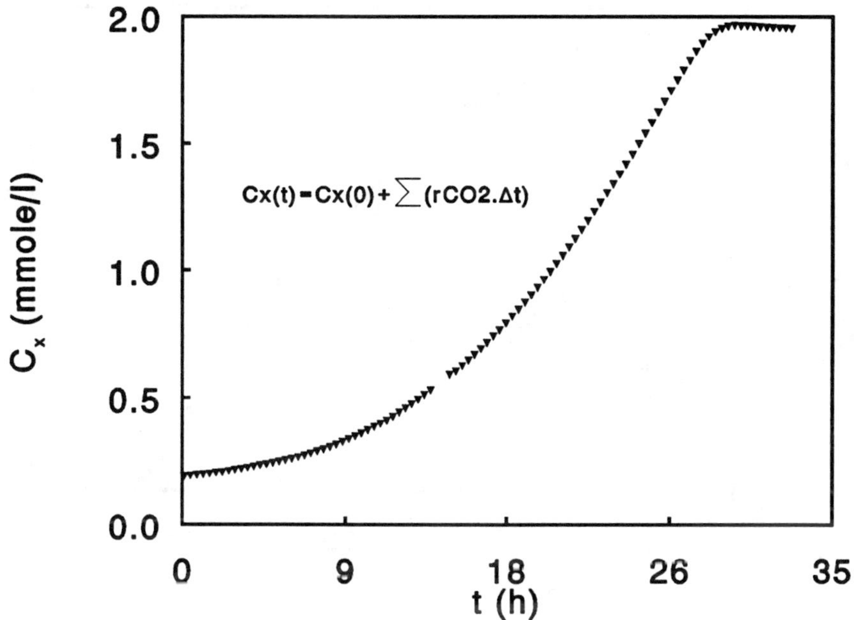

Figure 5: Biomass concentration from carbon dioxide consumption in batch culture on ferrous iron

the ferrous iron concentration from the oxygen and carbon dioxide consumption.

In Figure 7 both the specific oxygen consumption rate, q_{O2}, and the specific growth rate, μ, are plotted against time. These specific rates are derived from the measured oxygen and carbon dioxide consumption rates at time t; $r_{O2}(t)$ and $r_{CO2}(t)$ in Figure 4, divided by the biomass concentration at time t, $C_x(t)$ in Figure 5. It is expected that in the initial phase of the experiment maximum specific rates, q_{O2}^{max} and μ^{max}, will occur because ferrous iron is high and ferric iron is low. It appears however that up to 12 hours the specific growth rate, μ, and the specific oxygen consumption rate, q_{O2}, are uncoupled. The specific oxygen consumption rate is high from the start (3 to 4 times higher than the value of q_{O2} in the continuous culture). The specific growth rate starts at the value of μ in the continuous culture (D=0.033 h^{-1}) and increases to a maximum specific growth rate of about 0.1 h^{-1}. The specific rates of the biomass decrease at a certain ratio of the ferric and ferrous iron concentration. This is the kinetic behaviour that has to be described by the kinetic equations and will be discussed in Part II(Boon *et al.*, 1994b) of this paper. Note that the point were the decrease of the specific rates starts (after about 15 hours) does not coincide with the point were the oxygen and carbon dioxide consumption rate start to decrease (after about 27 hours): The increase of the oxygen and carbon dioxide consumption rates is

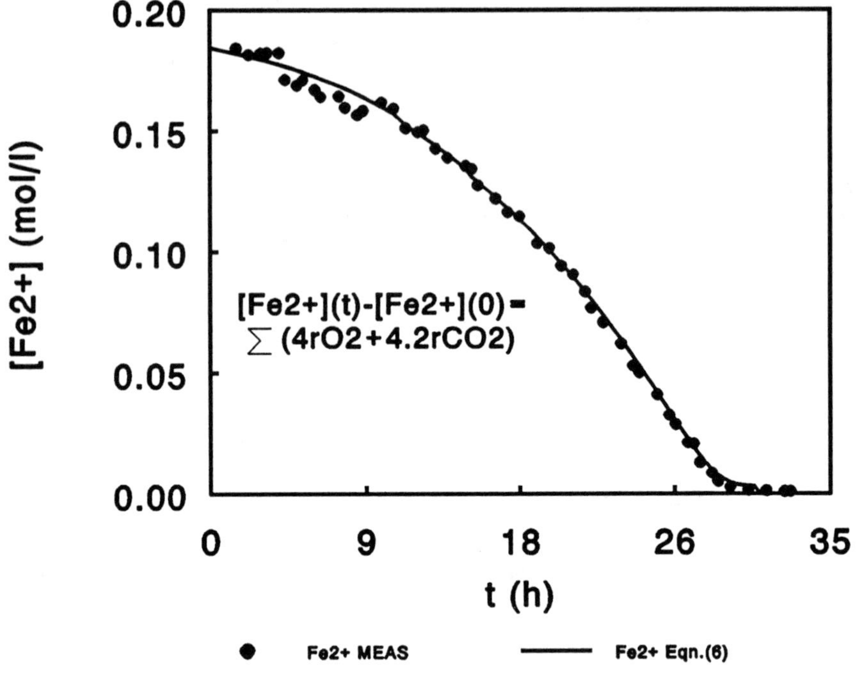

Figure 6: Ferrous iron concentration calculated from degree of reduction balance in batch culture on ferrous iron

mainly caused by the increase of the biomass concentration whereas changes in the specific rates are caused by the kinetic behaviour of the cells.

During the course of the batch culture experiment, BOM measurements were performed with samples from the batch culture to determine the maximum specific oxygen consumption rate of the cells. In Figure 8 both q_{O2} (Figure 7) and q_{O2}^{max} from BOM measurements at time t, are plotted against time. From this figure it can be seen that in the initial stage up to 15 hours of the experiment the specific oxygen consumption rate of the cells is equal to their maximum specific oxygen consumption rate. After 15 hours the value of q_{O2} starts to decrease. In Figure 6, a decrease of the specific oxygen consumption rate occurs if the ferrous iron concentration drops below 8 g/l and consequently the ferric iron concentration exceeds 4 g/l. This kinetic behaviour will be discussed further in Part II(Boon *et al.*, 1994b).

Examples on Bacterial Oxidation of Ferrous Iron in Continuous Cultures

Figures 9 to 13 give examples of measurements and calculations on a continuous culture experiment at dilution rates between 0.01 and 0.09 h^{-1} and a total iron

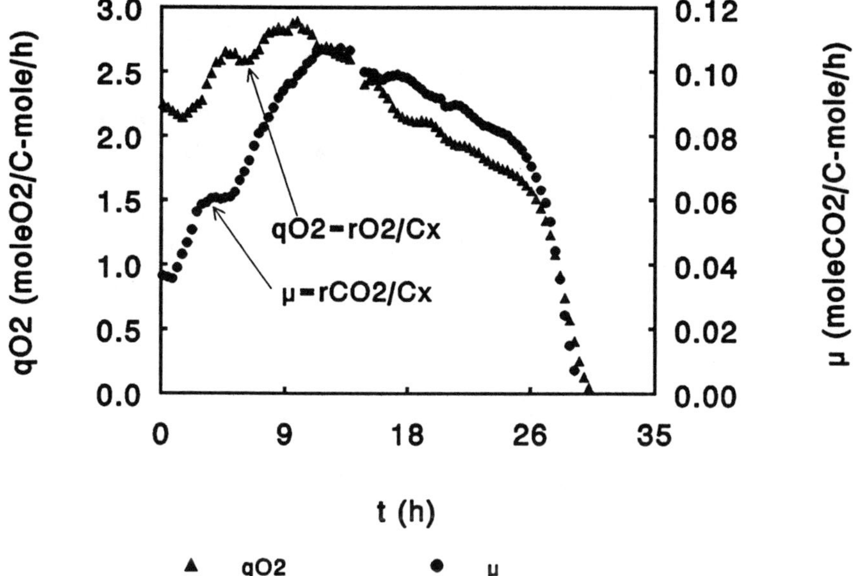

Figure 7: Specific oxygen consumption rate and specific growth rate in batch culture on ferrous iron

concentration of 0.21 mole/l.

In Figure 9 the measured oxygen and carbon dioxide consumption rates as calculated from the off-gas analyses are plotted against the dilution rate. Each data point shows the average oxygen and carbon dioxide consumption rate that is measured during a steady state. The plot shows the typical behaviour of a continuous culture experiment: the oxygen and carbon dioxide consumption rates increase with the dilution rate because more substrate (ferrous iron) is available per unit of time. Above a dilution rate of about 0.07 h^{-1} wash-out behaviour starts to occur.

In Figure 10 the biomass concentration as calculated from the carbon dioxide consumption rate is plotted as a function of the dilution rate. Also the biomass concentration as measured with TOC (total organic carbon) measurements is plotted. From this graph it can be seen that the biomass concentration calculated from the carbon dioxide consumption rate agrees with the measured concentration of organic carbon. From this graph it can also be concluded that the biomass concentration at low dilution rate increases with increasing dilution rates. In other words, the yield, $Y_{Fe,x}$, of biomass on ferrous iron at low dilution rates increases at increasing dilution rates or growth rates. This effect is caused by the fact that an increasing amount of the available substrate is used for maintenance requirements of the cells at increasing residence times or low growth rates where at dilution rates near to zero the major part of the substrate will be used as maintenance energy. The decrease of the

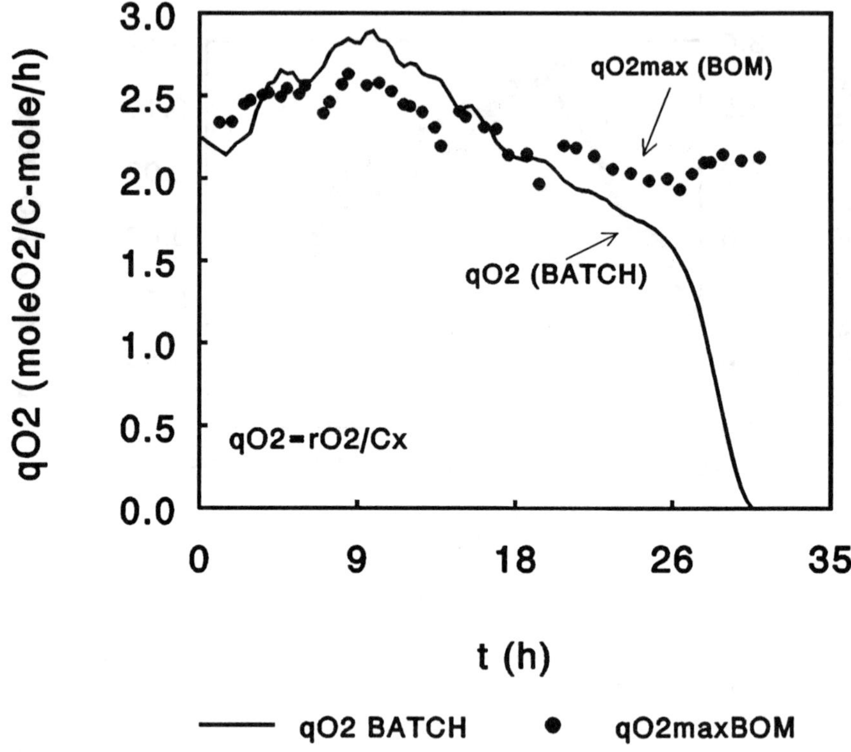

Figure 8: Specific oxygen consumption rate from off-gas analysis and maximum specific oxygen consumption rate from BOM measurement in batch culture on ferrous iron.

biomass concentration at high dilution rates is caused by wash-out behaviour.

Figure 11 shows the ferrous iron concentration as analyzed with an *o*-phenantroline analysis shown as data points. This plot shows that the ferrous iron concentration in the continuous culture increases steeply at high dilution rates, which is caused by wash-out behaviour. Therefore, at high dilution rates a significant part of the substrate is not oxidized, which is in agreement with the decrease of the biomass concentration in this range (Figure 10) and the decrease of the oxygen and carbon dioxide consumption rate (Figure 9).

In Figure 12 the specific oxygen consumption rate, q_{O2}, is plotted against the dilution rate. The specific oxygen consumption rate is derived from the measured oxygen consumption rates at a dilution rate, D, divided by the biomass concentration at that dilution rate (Figure 10). In Figure 12, the maximum specific oxygen consumption rate, q_{O2}^{max}, as measured in the BOM, is plotted. At dilution rates near wash-out, q_{O2} in the continuous culture, and q_{O2}^{max} become equal. The value of q_{O2}^{max} of cells in the

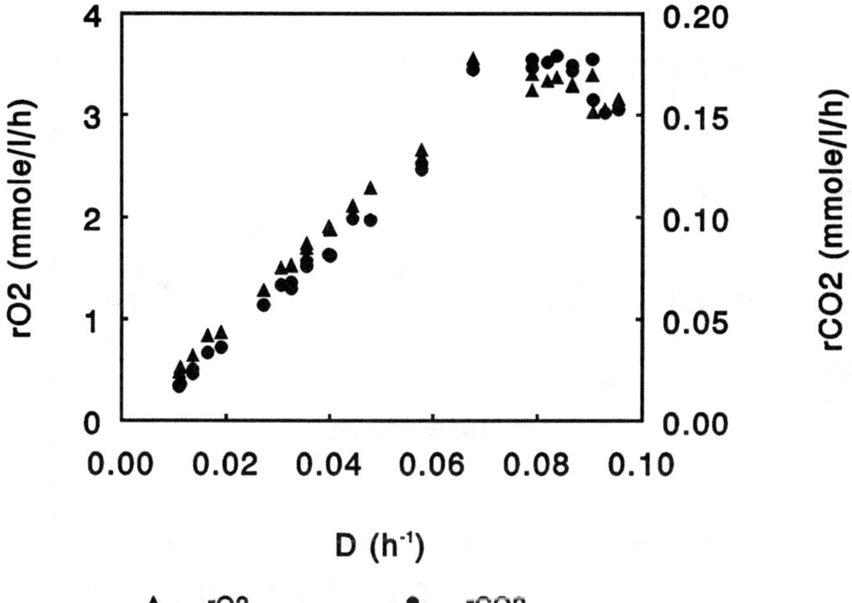

Figure 9: Oxygen and carbon dioxide consumption rates from off-gas analysis in continuous culture on ferrous iron.

continuous culture decreases slightly at high dilution rates.

According to the Pirt equation (Eqn. 6) q_{O2} and μ are related as $q_{O2} = \mu/Y_{ox}^{max} + m_o$. The slope of this line is equal to the reciprocal maximum yield, $1/Y_{ox}^{max}$, and the intercept on the Y-axis is equal to the maintenance coefficient, m_o. From Figure 12 it can be concluded that the specific oxygen consumption rate increases almost linearly with the specific growth rate.

The Pirt equation can also be rewritten to a relation between the ratio of the oxygen consumption rate and the growth rate ($1/Y^{ox}$), and the dilution rate (Eqn. 7). In Figure 13 the reciprocal of the yield on oxygen (r_{O2}/r_{CO2}) is plotted against the reciprocal of the dilution rate. According to equation 7 this plot should yield a straight line with a slope equal to the maintenance coefficient, m_o, and an intercept equal to the reciprocal value of the maximum yield, $1/Y_{ox}^{max}$.

From this graph it can be seen that the actual yield depends on the specific growth rate. The values of Y_{ox}^{max} and m_o are 0.054 C-mole/moleO_2 and 0.12 moleO_2/C-mole/h respectively.

Examples on bio-oxidation of pyrite in batch cultures

Figures 14 to 17 show plots of a typical batch culture experiment on the bacterial oxidation of pyrite and illustrate the usefulness of off-gas analysis. 20 g/l pyrite was

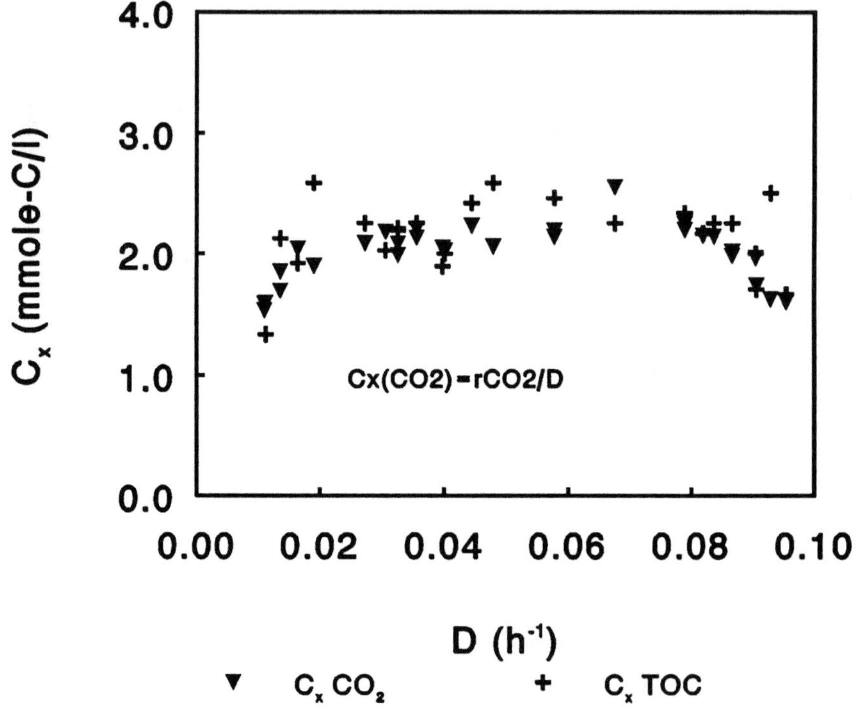

Figure 10: Biomass concentration in a continuous culture on ferrous iron.

inoculated with a small inoculum (about 2%). After 470 hours of the experiment half of the supernatant (without pyrite) was replaced with fresh iron free medium, and new pyrite was added.

In Figure 14 the oxygen and carbon dioxide consumption rates are plotted against time. Both curves show the same pattern, suggesting that the oxygen and carbon dioxide consumption rates are coupled, i.e. pyrite oxidation and growth are coupled. After the addition of fresh pyrite an immediate increase of the oxygen and carbon dioxide consumption was measured.

Figure 15 shows the concentration of biomass as calculated from the carbon dioxide consumption rate. The TOC in pyrite-free supernatant was also measured. It can be seen that there is no significant difference between the total biomass concentration in the fermenter (from CO_2 measurements) and the concentration of free bacteria in the supernatant.

The line in Figure 16 shows the dissolved iron concentration as calculated from the oxygen and carbon dioxide consumption, using the degree of reduction balance on

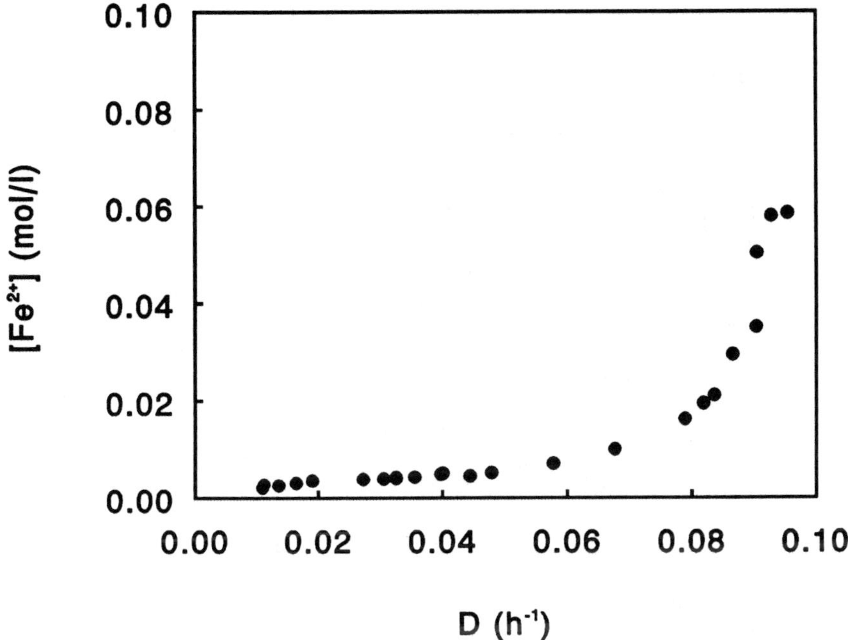

Figure 11: Ferrous iron concentration in continuous culture on ferrous iron.

pyrite. The measured iron concentrations are also plotted. It can be seen that the measured and the calculated values agree. This implies that the conversion of pyrite can be calculated from the oxygen and carbon dioxide consumption. Because the degree of reduction balance was derived from the assumption that pyrite is oxidized to ferric iron and sulphate it can also be concluded that this stoichiometry is correct (in case of the formation of elemental sulphur this degree of reduction balance would not apply).

In Figure 17 the specific oxygen consumption and growth rates are plotted against time.

LIST OF SYMBOLS

C_x	Biomass concentration (molesC/l)
D	Dilution rate (l/l/h)
$[Fe^{2+}]$	Ferrous iron concentration (moles/l)
$[Fe^{3+}]$	Ferric iron concentration (moles/l)
m_o	Maintenance coefficient on O_2 (molesO_2/molesC/s)
q_{O2}	Specific oxygen consumption rate (molesO_2/molesC/s)
$q_{O2,max}$	Maximum specific O_2 consumption rate (molesO_2/molesC/s)
μ	Specific growth rate on Fe^{2+} (h^{-1})
μ_{max}	Maximum specific growth rate on Fe^{2+} (h^{-1})
r_{CO2}	Carbon dioxide consumption rate (molesCO_2/l/s)
r_{O2}	Oxygen consumption rate (molesO_2/l/s)
r_{Fe2+}	Ferrous oxidation rate (molesFe^{2+}/l/s)
r_{FeS2}	Pyrite oxidation rate (molesFe^{3+}/l/s)
r_x	Growth rate on Fe^{2+} (mol C/l/s)
t	time (h)
$Y_{Fe,x}$	Yield of biomass on ferrous iron (molesC/molesFe^{2+})
Y_{ox}	Yield of biomass on oxygen (molesC/molesO_2)
Y_{ox}^{max}	Maximum yield coefficient (molesC/molesO_2)

REFERENCES

Boon,M, G.S.Hansford and J.J.Heijnen (1994), "Recent developments in modelling bio-oxidation kinetics and their implications in practice: Part II, Kinetic modelling of the bio-oxidation of sulphide minerals in terms of the critical sub-processes involved.",Minerals Bioprocessing II, Proceedings of Engineering Foundation Conference, Salt Lake City (July 1994), D.Holmes and R.Smith (Eds.) The Minerals, Metals and Materials Society, Warrendale, PA (in press).

Hansford,G.S., "Gold biohydrometallurgy: Current design and operation of bio-oxidation plants, new research tools and challenges",Minerals Bioprocessing II, Proceedings of Engineering Foundation Conference, Salt Lake City (July 1994), D.Holmes and R.Smith (Eds.) The Minerals, Metals and Materials Society, Warrendale, PA (in press).

Hansford,G.S. and D.M.Miller (1993), "Bio-oxidation of a gold-bearing pyrite-arsenopyrite concentrate". FEMS Microbiology Reviews, **11**, 175-182 .

Hansford,G.S. and A.D.Bailey (1992), "The logistic equation for modelling bio-oxidation kinetics", Minerals Engineering, **5**,1355-1364.

Jones,C.A. and D.P.Kelly, D.P. (1983),"Growth of *Thiobacillus ferrooxidans* on ferrous iron in chemostat culture: influence of product and substrate inhibition" J.Chem.Tech.Biotechnol., **33B**, 241-261.

Lizama,H.M. and I.Suzuki (1989),"Synergistic competitive inhibition of ferrous iron oxidation by *Thiobacillus ferrooxidans* by increasing concentrations of ferric iron and cells" Appl. and Environmental Microbiol., **55**, 2588-2591.

Pinches,A., J.T.Chapman, W.A.M.te Riele and M.van Staden (1988),"The performance of bacterial leach reactors for the pre-oxidation of refractory gold-bearing concentrates", Biohydrometallurgy, Proceedings of the International Symposium, Warwick 1987, Norris,P.R. and D.P.Kelly (Eds)STL, Kew, Surrey, UK.

Pirt,S.J. (1982),"Maintenance energy: A general model for energy limited and energy sufficient growth", Arch.Microbiol., **133**, 300-302.

Roels,J.A. (1983), Energetics and Kinetics in Biotechnology, Elsevier Biomedical Press, Amsterdam.

van Aswegen,P.C. (1993),"Bio-oxidation of refractory gold ores, the GENMIN experience", Proc. Biomine'93, Adelaide, March 1993, Australian Mining Foundation, Adelaide, Australia.

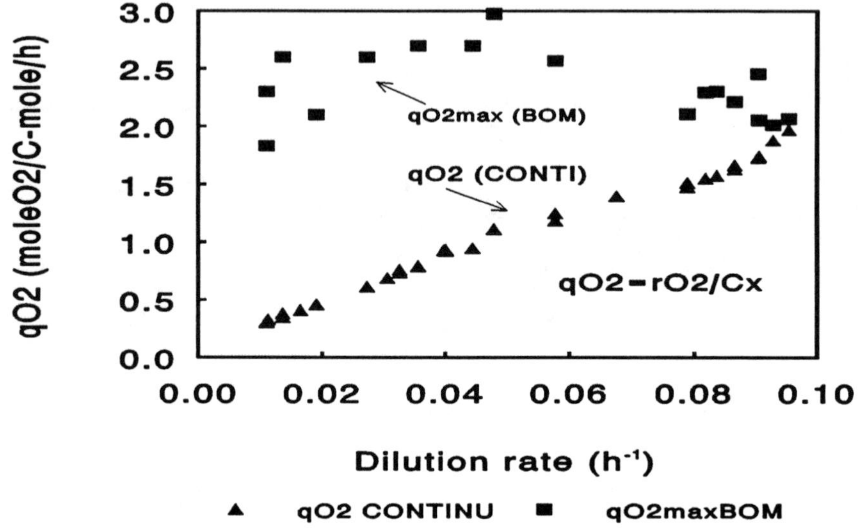

Figure 12: Specific oxygen consumption in a continuous culture on ferrous iron.

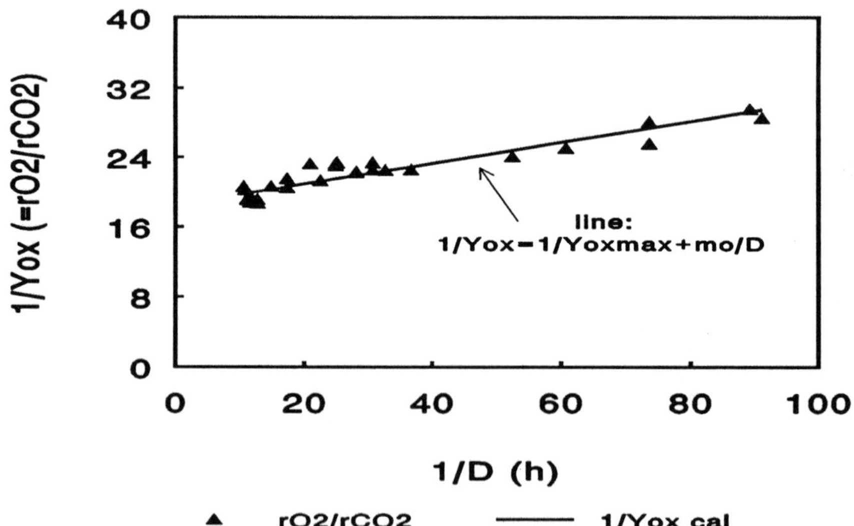

Figure 13: Reciprocal plot of yield on oxygen against dilution rate for a continuous culture on ferrous iron.

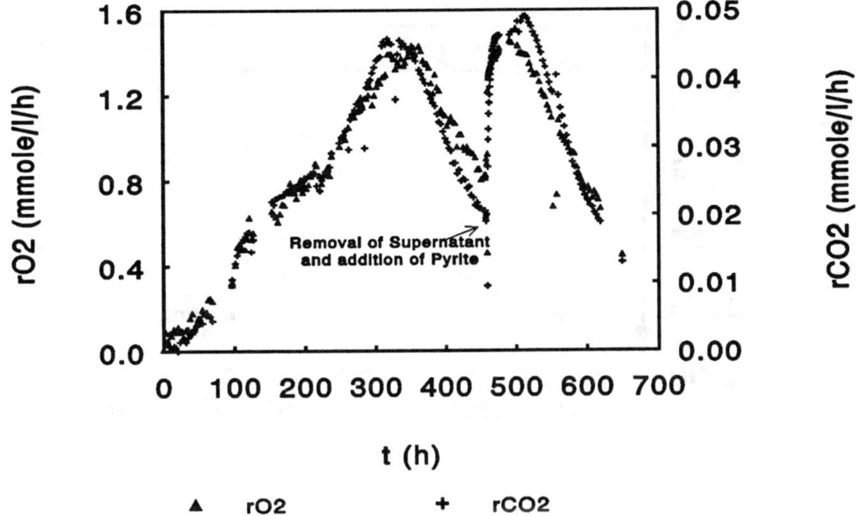

Figure 14: Oxygen and carbon dioxide consumption rates in batch culture on pyrite.

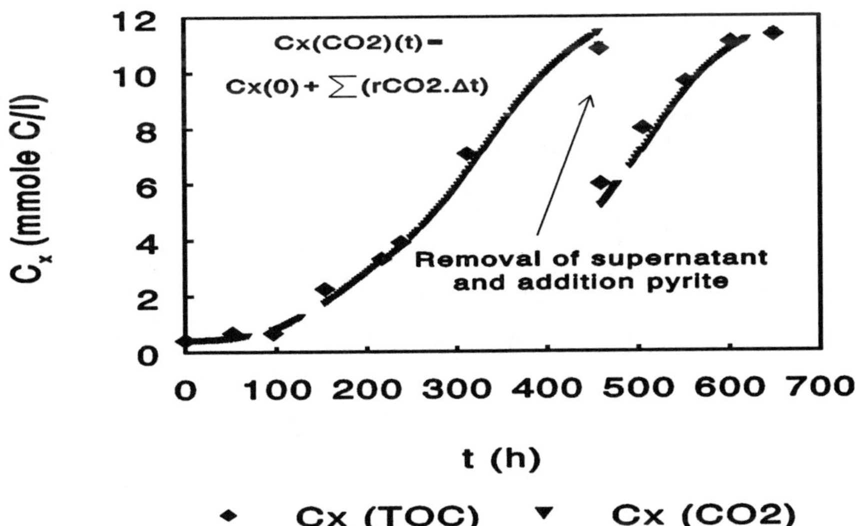

Figure 15: Biomass concentration in a batch culture on pyrite.

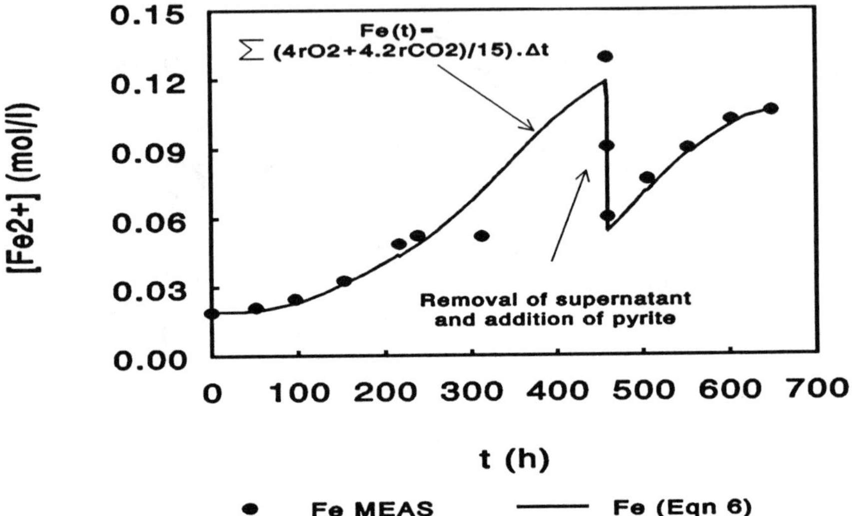

Figure 16: Total dissolved iron concentration versus time in a batch on pyrite calculated from degree of reduction balance and from analysis.

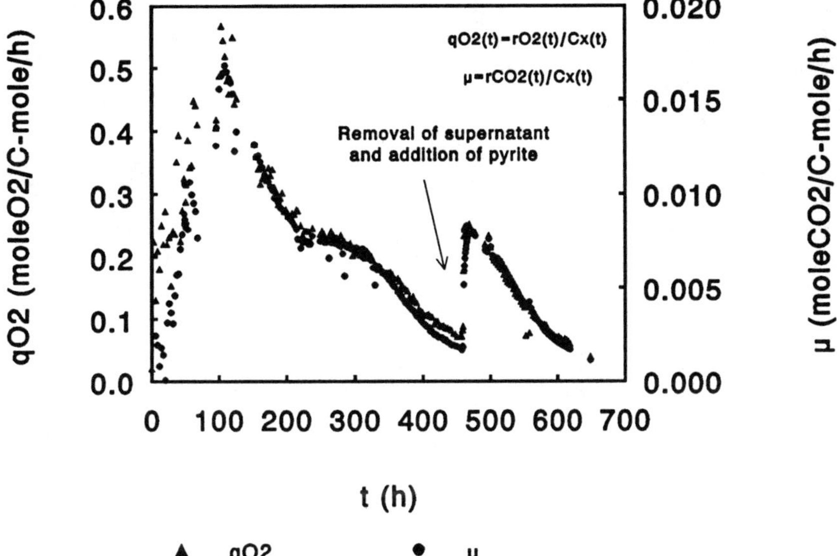

Figure 17: Specific oxygen consumption rate and growth rate in batch culture on pyrite

RECENT DEVELOPMENTS IN MODELLING BIO-OXIDATION KINETICS PART II: KINETIC MODELLING OF THE BIO-OXIDATION OF SULPHIDE MINERALS IN TERMS OF THE CRITICAL SUB-PROCESSES INVOLVED

M. Boon, G.S. Hansford[1] and J.J. Heijnen,
Department of Biochemical Engineering
Kluyver Laboratory of Biotechnology
Julianalaan 67, 2628 BC Delft, The Netherlands

SUMMARY

Kinetic models for bio-oxidation are presented. The bacterial oxidation of ferrous iron is shown to follow competitive inhibition kinetics. Good agreement is obtained between the prediction of the steady state data from continuous culture using off-gas analysis. However when the oxygen utilization rate is measured off-line in the biological oxygen monitor, BOM, different values of the parameters are necessary in order to fit the data. Batch data are not predicted very well by the model. But using values of the parameter, q_{O2}^{max}, from BOM measurements at high ferrous concentrations, it is possible to predict batch performance.

Kinetics based on an indirect mechanism were used to simulate the course of batch bacterial leaching of zinc sulphide in iron sulphate solution. The results of the simulation show that by means of off-gas measurement of oxygen and carbon dioxide utilization rates it should be possible to confirm the indirect mechanism.

A simulation of the bio-oxidation of pyrite was carried out based on a direct mechanism. The results show distinctive patterns for the oxygen and carbon dioxide utilization rates and for the specific oxygen utilization rate and growth rate. Similar trends were found in the data from a batch experiment.

INTRODUCTION

Bio-oxidation has been used for many years for the bioleaching of copper from low-grade sulphide ores and recently for the pretreatment of refractory sulphide and arsenical gold ores (Brierley and Luinstra, 1993) and concentrates (van Aswegen, 1993). Yet surprisingly the mechanism by which these processes occur has not been defined and proven rate expressions do not exist.

---

[1]On sabbatical leave from: Department of Chemical Engineering, University of Cape Town, Rondebosch 7700, South Africa

Mineral Bioprocessing II
Edited by David S. Holmes and Ross W. Smith
The Minerals, Metals & Materials Society, 1995

Most of the kinetic models on bio-oxidation of mineral sulphides presented in the literature are non-mechanistic (Hansford, 1994). A mechanistic kinetic model is one based on the mechanism that is supposed to be involved in the bio-oxidation of the mineral sulphide. Some authors support the *direct* mechanism, assuming that bacteria are capable of directly oxidizing the mineral surface. Others prefer the *indirect* mechanism, claiming that bio-oxidation takes place by means of a chemical reaction of the mineral surface with ferric iron; the role of the bacteria is to oxidize and recycle the ferrous iron thus formed to ferric iron.

Boon and Heijnen (1993) have proposed that the bio-oxidation of sulphide minerals (MeS) can be considered to consist of several sub-processes each of which can be described in terms of a mechanism and related rate expression. These rate expressions describe the rate of the sub-processas a function of the process variables such as, total iron concentration, ferrous to ferric ratio, pH and temperature. Furthermore it is assumed that each of the sub-processes can be examined seperately and that the overall bio-oxidation rate can be predicted from the kinetics of the critical or rate determining sub-processes.

Besides the chemical and/or bacterial oxidation reactions involved in the bio-oxidation process, other sub-processes might also play a role; e.g. oxygen and carbon dioxide transfer from the gas to the liquid phase, formation of precipitate layers such as jarosites on the mineral surface, mass transfer of reactants and products between the bulk liquid and the reaction surface through layers on the surface, attachment of bacteria to the mineral surface, etc.. In industrially used ores there are even more sub-processes such as mass transfer through the gangue, electrochemical reactions between different minerals, etc. which may be important. The possible sub-processes which can occur are shown in Figure 1.

However, in order to examine which sub-process is rate limiting in the overall bio-oxidation rate, each sub-process needs to be identified, and how its rate is affected by process conditions needs to be known. In this way mechanistic models provide a tool to predict what will happen to the overall bio-oxidation rate if certain process conditions (e.g. biomass concentration, particle diameter, ferrous and ferric iron concentration, pH etc.) are changed.

Accordingly, mechanistic modelling of bio-oxidation kinetics aims at the determination of rate limiting sub-processes involved, and at the mathematical description of these sub-processes as a function of the process conditions. This paper will focus on the mechanisms involved in bio-oxidation reactions of sulphide minerals and on the kinetic modelling of these reactions. It will show how the techniques described in Part I (Boon *et al.*, 1994) can be used to examine some of the sub-processes individually and to show

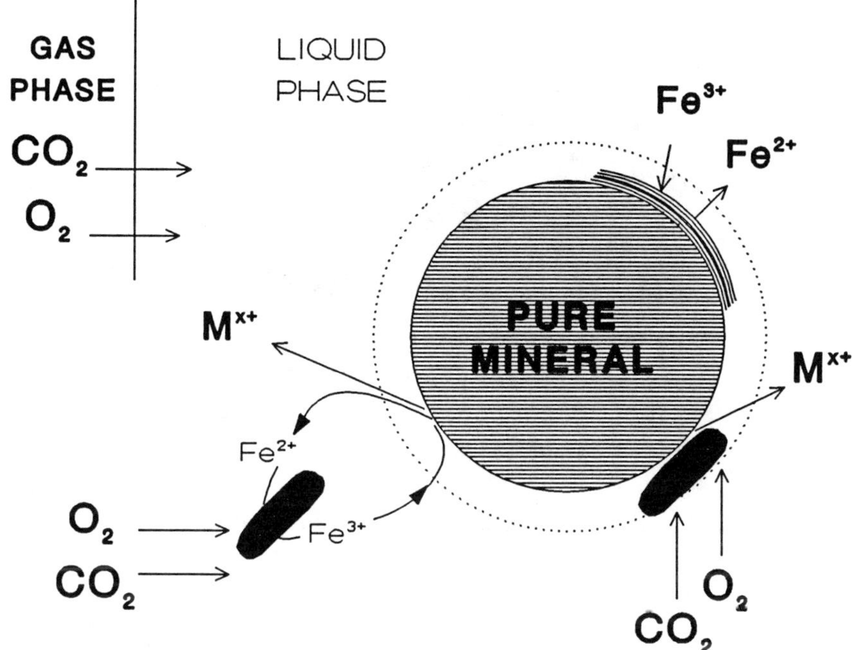

Figure 1 Sub-processes in the bio-oxidation of sulphide minerals.

that there are experimentally measureable differences in the kinetics depending on whether a direct or indirect mechanism applies. It will also show that in examining one sub-process it is assumed that all other rate limiting processes such as, gas-liquid mass transfer of oxygen and carbon dioxide, or diffusion through layers of precipitate are eliminated. Therefore, oxygen and carbon dioxide transfer from the gas to the liquid phase needs to be sufficiently high, and precipitate formation needs to be prevented in the kinetic experiments.

MECHANISMS IN BIO-OXIDATION REACTIONS

Direct versus Indirect Mechanism

The direct and indirect mechanisms for bio-oxidation of sulphide minerals have been presented and examined in detail elsewhere (Boon and Heijnen, 1993). In the direct mechanism, it is assumed that the bacteria associate closely with the sulphide mineral surface, and that direct participation of the bacteria in the oxidation reactions occurs. However, the specific mechanism, enzymatic or electrochemical, by which this occurs has not been described. It is generally assumed that attachment of the bacteria to the mineral surface is required in this mechanism, as shown in Figure 2.

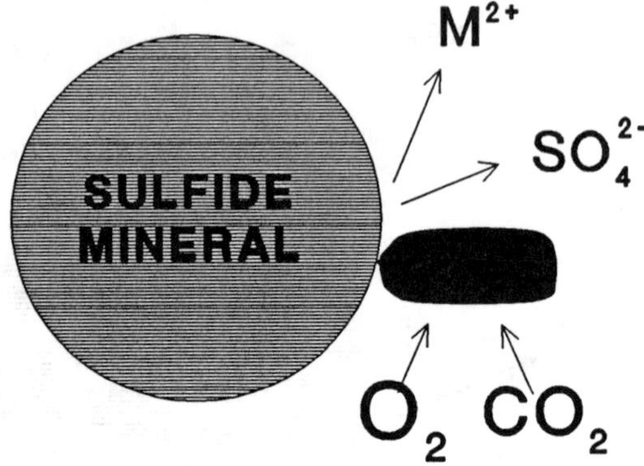

Figure 2: **The direct mechanism of bio-oxidation of sulphide minerals.**

On the other hand, in the indirect mechanism, it is assumed that the sulphide mineral is oxidized chemically by ferric iron leading to the dissolution of the metal cation. In this reaction, ferrous iron and elemental sulphur are produced. The role of the bacteria is to oxidize the ferrous iron to ferric and to oxidize the sulphur to sulphate. This does not require them to be closely associated with the mineral surface. The indirect mechanism is shown in Figure 3.

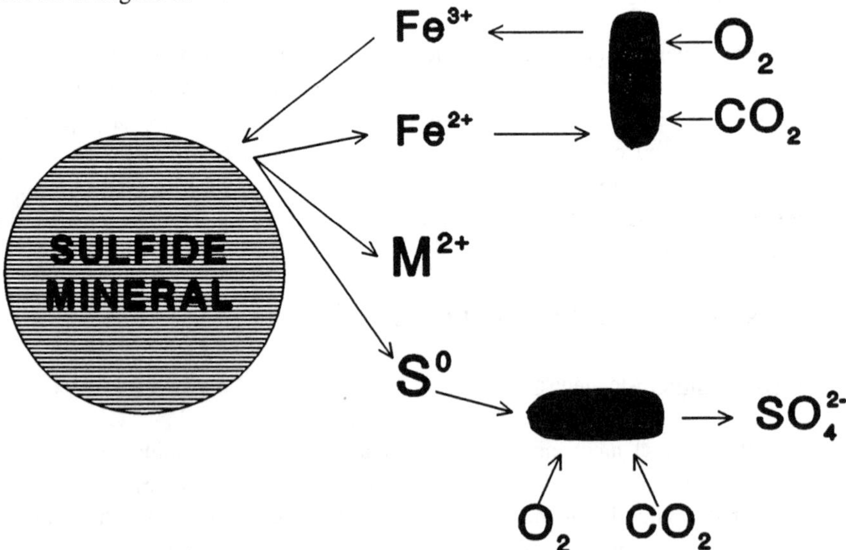

Figure 3: **The indirect mechanism for the bio-oxidation of sulphide minerals.**

Conclusions from Kinetic Data in the Chemical- and Bio-oxidation Literature

In order to distinguish between the direct and indirect mechanisms, the rates of oxidation in the presence of thiobacilli and under sterile conditions have been compared. If the rate under sterile conditions is lower than in the presence of bacteria, it is assumed that a direct mechanism applies. However, it is essential to ensure that the process conditions, particularly the total dissolved iron concentration and the ferric to ferrous ratio are the same in both cases. In many reported cases, the process conditions in the sterile experiment are not the same because ferric iron is consumed, whereas in the bio-oxidation experiment ferric iron is regenerated by bacteria, which will also occur in the case of an indirect mechanism

An extensive examination of kinetic data in the literature on both chemical and bacterial leaching of sphalerite, pyrite and chalcopyrite was performed (Boon and Heijnen, 1993). Specific oxidation rates of pure metal sulphides, k_1 (mole MeS/mole MeS/h) or k_A (MeS/m^2MeS/h), were derived from data in the literature, for both chemical oxidation with ferric iron and bio-oxidation at chosen "standard" conditions ([Fe^{3+}] = 0.01 or 0.1 or 1M, T=30^{0}C and different particle diameters). From this study, it was concluded that the bio-oxidation reaction of pure ZnS (and probably also other $Me^{2+}S^{2-}$ minerals) consists of chemical oxidation with ferric iron and bacterial oxidation of sulphur and of ferrous iron to regenerate ferric iron, whereas the bio-oxidation reaction of $CuFeS_2$ and FeS_2 consists of a direct bacterial attack. Thus, at the same process conditions the bio-oxidation rate of pyrite and chalcopyrite is significantly larger (5 to 20 times) in the presence of bacteria, whereas it is about the same with and without bacteria in the case of $Me^{2+}S^{2-}$.

KINETIC MODEL FOR THE INDIRECT MECHANISM

It is assumed that the bio-oxidation reaction of an indirect mechanism consists of three sub-processes that can be examined and kinetically modelled seperately:

(1) Chemical oxidation of the sulfide mineral with ferric iron. In this work ZnS was taken as a model substrate:

$$ZnS+2Fe^{3+} \rightarrow Zn^{2+}+S^0+2Fe^{2+} \quad (1)$$

(2) Bacterial oxidation of ferrous to ferric iron.

$$2Fe^{2+}+0.5O_2+2H^+ \rightarrow 2Fe^{3+}+H_2O \quad (2)$$

(3) Bacterial oxidation of elemental sulphur.

$$S^0 + 1.5O_2 + H_2O \rightarrow SO_4^{2-} + 2H^+ \tag{3}$$

The modelling of chemical ZnS oxidation with ferric iron is only discussed briefly because excellent work has been done by others. Bacterial oxidation kinetics of ferrous iron will be discussed more extensively. The bacterial oxidation of elemental sulphur will not be discussed. From kinetic data on the chemical oxidation of pure ZnS it was concluded that the sulphur layer formed on the mineral surface does not cause mass transfer limitation (Boon and Heijnen, 1993) and therefore the bacterial oxidation of the chemically formed sulphur should not have an significant effect on the overall bio-oxidation rate.

Chemical Oxidation of ZnS with Ferric Iron

Most authors agree that the chemical oxidation of sphalerite with ferric iron can be described with a simple shrinking core model. The kinetic model derived by Crundwell (1987) describes the chemical oxidation rate of zinc sulphide in ferric sulphate solution as shown below:

$$\frac{d[ZnS]}{dt} = k_{chem} * A_0 * [Fe^{3+}]_{active}^{0.5} * (\frac{[ZnS]_0 - [ZnS]}{[ZnS]_0})^{2/3}) \tag{4}$$

In this equation $[Fe^{3+}]_{active}$ is the concentration of active ferric iron species, being Fe^{3+} and $FeHSO_4^{2+}$, whereas $FeSO_4^+$ is not active. In a sulphate solution the major part of ferric iron exists however as $FeSO_4^+$ ions. Increasing concentrations of ferrous sulphate causes a dramatic decrease of the concentration of active ferric iron species. This effect explains the apparent inhibition of the chemical oxidation rate of zinc sulphide caused by ferrous iron as observed by Verbaan and Crundwell (1986).

Competitive Inhibition of Bio-oxidation of Ferrous Iron

In the literature (Lacey and Lawson, 1970; Kelly and Jones, 1978 and Jones and Kelly, 1983) it is assumed that ferric iron inhibits the bacterial oxidation of ferrous iron according to competitive inhibition kinetics. This can be written in terms of the specific oxygen consumption rate as:

$$-q_{O_2} = \frac{q_{O_2,\max}}{1 + \frac{K_s}{[Fe^{2+}]} + \frac{K_s}{K_i} * \frac{[Fe^{3+}]}{[Fe^{2+}]}} \tag{5}$$

In this equation $q_{O2,max}$ is the maximum specific oxygen consumption rate, indicating the specific oxygen consumption rate at an excess of ferrous iron and no inhibition effects from ferric iron. K_s is the saturation constant for ferrous iron. K_i is the inhibition constant of ferric iron. As shown in Part I (Boon *et al., 1994)* the energy generated from the bio-oxidation of ferrous iron is used by the bacteria for both growth and maintenance. This is described in the Pirt equation which can be rewritten as a relationship between the specific oxygen consumption rate and the specific growth rate:

$$q_{O_2} = \frac{\mu}{Y_{ox}^{max}} + m_o \tag{6}$$

and, therefore, the following relation between q_{O2max} and μ_{max} holds:

$$q_{O_2,max} = \frac{\mu_{max}}{Y_{ox}^{max}} + m_o \tag{7}$$

from which it can be shown that for the inhibition of the specific growth rate:

$$\mu = \frac{\mu_{max} + m_o * Y_{ox}^{max}}{1 + \frac{K_s}{[Fe^{2+}]} + \frac{K_s}{K_i} * \frac{[Fe^{3+}]}{[Fe^{2+}]}} - m_o * Y_{ox}^{max} \tag{8}$$

In the literature on bacterial ferrous iron kinetics different experimental systems have been used to determine model parameters (e.g. continuous cultures, batch cultures and respiration experiments). Table I shows which parameters can be obtained from each experimental method. However, to date no attempt has been made to examine whether the kinetic parameters from these different systems and methods agree.

Rate Limiting Regimes

When the kinetic models together with the stoichiometric and kinetic parameters on the two sub-processes are available, then it is possible to determine whether the overall bio-oxidation kinetics of the metal sulphide can be described by combination of these sub-processes. This requires that there are no mass transfer limitations or iron sulphate precipitate formation. It is then possible to determine which factors control the overall bio-oxidation rate; chemical reaction or bacterial activity.

Table I	Stoichiometric Parameters		Kinetic Parameters		Others
	Y_{sx}	Y_{ox}	μ	q_{O2}	
Continuous culture	Y_{sx}^{max}, m_s	Y_{ox}^{max}, m_o	μ_{max}, K_s, K_i	K_s, K_i	
Batch culture	Y_{sx}^{max}, m_s	Y_{ox}^{max}, m_o	μ_{max}, K_s, K_i	$q_{O2,max}$, K_s, K_i	
BOM measurement				$q_{O2,max}$, K_s, K_i	K_{O2}, pH effect

KINETIC MODEL FOR THE DIRECT MECHANISM

In the literature kinetic models on direct oxidation of a sulphide mineral have been proposed. In these models it was assumed that only attached bacteria oxidize the mineral surface. If this is the case, then the bacterial pyrite oxidation rate is determined by three numbers: the concentration of attached bacteria, (C-mole/m^2_{pyrite}), the specific growth rate of bacteria on pyrite, μ_{FeS2}, and the yield of bacteria on pyrite, $Y_{FeS2,x}$ (C-mole/moleFeS_2):

$$r_{FeS_2} = \frac{\mu_{FeS_2}}{Y_{FeS_2}} . C_x^* . a_{FeS_2} \qquad (9)$$

In this equation a_{FeS2} is the pyrite surface per volume (m^2_{FeS2}/m^3_{slurry}).

<u>Rate Limiting Regimes</u>

If this kinetic model holds and the kinetic and the parameters in the model are determined, then it can be examined which factors determine the bio-oxidation rate of pyrite: would more bacteria or would smaller size fraction improve the rate.

EXPERIMENTAL METHODS

<u>Bacterial Culture</u>

Kinetic experiments on ferrous iron were performed with a pure strain of *Thiobacillus ferrooxidans* in ferrous sulphate solution at different concentrations. For the kinetic experiments on pyrite, a pyrite concentrate, with particle diameters of -75+53 μm, from Prieska Copper Mine, South Africa and a wild culture of thiobacilli from the Gamsberg deposit, Namaqualand, South Africa were used. The equipment as described in Part I was used for all experiments.

Continuous Culture Experiments with Ferrous Iron

The continuous culture experiments on ferrous iron were performed as follows. For each dilution rate at least three times the residence time (1/D) was allowed to elapse before steady state was attained. A steady state was assumed when the oxygen and carbon dioxide consumption rates and the ferrous iron concentration are constant. The concentrations of oxygen and carbon dioxide in air and in the off-gas (on line), and the concentrations of total and ferrous iron (off line, with the colorometric ortho-phenantroline method, according to ASTM D1068), and the total organic carbon concentration in the influent and effluent (TOC, off line), were measured over two to three days. Also BOM measurements were performed to determine the maximum specific oxygen consumption rate, $q_{O2,max}$. For these BOM measurements samples from the continuous culture in steady state were diluted with an excess of ferrous iron solution at the same total iron concentration as the continuous culture suspension. Continuous culture experiments were performed at total iron concentrations of 0.05, 0.11, 0.16, 0.21, 0.26 and 0.36 mole/l, and dilution rates between 0.01 and 0.09 h^{-1}.

Batch Culture Experiments with Ferrous Iron

The procedure of batch culture experiments with ferrous iron was as follows. A ferrous iron solution with 0.21 mole ferrous iron per litre was prepared at pH 1.8-1.9. The inoculum was taken from the continuous culture in steady state, thus having a defined initial concentration of biomass and specific growth rate in the batch culture. Oxygen and carbon dioxide concentrations in air and the off-gas were measured on line every 20 minutes. A TOC analysis was only performed at the end of an experiment. Ferrous iron concentrations were analyzed every hour in the initial phase and at intervals decreasing from 30 to 8 minutes in the later and end phase. The pH of every sample was measured and the pH in the batch culture was controlled by hand between 1.8 and 1.9 by addition of 2N H_2SO_4. BOM measurements were performed with samples from the batch culture to determine the maximum specific oxygen consumption rate of the cells, $q_{O2,max}$.

Biological Oxygen Monitor Experiments with Ferrous and Ferric Iron

BOM experiments were performed to determine the specific oxygen consumption rate of the bacteria at different ferric and ferrous iron concentrations. A BOM experiment is performed as follows. Cell suspension from the continuous culture in steady state (thus having a known biomass concentration) was used. For each BOM experiment 14 samples of cell suspension at different ferrous and ferric iron concentrations and ratios were prepared. These samples were derived by diluting the cell suspension with iron free medium, ferrous and ferric iron solutions at pH 1.8-1.9. The decrease rate of the dissolved oxygen activity in each sample is measured in the BOM.

Batch Culture Experiments on Pyrite

Batch culture experiments on pyrite were performed as follows. Iron free medium (pH 1.6-1.7) and pyrite (20 - 60 gram/l) were inoculated with Gamsberg inoculum. Oxygen and carbon dioxide concentrations in air and in the off-gas were measured on line every hour. Total iron, pH and TOC in pyrite free cell suspension was measured once or twice a day. The pH in the batch culture was controlled hand wise at 1.6-1.7 by addition of 2 N NaOH.

RESULTS and DISCUSSION

The steady state results of continuous culture on ferrous iron at a number of feed concentrations and dilution rates will be discussed. Some of the results from the 12 gl^{-1} feed concentration have been presented in Part I (Boon *et al.*,1994). In Figure 10 of that paper it can be seen that the oxygen and carbon dioxide utilization rates, r_{O2} and r_{CO2} increase linearly with dilution rate up to about D = 0.08 h^{-1} and then drop off due to wash-out. In Figure 11 of Part I, the substrate ferrous iron concentration is plotted against dilution rate and shows typical chemostat behaviour. The substrate ferrous iron concentration is plotted versus dilution rate for all the feed concentrations in Figure 4. In order to test competitive inhibition kinetics use is made of Equation 5 for specific oxygen utilization rate, q_{O2}. This is shown in Figures 5 and 6 where it can be seen that there is fairly good agreement using the following values of the parameters: q_{O2}^{max} = 2.2 mole$(moleC)^{-1}h^{-1}$ and K_s/K_I = 0.03.

Oxygen utilization rate measurements were also made in the biological oxygen monitor with samples taken from the continuous culture at steady state ($[Fe^{2+}]_{feed}$ = 12 g/l, D = 0.036 h^{-1}) and the $[Fe^{3+}]/[Fe^{2+}]$ ratio adjusted to that shown. The results are plotted in Figure 7 and 8 where it can be seen that the competitive inhibition kinetics, Equation 5, fit the data, but give slightly different values for the kinetic parameters q_{O2}^{max} = 2.85 mole$(moleC)^{-1}h^{-1}$ and K_s/K_I = 0.10).
The difference is probably due to the fact that the bacteria need time to adjust to the changed ferric/ferrous ratio of the BOM than that used in the continuous culture.

experiments on pyrite, a pyrite concentrate, with particle diameters of -75+53 μm, from Prieska Copper Mine, South Africa and a wild culture of thiobacilli from the Gamsberg deposit, Namaqualand, South Africa were used. The equipment as described in Part I was used for all experiments.

Continuous Culture Experiments with Ferrous Iron

The continuous culture experiments on ferrous iron were performed as follows. For each dilution rate at least three times the residence time (1/D) was allowed to elapse before steady state was attained. A steady state was assumed when the oxygen and carbon dioxide consumption rates and the ferrous iron concentration are constant. The concentrations of oxygen and carbon dioxide in air and in the off-gas (on line), and the concentrations of total and ferrous iron (off line, with the colorometric ortho-phenantroline method, according to ASTM D1068), and the total organic carbon concentration in the influent and effluent (TOC, off line), were measured over two to three days. Also BOM measurements were performed to determine the maximum specific oxygen consumption rate, $q_{O2,max}$. For these BOM measurements samples from the continuous culture in steady state were diluted with an excess of ferrous iron solution at the same total iron concentration as the continuous culture suspension. Continuous culture experiments were performed at total iron concentrations of 0.05, 0.11, 0.16, 0.21, 0.26 and 0.36 mole/l, and dilution rates between 0.01 and 0.09 h^{-1}.

Batch Culture Experiments with Ferrous Iron

The procedure of batch culture experiments with ferrous iron was as follows. A ferrous iron solution with 0.21 mole ferrous iron per litre was prepared at pH 1.8-1.9. The inoculum was taken from the continuous culture in steady state, thus having a defined initial concentration of biomass and specific growth rate in the batch culture. Oxygen and carbon dioxide concentrations in air and the off-gas were measured on line every 20 minutes. A TOC analysis was only performed at the end of an experiment. Ferrous iron concentrations were analyzed every hour in the initial phase and at intervals decreasing from 30 to 8 minutes in the later and end phase. The pH of every sample was measured and the pH in the batch culture was controlled by hand between 1.8 and 1.9 by addition of 2N H_2SO_4. BOM measurements were performed with samples from the batch culture to determine the maximum specific oxygen consumption rate of the cells, $q_{O2,max}$.

Biological Oxygen Monitor Experiments with Ferrous and Ferric Iron

BOM experiments were performed to determine the specific oxygen consumption rate of the bacteria at different ferric and ferrous iron concentrations. A BOM experiment is performed as follows. Cell suspension from the continuous culture in steady state (thus

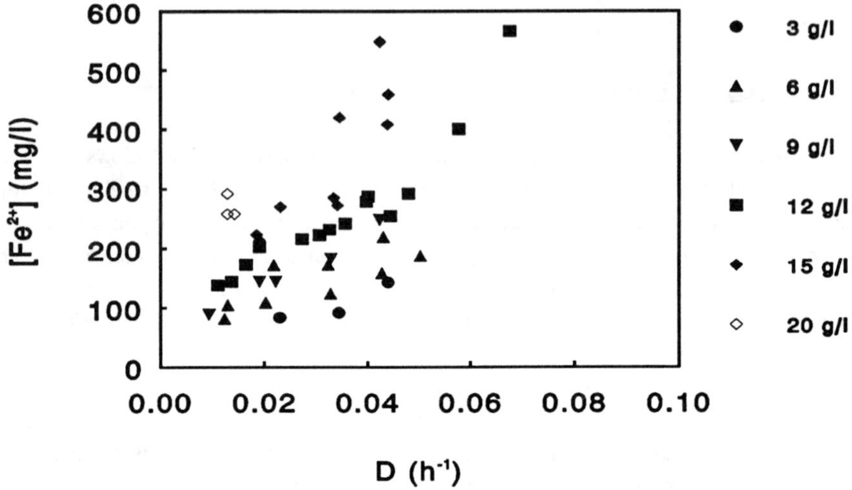

Figure 4: Ferrous iron concentration as a function of dilution rate in continuous on ferrous iron.

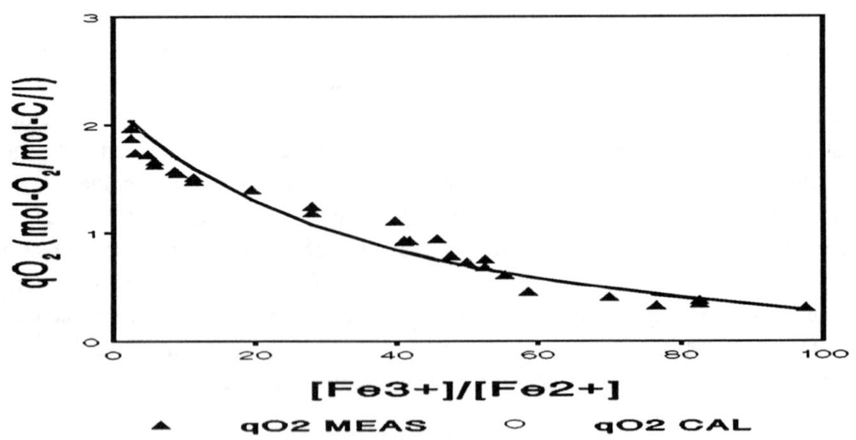

Figure 5: Specific oxygen utilization rate as a function of ferric:ferrous ratio in continuous culture on ferrous iron.

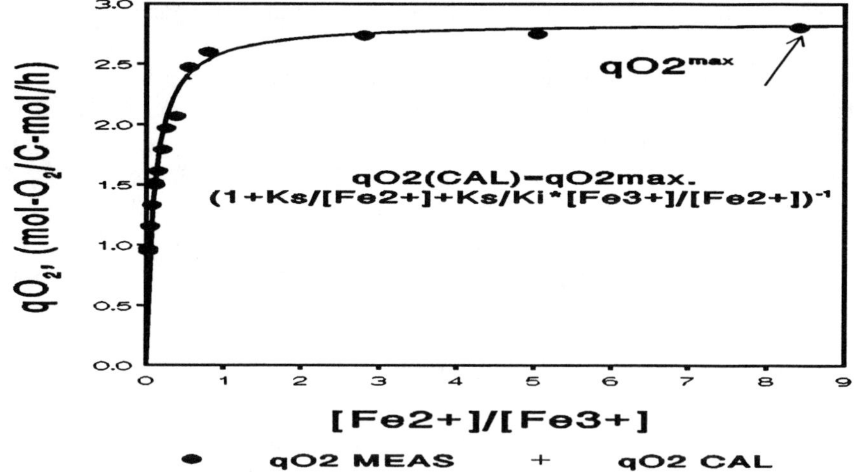

Figure 6: Oxygen utilization rates from BOM measurements in continuous culture on ferrous iron

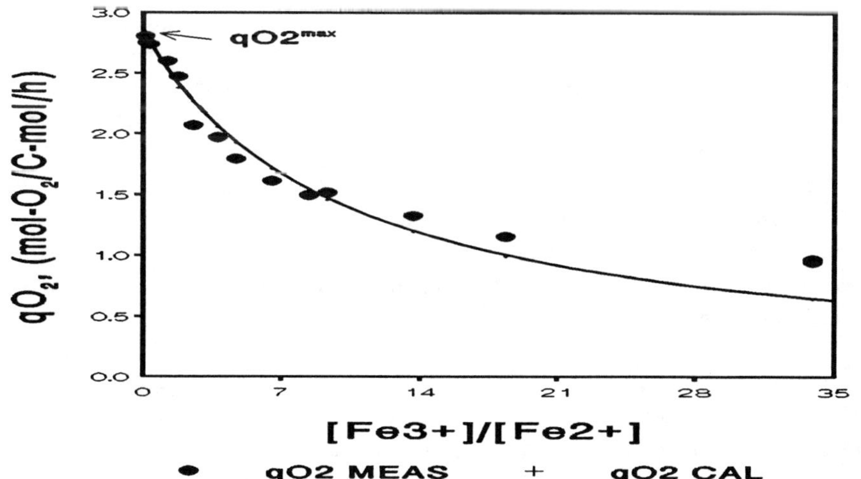

Figure 7: Oxygen utilization rates from BOM measurements in continuous culture on ferrous iron.

This adjustment period was also observed in cells taken from the continuous culture and grown in batch culture. This has been shown in Figure 8 of Part I (Boon *et al.*,1994). From Figure 9 where q_{O2} measured by off-gas analysis in the batch is plotted against ferric/ferrous ratio, it appears that competitive inhibition kinetics, Equation 5 are a good fit of the results. However, when plotted against the reciprocal, ferrous/ferric ratio, in Figure 10, it can be seen that the fit is not very good using a constant value for q_{O2}^{max} of 2.2 mole(moleC)$^{-1}$h^{-1} and K_s/K_I = 0.05. BOM measurements were made with samples from the batch and q_{O2} values were calculated using these q_{O2}^{max} values and K_s/K_I = 0.05. These are also plotted in Figure 10 and show better agreement with the q_{O2} values measured directly in the batch reactor.

IMPLICATIONS OF MECHANISTIC MODELS

It is of interest to see what differences are predicted for the batch bio-oxidation of solid sulphides by the direct and indirect mechanisms. What is attempted here is to show by simulations using models based on the direct and indirect mechanisms is that there are differences in oxygen utilization rates that can be detected from the off-gas measurements described in Part I.

An example of a system where the indirect mechanism is thought to apply is the bioleaching of zinc sulphide as described in Equations 4 and 5. Results of a simulation using PSIe (van den Bosch *et al.*, 1990) are shown in Figure 11. It can be seen that there is initially rapid growth and consumption of ferrous iron and an increase in oxygen and carbon dioxide consumption until all of the ferrous iron has been used up. During this time, there is only a small amount of zinc leached. From about 17 hours, when the ferrous iron has been converted to the ferric form oxygen and growth rate drops rapidly and the zinc leach rate increases. Oxygen consumption appears to be used to maintain a low ferrous and high ferric iron concentration. From the simulation results it can be seen that r_{O2} and r_{CO2} are the most sensitive parameters and as shown in Part I they can be measured using off-gas analysis.

As an example of the direct mechanism, the bio-oxidation of pyrite is proposed. Using Equations 6 and 9, a batch bio-oxidation of pyrite was simulated. The results are shown in Figure 12. Here, because of the direct mechanism, a different behaviour is observed. The concentration of active bacteria, those attached to the pyrite surface, starts at a value close to its maximum and after a small increase stays at that value per unit surface. At the same time, due to oxidation according to an assumed shrinking particle model, the volumetric concentration of active attached bacteria decreases. This can also be seen by the steady decrease in r_{O2} and r_{CO2}. Because the pyrite surface is saturated, bacterial growth leads to an increase in the total volumetric concentration of bacteria. Most of these are not attached to the pyrite substrate and are therefore not active. The growth rate, μ, and the specific oxygen consumption rate, q_{O2}, are calculated by dividing the rates by the total, free and attached, bacterial concentration and exhibit a rapid decrease from the start of the batch. Once again r_{O2} and r_{CO2} and the specific rates show distinctive patterns which can be measured using off-gas analysis.

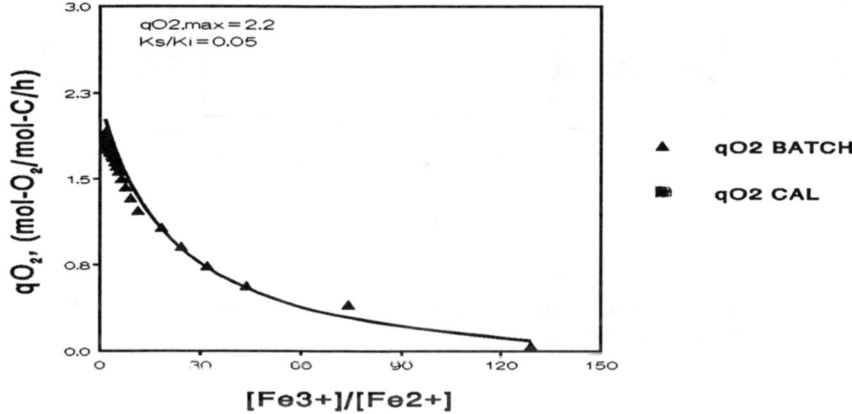

Figure 8: **Specific oxygen utilization rates from off-gas analysis in batch culture on ferrous iron as a function of ferric:ferrous ratio.**

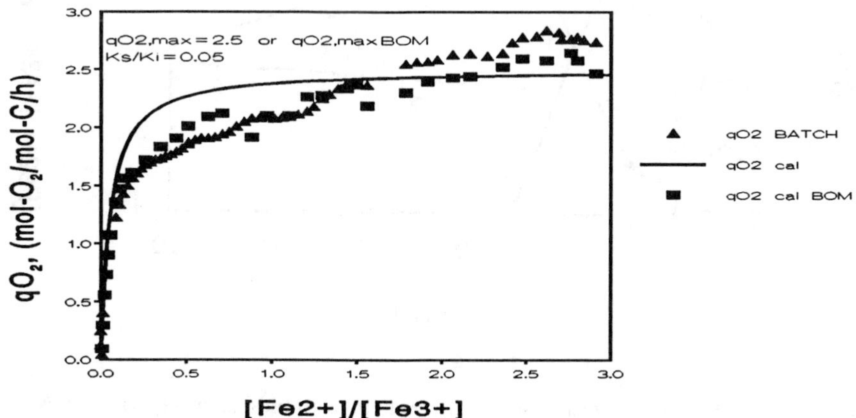

Figure 9: **Specific oxygen utilization rates from off-gas measurements in batch culture on ferrous iron as a function of ferrous:ferric ratio.**

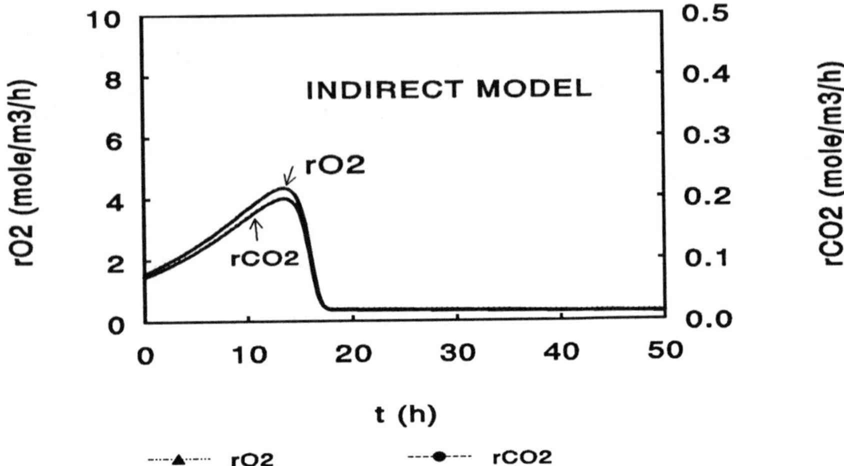

Figure 10: Simulation of the bio-oxidation of zinc sulphide using equations based on the indirect mechanism.

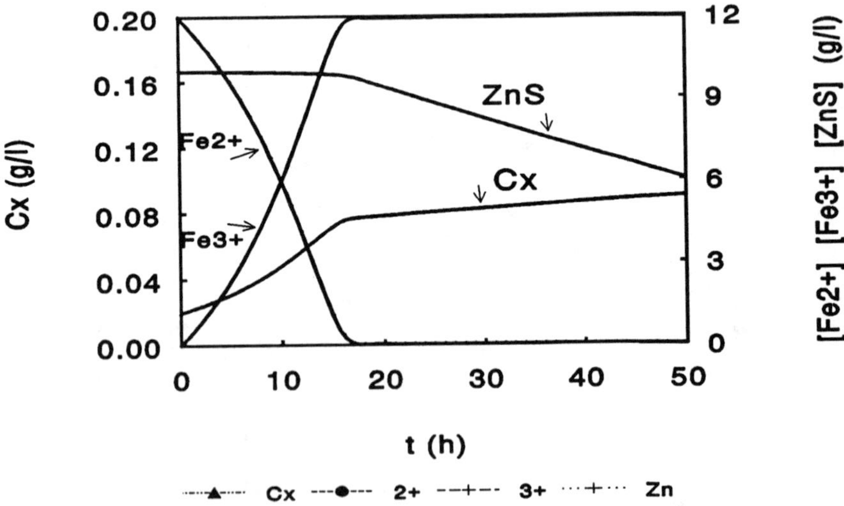

Figure 11: Simulation of the bio-oxidation of zinc sulphide using equations based on the indirect mechanism

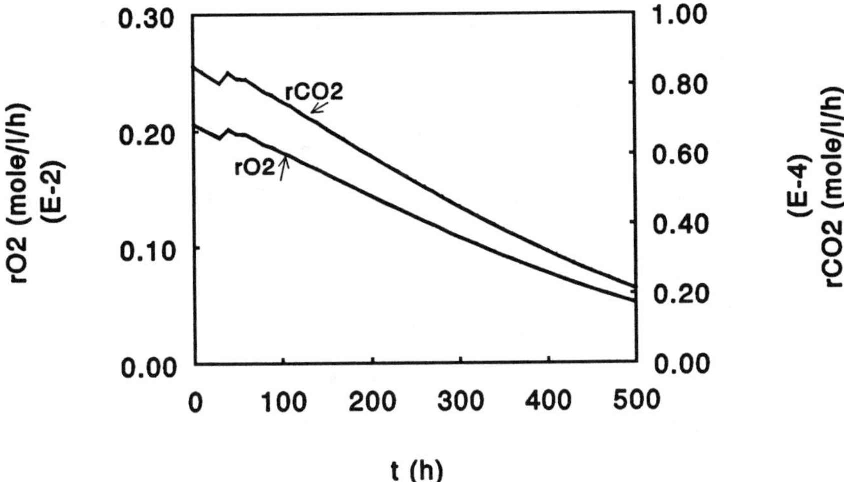

Figure 12: Simulation of the batch bio-oxidation of pyrite using equations based on the direct mechanism.

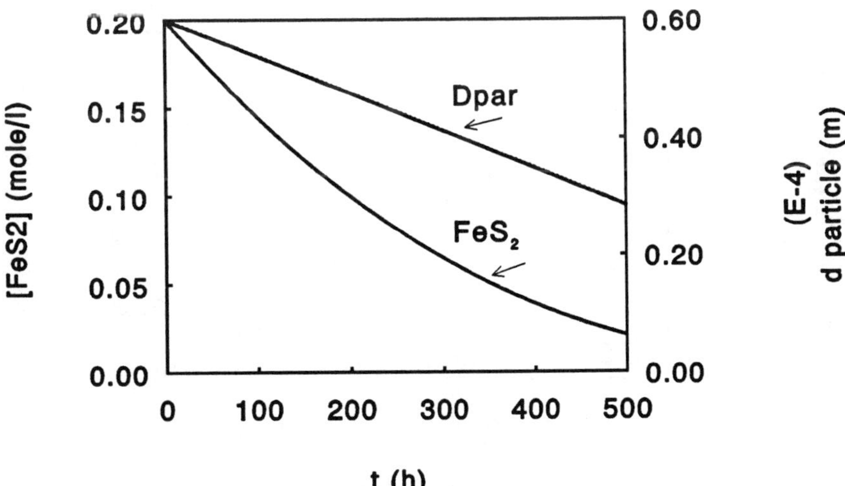

Figure 13: Simulation of batch bio-oxidation of pyrite using equations based on the direct mechanism.

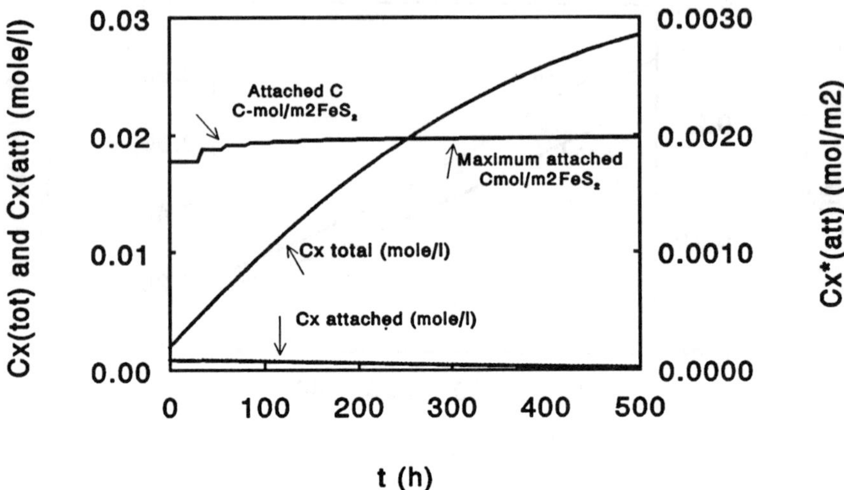

Figure 14: Simulation of the batch bio-oxidation of pyrite using equations based on the direct mechanism.

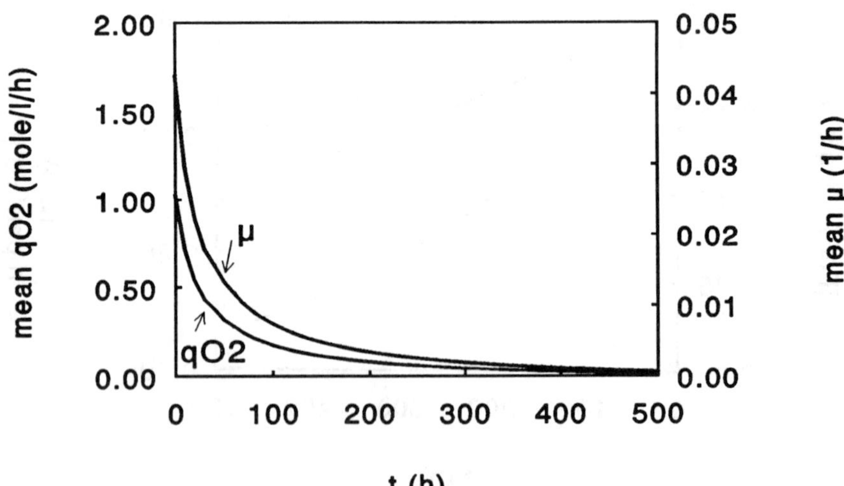

Figure 15: Simulation of the batch bio-oxidation of pyrite using equations based on the direct mechanism.

LIST OF SYMBOLS

A_0	Initial BET surface of the mineral surface in the slurry (m^2/m^3_{slurry})
a_{FeS2}	Pyrite surface per volume (m^2_{FeS2}/m^3_{slurry})
$C_x^{\bullet}$	Concentration of biomass on surface (C-mole/m^2)
D	Dilution rate (l/l/h)
$[Fe^{2+}]$	Ferrous iron concentration (moles/l)
$[Fe^{3+}]$	Ferric iron concentration (moles/l)
$[Fe^{3+}]_{active}$	Concentration of active ferric iron species (only Fe^{3+} and $FeHSO_{42+}$)
k_{chem}	Rate constant in chemical oxidation reaction ($(moles.m^3)^{1/2}/m^2s$)
k_1	First order rate constant in concentration (moles MeS/moles MeS/h)
k_A	First order rate constant in mineral surface (moles MeS/m^2 MeS/h)
K_s	Monod constant (molesFe^{2+}/l)
K_i	Inhibition constant (molesFe^{3+}/l)
m_s	Maintenance coefficient on substrate (moles substrate/molesC/s)
m_o	Maintenance coefficient on O_2 (molesO_2/molesC/s)
q_{O2}	Specific oxygen consumption rate (molesO_2/molesC/s)
$q_{O2,max}$	Maximum specific O_2 consumption rate (molesO_2/molesC/s)
μ	Specific growth rate on (h^{-1})
μ_{max}	Maximum specific growth rate on (h^{-1})
r_{CO2}	Carbon dioxide consumption rate (molesCO_2/l/s)
r_{O2}	Oxygen consumption rate (molesO_2/l/s)
r_{FeS2}	Pyrite oxidation rate (molesFeS_2/l/s)
t	time (h)
T	Temperature, (K)
$Y_{FeS2,x}$	Yield of biomass on ferrous iron (molesC/molesFe^{2+})
Y_{ox}	Yield of biomass on oxygen (molesC/molesO_2)
$Y_{FeS2,x}^{max}$	Maximum yield coefficient on ferrous iron (molesC/molesFe^{2+})
Y_{ox}^{max}	Maximum yield coefficient (molesC/molesO_2)

ACKNOWLEDGEMENTS

The authors acknowledge Cor Ras, Miranda Snijder, Christian Thone, Heleen Brasser and Frank Crundwell who contributed to this work.

REFERENCES

Boon,M., J.J.Heijnen and G.S.Hansford (1994), "Recent developments in modelling bio-oxidation kinetics and their implications in practice: Part I, Measurement methods.",Minerals Bioprocessing II, Proceedings of Engineering Foundation Conference, Salt Lake City (July 1994), D.Holmes and R.Smith (Eds.) The Minerals, Metals and Materials Society, Warrendale, PA (in press).

Boon M. and J.J.Heijnen (1993), "Mechanisms and rate limiting steps in bioleaching of sphalerite, chalcopyrite and pyrite with *Thiobacillus ferrooxidans*", Biohydrometallurgical Technologies Vol I, 469-478, A.E.Torma, J.E.Wey and V.L.Lakshmanan (Eds.) The Minerals, Metals and Materials Society, Warrendale, PA (1993).Proc. Int. Biohydrometallurgy Symposium, Jackson, Wyoming, USA, (August 1993), 217-236.

Crundwell F.K. (1987), "Kinetics and mechanisms of the oxidative dissolution of a zinc sulphide concentrate in ferric sulphate solutions", Hydrometallurgy, **19**, 227-242.

Hansford,G.S. (1994), "Gold biohydrometallurgy: Current design and operation of bio-oxidation plants, new research tools and challenges",Minerals Bioprocessing II, Processing of Engineering Foundation Conference, Salt Lake City (July 1994), D.Holmes and R.Smith (Eds.) The Minerals, Metals and Materials Society, Warrendale, PA (in press).

Jones C.A. and D.P.Kelly (1983), "Growth of *Thiobacillus ferrooxidans* on ferrous iron in chemostat culture: influence of product and substrate inhibition", J.Chem.Tech.Biotechnol. **33B**, 241-261.

Kelly D.P. and C.A.Jones (1978), "Factors affecting metabolism and ferrous iron oxidation in suspensions and batch cultures of *Thiobacillus ferrooxidans*: relevance to ferric iron leach regeneration", Metallurgical Applications of Bacterial Leaching and Related Microbiological Phenomena, L.E. Murr, A.E.Torma; J.A. Brierley (Eds), Academic Press, New York, 19-44.

Lacey D.T. and F.Lawson (1970), "Kinetics of the liquid-phase oxidation of acid ferrous sulphate by the bacterium *Thiobacillus ferroxidans*", Biotechnol.Bioeng., **12**, 29-50.

van Aswegen,P.C. (1993),"Bio-oxidation of refractory gold ores, the GENMIN experience", Proc. Biomine'93, Adelaide, March 1993, Australian Mining Foundation, Adelaide, Australia.

van den Bosch,P.P.J., H.Butler and A.R.M.Soeterboek (1990), Modelling and Simulation with PSI/e. BOZA Automatiseering BV and TUDelft, Pijnacker and Delft, The Netherlands.

Verbaan B. and F.K.Crundwell (1986), "An electrochemical model for the leaching of a sphalerite (ZnS) concentrate", Hydrometallurgy, **16**, 345-359.

II.

GENETIC STUDIES OF MICROORGANISMS OF INTEREST IN MINERALS BIOPROCESSING

GENETIC AND PHENOTYPIC INSTABILITY IN *THIOBACILLUS FERROOXIDANS* AND WHY THESE PHENOMENA ARE IMPORTANT TO UNDERSTAND.

David S. Holmes, Depto. Biología, Facultad de Ciencias, Universidad de Chile,

Las Palmeras 3425, Santiago, Chile.

ABSTRACT

A description of the physical characteristics of insertion sequences and their role in altering the genetic and phenotypic properties of bacteria is provided for the non-specialist. Special reference is made to the insertion sequences present in *Thiobacillus ferrooxidans* and a discussion is presented as to why it is important, both for fundamental and practical reasons, to understand their structure and function.

Mineral Bioprocessing II
Edited by David S. Holmes and Ross W. Smith
The Minerals, Metals & Materials Society, 1995

INTRODUCTION

Thiobacillus ferrooxidans in an autotrophic, chemolithotrophic, acidophilic, Gram negative bacterium that participates in the bioleaching of minerals from ores. It derives the energy required for growth from the oxidation of ferrous ions or reduced sulfur compounds (7).

The majority of the genetic information in the chromosome of a typical bacterium is thought to consist of genes that encode the structural proteins and enzymes that constitute the bacterium and permit it to grow and reproduce. Normally, these genes are quite stable, not only with regard to their DNA sequence but also in the position that they occupy in the chromosome. Only rarely do such genes undergo mutation or change position in the chromosome by rearrangement.

However, an increasing number of bacteria are being described that contain, in addition to the typical array of genes, DNA sequences, called insertion sequences and transposons, that can move around the genome (4). In some instances such sequences may promote genetic rearrangements and mutations of nearby genes with a frequency that can be far above the normal level for such events.

The bulk of this paper will describe the characteristics of insertion sequences with special reference to those in *T. ferrooxidans*. Transposons have not been described in *T. ferrooxidans* but because of their potential significance in the genetic engineering of these microorganisms they will be discussed towards the end of this paper.

BACTERIAL INSERTION SEQUENCES

Figure 1 shows the characteristic features of a typical bacterial insertion sequence (See 3,4 for review). It is normally about 1 to 3 kilobases (kb) in length and contains a gene that encodes a transposase enzyme that is involved in the ability of the insertion sequence to undergo movement (transposition) within the chromosome, or between the chromosome and any plasmid that the organism might have. Also, the typical insertion sequence contains, at one end, a short DNA sequence of about 20 nucleotides that is repeated, almost perfectly, at the other end, but in reverse order. These are called " terminal inverted repeats" or TIRs. They have the ability to form a folded structure by virtue of their reverse complementarity. Such a structure is thought to form when the insertion sequence enters or leaves the chromosome. When an insertion sequence enters into a new region of the chromosome, called the target site, it duplicates part of the chromosome (normally about 5-8 base pairs). This is called a target site duplication or "TSD" (see fig. 1).

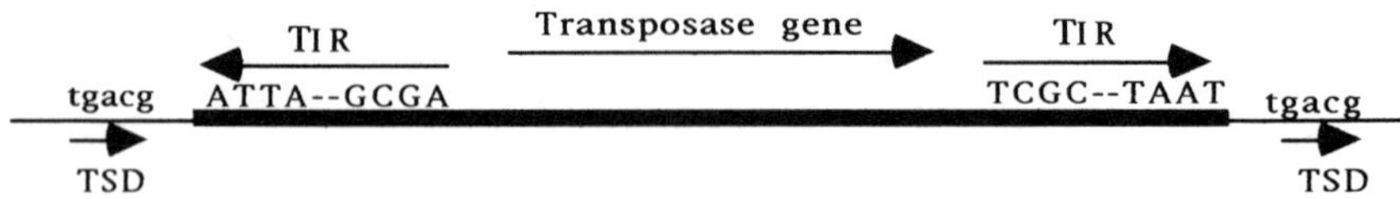

FIGURE 1. A typical bacterial insertion sequence. TSD=target site duplication; TIR= terminal inverted repeat; The thin line represents the chromosomal DNA and the thick line represents the insertion sequence. Only one of the two DNA strands is shown.

At times, usually during replication of the bacterial chromosome, a copy of the insertion sequence can be made and the new copy can move into another region of the chromosome, leaving the original insertion sequence in place. Occasionally, an insertion sequence excises from the chromosome and can either be lost or, at times, it can reintegrate into the chromosome in another position. When it leaves the chromosome it usually leaves behind the duplicated copy of the target site, creating a permanent mutation in that region.

Insertion sequences generally insert at random into the chromosome, although there may be a preference for certain types of DNA sequence in some cases. Since the genes of a bacterium are aligned along the chromosome with very little space between them the probability that an insertion sequence will integrate into a gene when it transposes is very high. Insertion sequences generally contain stop codons in all frames of reference with respect to the reading of the triplet code. Therefore, they interfere with proper translation of any gene into which they integrate and generally cause a lethal mutation. Thus, the only viable examples of integration are when the insertion sequence (i) integrates into a non-essential region of the chromosome such as a non-essential gene or into an inert or inoperative segment of DNA or (ii) when the integration results in the synthesis of a novel, but non-lethal, protein. Because the majority of insertion events kill the host bacterium, the frequency of transposition can only be estimated by indirect methods.

Clearly, insertion sequences cannot multiply and spread without check in the chromosome of a bacterium because of their potential lethality. In a sense, insertion sequences are parasitic, using bacteria as hosts to replicate and maintain themselves and, like many parasites, they use feedback control to prevent their undue proliferation and the death of their hosts. On the other hand insertion sequences are common in bacteria in multiple copies, and there may be advantages to the host bacteria in having them. Such advantages could include an increased ability to undergo genetic change.

One is left, anthropomorphically speaking, with a picture of an evolutionary balance between the advantages and disadvantages of carrying a system that can lead to an increase in the rate of beneficial genetic changes as well as, perhaps, a very high rate of increase in the number of lethal mutations.

INSERTION SEQUENCES IN THIOBACILLUS FERROOXIDANS

General Description

Insertion sequences have been found in many strains of *T. ferrooxidans*, but have only been described in detail in *T. ferrooxidans* ATCC19859. In this strain two different insertion sequences, termed IST1 and IST2, have been characterized (10-12). Both are about the same size, approximately 1.2-1.5 kb in length, and both have the characteristic features of classic prokaryotic insertion sequences as described above. There are about 20 copies per chromosome of IST1 and about 10 copies of IST2. In addition, IST1, but not IST2, is present in about 2 copies on one of the plasmids of this strain. Both types of insertion sequence are capable of transposition within the chromosome but IST1 appears to move more frequently than IST2. A diagrammatic representation of the distribution of IST1 and IST2 on the chromosome and plasmid of this strain of *T. ferrooxidans* is shown in figure 2.

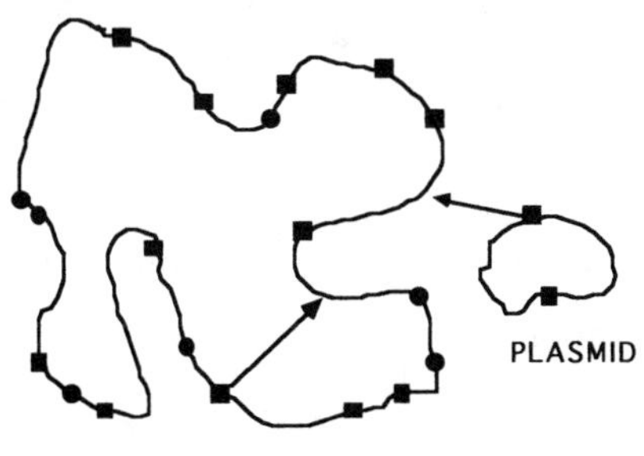

FIGURE 2.

Diagrammatic representation of the distribution of the insertion sequences IST1 and IST2 on the chromosome and plasmid of T. ferrooxidans ATCC19859. The arrows indicate that the insertion sequences are capable of transposition both within the chromosome and between the chromosome and the plasmid.

■ = IST1 ● = IST2

IST1 and IST2 insertion sequences are found in many strains of *T. ferrooxidans* from various regions of the world (5,6). Therefore, they are thought to be important in the competitive survival of the species. However, because they are not found in all strains of *T. ferrooxidans* they cannot be essential to its existence. Preliminary evidence indicates that sequences related to IST1 and IST2 exist in *T. thiooxidans* but not in several heterotrophic, acidophilic bacteria that have been examined.

Alteration of Genotype and Phenotype

Insertion sequences have been shown to induce specific mutations in several microorganisms either by interfering with the proper translation of genes or by bringing genes under the control of promoters present within the insertion sequence. Both of these effects are suspected to occur in *T. ferrooxidans.*

Figure 3 shows a diagrammatic representation of the insertion of IST-1 into an open reading frame (ORF) in *T. ferrooxidans* ATCC19859 (12). The insertion of the IST-1 into this position prevents normal translation of the gene because IST-1 contains stop codons in all three reading frames. However, because IST1 may contain its own genetic control elements within the terminal inverted repeats, it could, theoretically, initiate transcription and translation of a novel protein. This possibility remains to be investigated.

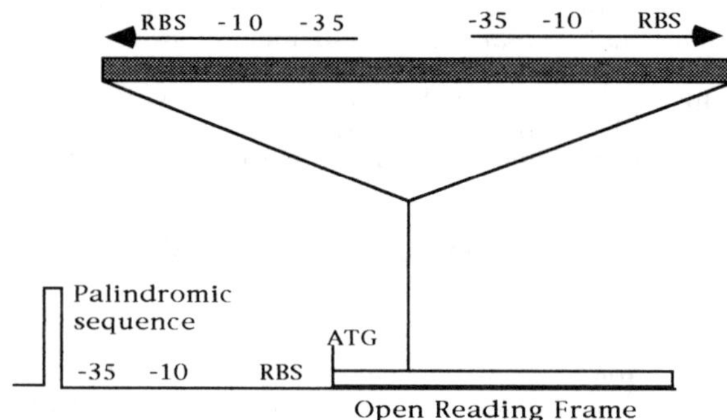

Figure 3. Insertion of IST1 into an open reading frame (ORF) that may encode a protein involved in iron oxidation in *T. ferrooxidans.* -10 and -35 refer to potential genetic control regions; RBS = ribosome binding site; the palindromic sequence may be involved in the regulation of the ORF. The insertion of IST1 could prevent the expression of the ORF or, by virtue of the potential control regions found on its inverted terminal repeats it could result in the synthesis of a novel protein from the ORF.

What is particularly interesting is that insertion of IST1 into this ORF is correlated with the loss of the ability of *T. ferrooxidans* to oxidize iron. Therefore, it is suspected that the ORF encodes a protein that is used either directly or indirectly in the oxidation of iron. A comparison of the potential protein encoded by the ORF with the sequences present in the data banks reveals no strong homologies with any known protein. In addition to containing canonical genetic regulatory regions, termed -10 and -35 regions, the ORF also contains a canonical ribosome binding site and a palindromic sequence of 19 base pairs upstream from the putative -35 region. Palindromic sequences such as the one observed are frequently found in regulated genes. Therefore, although it remains to be proved that the ORF is translated into a protein, it contains all the essential elements of a gene.

A second point that is interesting about this particular insertion event is that it appears to be reversible, that is to say the insertion sequence can be excised from the ORF, restoring the ability to oxidize iron. This phenomenon has been termed phenotypic switching (8). The implication is that the target site duplication, that is generated on insertion of the IST-1, is removed on excision of the IST-1. Removal of the target site duplication has only been observed in one other case.

IST-2 has also been shown to insert into ORFs, potentially inactivating them or altering the amino acid sequence of proteins (1). An example of this type of activity is shown in figure 4. In this example, the level of expression and the number of base pairs of the ORF could be modulated by the presence of IST2 insertion sequences numbered 3 and 4. For example, in *T. ferrooxidans* ATCC19859 (grown in iron) the ORF could be initiated by IST2 #3 and, therefore, be under its control, and it could terminate 34 nucleotides inside ORF #4 where it encounters a stop codon.

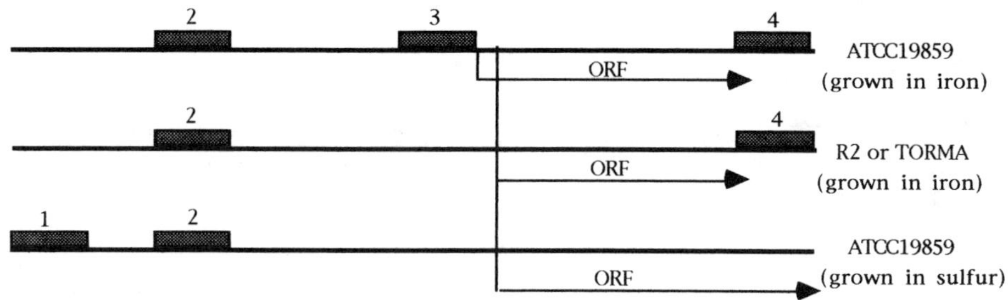

FIGURE 4. Examples of the position of several IST2 insertion sequences in different strains of *T. ferrooxidans* grown in either iron or sulfur medium. The position of one particular ORF whose size and activity are potentially modulated by the IST2 insertion sequences # 3 and #4 is shown. Other ORFs whose expression could be modulated by IST2 #1, #2 and #3 are not shown, but further details are available in (1).

In contrast, *T. ferrooxidans* ATCC19859 (grown in sulfur) contains neither IST2 #3 nor #4 and, consequently, the equivalent ORF could be initiated from its own control sequences and terminate at some, as yet undetermined position in the chromosome. Finally, *T. ferrooxidans* strains Torma and R2 contain IST2 #4 but not #3 and so the equivalent ORF in these strains could be initiated from the chromosome (like sulfur grown ATCC19859) but terminated in IST2 #4 (like iron grown ATCC19859). It remains to be determined whether the relative growth conditions affect the positioning of the IST2 insertion sequences or whether it is merely coincidental. However, the significant point is that the chromosomal locations of the insertion sequences vary and that these variations have the potential to modulate the expression of genes.

Genomic Rearrangements

In addition to directly effecting gene expiation, insertion sequences have been implicated in such genomic rearrangements as deletions, inversions and replicon fusions (3). They can also promote the recruitment of foreign genes, creating new catabolic pathways (9). As yet, none of these phenomena have been detected in *T. ferrooxidans* but because they are widespread and because of their potential importance in the evolution of new genes a brief description of their activities will be given (see figure 5A-C).

Figure 5A illustrates how a plasmid, carrying a copy of an insertion sequence, can integrate into the chromosome of a bacterium. This permits a plasmid, together with any genetically engineered genes that it might carry, to become incorporated into the chromosome, stabilizing it to some extent. The process can reverse, permitting a plasmid that is already integrated to leave the chromosome and lead an independent existence. If, during the tenure of the plasmid in the chromosome, genetic recombination takes place, then the plasmid leaves the chromosome with different genes than it arrived with.

DNA does not have to be in the form of a plasmid to become integrated into the chromosome. In principle, any DNA that carries two insertion sequences can enter the chromosome by the type of recombination event shown in figure 5B.

Figure 5C illustrates how crossing over between four copies of an insertion sequence can lead to rearrangement of genes within a bacterial chromosome.

Rearrangement of genes is important because it can lead to new combinations of genetic information and it can alter the regulation or expression of existing genes. This type of activity, together with the integration events illustrated in figure 5A and 5B, are major forces, in generating genetic diversity.

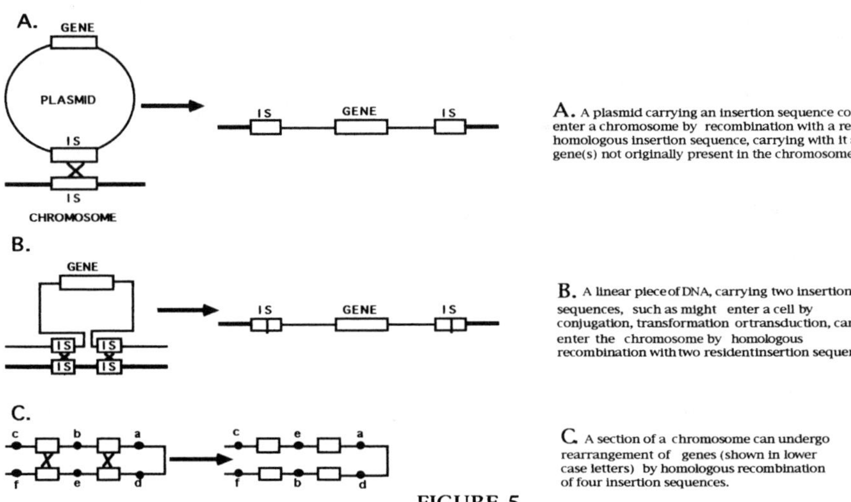

A. A plasmid carrying an insertion sequence could enter a chromosome by recombination with a resident homologous insertion sequence, carrying with it a gene(s) not originally present in the chromosome.

B. A linear piece of DNA, carrying two insertion sequences, such as might enter a cell by conjugation, transformation or transduction, can enter the chromosome by homologous recombination with two resident insertion sequences.

C. A section of a chromosome can undergo rearrangement of genes (shown in lower case letters) by homologous recombination of four insertion sequences.

FIGURE 5.

Recombination, chromosomal rearrangement and plasmid integration can take place in the absence of insertion sequences, but occur much more frequently in their presence. There are two reasons for this. First, insertion sequences encode a transposase enzyme that facilitates the integration and excision of the insertion sequences into the chromosome and, because integration and excision events can be imprecise, they promote the recombination and rearrangement of genes in the chromosome. Second, the majority of gene recombinations and rearrangements occur by a mechanism that involves the recognition of identical or homologous sequences prior to the crossing over of DNA If there are multiple copies of an insertion sequence in the chromosome, as there usually are, then there are more opportunities for homologous recognition and crossing over. Moreover, when there are two or more insertion sequences, crossing over can exchange or rearrange genes that lie between them, as shown in figure 5C.

When a bacterium takes up DNA into its cell by mating with another bacterium (conjugation), by uptake directly from the environment (transformation) or via the agency of a virus (transduction), then the chances of incorporation of the DNA into the chromosome are increased if such DNA contains an insertion sequence(s), especially one shared by the host bacterium. Without such incorporation the chances are that the incoming DNA would be destroyed by cellular enzymes.

Transposons

The discussion above stresses the important role played by insertion sequences in the evolution of bacteria. So important are they, that a variation of the insertion sequence, called a transposon, has been evolved naturally.

A transposon is a segment of DNA that consists essentially of two insertion sequences with a specific gene between them. Usually this gene is one that can confer drug resistance, so the transposon represents a mechanism that promotes the spread of drug resistance.

Transposons are useful because they can be introduced into a bacterium and then used to generate and isolate mutants. For example, if the transposon carries a drug resistant gene, then the incorporation of the transposon into the chromosome can be selected for because only those bacteria that take up and incorporate the transposon will survive when exposed to the drug. Since its introduction into the chromosome generally results in mutation, a transposon represents a way to generate mutations. A transposon has the additional advantage over other mechanisms for generating mutations (such as chemical mutation) in that it can be used to recover the gene into which it has inserted. This can be achieved by using restriction enzymes to cut the DNA, followed by the identification of the specific fragment carrying the transposon by hybridization with a copy of the transposon or by other tricks in the armamenterium of the geneticist.

Additionally, artificial transposons can be made in the laboratory by introducing a gene of interest between two insertion sequences. This creates a structure that is equivalent to the linear fragment of DNA shown in figure 5B. Since it appears to be difficult to introduce genes into *T. ferrooxidans,* an artificial transposon might expedite the insertion of a gene of interest into the chromosome of a strain of *T. ferrooxidans* that carries insertion sequences complementary to those found on the artificial transposon.

Some General Questions of Interest

A considerable amount of information remains to be elucidated regarding the role of IST1 and IST2, and other possible insertion sequences, in *T. ferrooxidans.* Some of the questions to be investigated include the following:

- It is not known whether insertion sequences other than IST1 and IST2 remain to be discovered in *T. ferrooxidans.* DNA reassociation experiments indicate that there might be other insertion sequences in *T. ferrooxidans* ATCC19859 but no data exists regarding other *T. ferrooxidans* strains. It is important to know the answer to this question for the following fundamental and practical reasons. Fundamentally, if strains of *T. ferrooxidans* exist without any insertion sequences then clearly insertion sequences in general are not essential to the life of *T. ferrooxidans,* as was mentioned above for IST1 and IST2 specifically. From a practical point of view there are two relevant considerations. First, insertion sequences seem be involved in the generation of new genotypes and phenotypes of *T. ferrooxidans* (5,6 and 8). Therefore, a knowledge of their structure and function would be of value as an aid to the development of new strains of *T. ferrooxidans* of industrial significance. Second, if strains of *T. ferrooxidans* are developed by genetic engineering for commercial applications then, perhaps, the genetic alterations should be done in a strain devoid of insertion sequences. This would reduce the frequency of unacceptable genetic changes to the genetically engineered characteristics, resulting from the types of activities of the insertion sequences described above.
- Although the observed frequency of transposition of IST1 and IST2 within the genome of *T. ferrooxidans* ATCC19859 is high (with that of IST1 being substantially higher), the actual frequency of transposition is not known and could be a good deal higher than the observed frequency. For reasons discussed above, it is presumed that many transposition events lead to the generation of lethal mutations and, because the resulting bacteria die, such transpositions, which may be the majority, are not observed or scored by the techniques that have been used to date.
- It is not known whether transposition occurs at random or whether there are preferred sites of integration of the insertion sequences in the genome. A recent study, illustrated in part in figure 4, describes the presence of several copies of IST2 in one particular region of the genome (1). This might seem to imply that this region represents a "hot spot" for integration. However, the techniques used in this analysis bias the data in favor of observing tightly clustered groups of insertion sequences. Thus, although clustering of insertion sequences clearly

takes place, the present data cannot be used to make generalizations about their random or non-random distribution.

One study, referred to in figure 3, demonstrates that IST1 frequently integrates into one particular region of the genome and its transposition into this region is associated with a loss of iron-oxidizing ability (12). This data may indicate that there is a preference of integration of IST1 in this region. However, it remains to be demonstrated whether the data is biased because this type of insertion leads to a readily detectable phenotype.

It is important to know whether transposition is random or whether it is restricted to certain regions of the genome because this will govern the utility of the insertion sequences as a general means of generating new strains of *T. ferrooxidans* both in the wild and by genetic selection in the laboratory.

- Recent work has shown that metabolic stress can lead to an increased frequency of transposition of insertion sequences in several different types of bacteria (2). However, it is not known whether stress can provoke a higher than "normal" rate of transposition of IST1 or IST2 in *T. ferrooxidans.* Holmes and Haq (6) observed a specific change in the location of an IST1 element when *T. ferrooxidans* was stressed with increasing concentrations of Cu^{+2} and Cádiz et al., (1) observed different patterns of insertion of IST2 depending on whether the *T. ferrooxidans* cells were grown in iron or sulfur medium (figure 4). However, in neither case were data provided that unequivocally addresses the issue of whether the changes in the metabolic growth of the microorganisms provoked an increased rate in the transposition of either IST1 or IST2.

Transposition of insertion sequences might represent a significant way for *T. ferrooxidans* to generate genetic diversity and, if this is true, then stress-induced transposition might represent a way for the species to adapt and survive changing environmental conditions. From a practical point of view it is important to determine whether stress-induced transposition occurs because, if it does, then a careful analysis of the phenomenon might suggest ways to promote the rate of adaptation of *T. ferrooxidans* to industrial conditions.

References

R. Cádiz, Gaete L., Jedlicki E., Yates J., Holmes D.S. and Orellana O, “Transposition of IST-2 in *T. ferrooxidans,*” Mol. Microbiol., 12, (1994), 165-170.

C. A. Cullis, "DNA Rearrangements in Response to Environmental Stress". Advances in Genetics 28, (1990) 73-97.

D. J. Galas and Chandler, M., "Bacterial Insertion Sequences". p: 109-162. In Berg, D. E. and Howe, M. M. (ed.), Mobile DNA. American Society for Microbiology, Washington, D. C., U.S.A (1989).

N. D. F. Grindley and Reed R. R., "Transpositional Recombination in Prokaryotes". Ann. Rev. Biochem. 54 (1985) 863-896.

D. S. Holmes, Yates J.R. and Schrader J., "Mobile Repeated DNA Sequences in *Thiobacillus ferrooxidans* and their Significance for Biomining", In: Biohydrometallurgy, P.R. Norris and D.P. Kelley eds., pp. 153-160, Science and Tech. Letters (1988).

D. S. Holmes and Haq, R. U., "Adaptation of *Thiobacillus ferrooxidans* for Industrial Applications". In Biohydrometallurgy 1989, Symposium Proceedings, Jackson Hole, Wyoming, U.S.A. (Salley, J., McCready, R. L. G. and Wichlacz, P. L., eds). pp 115-127 (1990).

J. G. Holt, ed, Bergey's Manual of Determinative Bacteriology. 9th edition. Williams and Wilkins, Baltimore, MD (1993).

J. A. Schrader and Holmes, D. S., "Phenotypic Switching of *Thiobacillus ferrooxidans*. J. Bacteriol. 170, (1988) 3915-3923.

S. Silver, A. M. Chakrabarty, B. Iglewski and S. Kaplan. "Pseudomonas: Biotransformations, Pathogenesis, and Evolving Biotechnology". Eds. Int. Symp. Molecular Biology of Pseudomonas. Chicago, IL (1990).

J. R. Yates and Holmes, D. S., "Two Families of Repeated DNA Sequences in *Thiobacillus ferrooxidans* ". J. Bacteriol. 169, (1987) 1861-1870.

J. R.Yates, Cunningham R.R. and Holmes D.S. "IST2: A New Insertion Sequence in *Thiobacillus ferrooxidans* ", Proc. Natl. Acad. Sci. USA. 85, (1988) 7284-7287 .

H. Zhao and Holmes D.S., "Insertion Sequence ISTI and Associated Phenotypic Switching in *Thiobacillus ferrooxidans* ." pp 667-671 in Biohydrometallurgical Technologies Vol. II eds A.E. Torma, M. L. Apel and C. L. Brierley, TMS, Warrendale, PA, USA (1993).

GENETICS OF *Thiobacillus ferrooxidans*: ADVANCEMENT AND PROJECTS: SEQUENCE AND ANALYSIS OF *rus* GENE ENCODING RUSTICYANIN AND *alaS* GENE ENCODING ALANYL-tRNA SYNTHETASE.

Nicolas Guiliani, Abderrahmane Bengrine, Francoise Borne
Marc Chippaux and Violaine Bonnefoy.
Laboratoire de Chimie Bactérienne, Marseille, France.

Abstract

As a first approach to the genetics of *Thiobacillus ferrooxidans,* we have decided to undertake a marker exchange mutagenesis program (reverse genetics). This program requires several conditions to be met, the essential ones being a genetic transfer technique and a non essential *T. ferrooxidans* gene.

One of the most investigated proteins in *T. ferrooxidans* is rusticyanin, though its exact role in the transfer of electrons between ferrous iron and oxygen remains to be elucidated. Construction of a mutant in which rusticyanin is no longer synthesized, could help in the elucidation of its function. Therefore, we decided to clone and study the *rus* gene encoding this protein.

The sequence of this gene and that of *alaS*, a gene encoding an alanyl-tRNA synthetase will be reported as well as a preliminary analysis of its regulation.

Electropermeabilization of ATCC33020 cells was demonstrated by the efflux of ATP from the cells. The possibility of using this technique to transfer DNA in this strain will be discussed.

Mineral Bioprocessing II
Edited by David S. Holmes and Ross W. Smith
The Minerals, Metals & Materials Society, 1995

INTRODUCTION

Thiobacillus ferrooxidans is one of the most important microorganisms involved in bioleaching. It obtains energy by the oxidation of iron and sulfur in ores and it obtains carbon by carbon dioxide fixation. One of the most interesting aspects of its unusual physiology concerns the respiratory processes it has developed to gain the energy required for growth, i.e. the oxidation by molecular oxygen of reduced sulfur compounds and ferrous iron, two of the most abundant elements in ores (pyrite, chalcopyrite, etc.). Reduced sulfur compounds are eventually oxidized to sulfuric acid, which acidifies the medium, while ferrous iron is oxidized to ferric iron which is a powerful oxidant.

The pH of the environment of *T. ferrooxidans* is very low, below pH2, while the internal pH is neutral. This pH gradient across the cytoplasmic membrane leads to a proton flux inside the cell. The protons which enter the cells are used in respiration processes to reduce oxygen to water while ferrous iron is oxidized to ferric iron, and reduced sulfur compounds are oxidized to sulfuric acid. This energy of protons is coupled to the synthesis of ATP in a classical chemiosmotic ATPase reaction. These physiological characteristics made *T. ferrooxidans* an attractive microorganism to study and several laboratories in the world are interested by the biochemistry and the molecular biology of the bioenergenetics of iron oxidation.

Different proteins involved in the electron transport between ferrous iron and oxygen have been already characterized: the FeII oxidase (1-3), which receives electrons direcly from ferrous iron, the cytochrome oxidase (3, 4), which gives electrons directly to oxygen, the rusticyanin (3, 5-9) and the cytochrome $^{c}552$ (3, 10). However, the number, the exact role and the order of the different components in the respiratory chain are not precisely known (11; Figure 1).

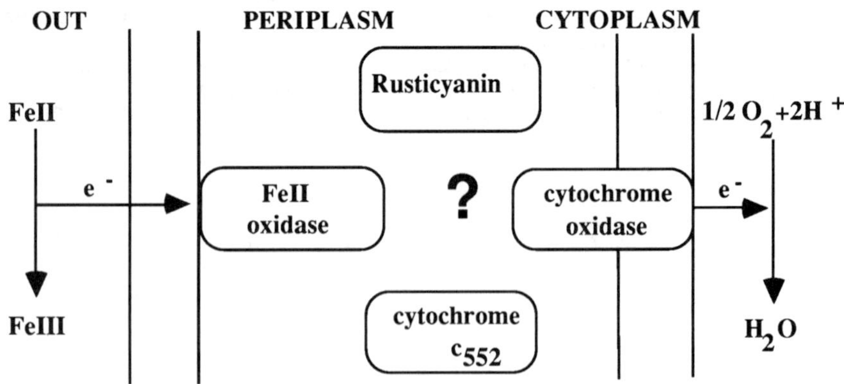

Figure 1: Electron transport between ferrous iron and oxygen.

One way to clarify these points is to utilize mutants in which one or several proteins of interest are no longer present. However, to construct strains in which a specific phenotype has been eliminated by conventional mutagenesis (spontaneous mutants, chemical mutagenic treatment, etc.) is not convenient. Genes involved in maturation of the protein (folding, cofactor incorporation, transport) or in regulation of its expression could be mutagenised in addition to the gene encoding the rusticyanin enzyme (*rus* gene). Therefore, we are interested in a reverse genetics approach (marker exchange mutagenesis) to construct a strain mutant in the rusticyanin encoding gene.

In reverse genetics, the gene of interest is cloned in *Escherichia coli*, mutagenised by insertion of an transposon containing transcription termination signal and a selectionable marker which can be used in both *E. coli* and *T. ferrooxidans*. This allows a direct selection of the clones with the mutagenised copy of the gene. This construction is then cloned on a broad host range plasmid able to replicate in *E. coli* and *T. ferrooxidans*. After introduction of this recombinant plasmid into *T. ferrooxidans*, homologous recombinations between the plasmid borne mutagenised copy and the chromosomal wild type copy may occur. The expected null mutants can be screened as having lost the plasmid born resistance marker but retained that carried by the transposon.

This paper will focus on the cloning and sequencing of the *rus* and the *alaS* genes of *T. ferrooxidans* encoding the rusticyanin and the alanyl-tRNA synthetase, respectively. Preliminary results on DNA transfer techniques from *E. coli* to *T. ferrooxidans* will be also reported.

MATERIALS AND METHODS

Strains, plasmids and growth conditions.

Thiobacillus ferrooxidans ATCC33020 was obtained from the American Type Collection.

Initial screening of the recombinant *T. ferrooxidans* chromosomal library, the propagation of plasmids and complementation analysis of the *recA* character were performed on *E. coli* HB101 (Δ(gpt-*proA*)62 *leuB6 thi-1 lacY1* hsd_B*20 recA rpsL20 ara-14 galK2 xyl-5 mtl-1 supE44* $mcrB_B$) strain. TG1 strain (supE hsdΔ5 thi Δ(lac-proAB) F': *traD36 proAB* $lacI^q$ lacZΔM15) was used for phagemids propagation. Complementation analysis of the *alaS* character was performed on Ab4132 strain (*alaS4 thi-1 metC56 lacY1 galK2 xyl-5 or xyl-7 ara-14 tfr-15 tsx-57 supE44*).

Phagemid Bluescript (Stratagéne) was used for sequencing.

Plasmids pNG1 and pNG2 construction has been already described (12). *E. coli* growth conditions have been described in Miller (13). $HgCl_2$ was used at 0.25 μg/ml. For analyzing *recA* phenotype, methyl methanesulfonate (MMS) was used at 0.01% (w/v).

Thiobacillus ferrooxidans growth conditions have been already published (12).

DNA manipulation

General DNA techniques were performed according to Ausubel et al. (14). Plasmid DNA preparations were obtained using Qiagen plasmid kit. Digestion with restriction endonucleases were carried out as recommended by the manufacturer. Preparation of competent *E. coli* cells and transformation were performed according to Chun et al. (15). The dideoxy chain termination method was used to sequence DNA using $\alpha(35_S)$dATP and the T^7 Sequence kit purchased from Pharmacia.

PCR reactions have been perfomed with Taq polymerase according to the manufacturer recommendations. The strategies used, the sequences of the different oligonucleotides and the amplification conditions, will be described elsewhere.

The ECL 3'-oligolabeling system and ECL detection reagents purchased from Amersham have been used for non radioactive labeling of oligonucleotides and detection.

Non radioactive labeling of DNA was performed by incorporation of Digoxigenin 11-dUTP with Taq polymerase during an amplification reaction (PCR) according to Boehringer rcommendations.

The DNA sequences were compiled, analysed and compared with the EMBL data bank sequences with the UWGCG package (FastA, BestFit: 16).

Electropermeabilization.

Electropermeabilization of ATCC33020 cells and the luciferase-luciferin assay were as described in Sizou et al. (17).

RNA techniques.

RNAs were prepared by the hot phenol method (13). RNAs were analysed by Northern hybridization according to the protocol described in Ausubel et al. (14).

RESULTS

Thiobacillus ferrooxidans genes.

Since our main interest concerns the study of the different componenets involved in ferrous iron oxidation, we decided to clone and sequence the *rus* gene encoding rusticyanin. This small periplasmic blue copper protein represents up to 5% of the proteins of *T. ferrooxidans* grown on ferrous iron (5, 18, 19). As indicated in the Introduction, its exact function in ferrous iron oxidation is still controversal.

We utilized *T. ferrooxidans* strain ATCC33020 although the amino-acid sequence of its *rus*ticyanin protein has not been determined. Degenerated oligonucleotide sequences were determined from three known amino acid sequences derived from other strains of *T. ferrooxidans* (7, 8, 9). A genomic bank of *T. ferrooxidans* ATCC33020 [kindly provided by Dr. Rawlings], was screened with one of these oligonucleotides. Nineteen clones responding positively were subjected to restriction analysis and Southern blotting experiments. A common 1,9 kb *Pst*I fragment hybridised with the probe. After sequencing no ORF corresponding to the rusticyanin gene could be found whereas the 3' end of gene *recA* (already cloned and sequenced by Dr. Rawling's group (20)) and the 5' end of a yet not reported gene were pesent (12). We have determined the entire sequence of this gene, termed *alaS*, since it encodes an alanyl-tRNA synthetase, a protein which charges the Ala-tRNA with alanine.

The deduced amino-acid sequence has a high degree of homology to *E. coli* AlaS sequence (21) as well as with the known amino terminal sequences of *Rhizobium meliloti* and *Rhizobium leguminosarum* biovar *viciae* AlaS (22; Figure 2).

```
AlaS T.f.    1 MQAQ A IRQAFLEYFVE QGHQIVPSSPLIPRNDPTLLFTNAGMVPFKD
AlaS E.c.    1 -SKST-E------DF-HSK ---V-A--S-V-H------------NQ---
AlaS R.l.    1 -SGVND --ST--D--KKN --E-------V-------M-------Q--N
AlaS R.m.    1 -SGVN E--SM--D--RKN --E--S----V-------M-------Q--N

AlaS T.f.   48 VFLGLETRPYRRAVSSQRCMRAGGKHNDLENVGYTARHHTFFEMLGNFSF
AlaS E.c.   50 -----DK-N-S--TT----V------------------------------
AlaS R.l.   49 --T---K-S-ST-TT--K-V-V-------D--------L-----------
AlaS R.m.   49 --T---Q-P-ST-ATA-K-V---------D--------------------

AlaS T.f.   98 GDYFKREVIGFAWRFLT ERLG  LPPEKLWITVYEEDDEAADIWMNEIG
AlaS E.c.  100 -----HDA-Q---EL--S-KWFA --K-R--V----S----YE--EK-V-
AlaS R.l.   99 -----ENA-EL--KLV- -   -FD--KHR-LV---S--E---TL-KKIA-
AlaS R.m.   99 -----ERA-EL--NLI-K-Y -   -DAKR-LV---HT--DTFGL-KKIA-

AlaS T.f.  145 IDPARLSRCG EK         DNFWSMGDTGPCGPCSEIFYDHGPHIPGGPP
AlaS E.c.  149 -PRE-II-I-DN-GAPYAS----Q----------T-------D--W----
AlaS R.l.  146 FSDDKII-IS          TS----Q---------------I-Q-ENVW----
AlaS R.m.  146 LSDDHII-I      A   TS----A------------------D--W----

AlaS T.f.  188 GSPEEDGDRYIEIWNLVFMQFDRDSSGTLTPLPKPSVDTGMGLERLAAVL
AlaS E.c.  199 ---------------I-----N-QAD--ME---------------I----
AlaS R.l   189 ---------FL-F 201
AlaS R.m.  189 --A------F--- 201

AlaS T.f.  238 QGVHNNYDTDLFKPLIAAAAAISGK TYGSDAATDI   SLRVLADHIRA
AlaS E.c.  249 -H-NS---I---RT--Q-V-     -V- -    ---LSNK----I-----S

AlaS T.f.  284 CSFLITDGVLPANEGRGYVLRRIIRRAVRHGRKLGM ETVFFHQLVAPLV
AlaS E.c.  291 -A---A---M-S--N----------------NM--AK-- --YK--G--I

AlaS T.f.  333 AEMGSAFPE LTRAQREVERALEREETRFRETLERGLSLL EEAIADLAA
AlaS E.c.  340 DV----G -D-K-Q-AQ--QV-KT--EQ-AR------A--D-- L-K-S

AlaS T.f.  381 GAAIPGEIIFRLADTFGFPVDLTADIARERDLIMDMEGYEAAMADERSPL
AlaS E.c.  387 -DTLD--TA---Y--Y---------VC---NIKV-EA-F----EEQ-RRA

AlaS T.f.  431 PCRLG GSGEVKTERVYHDLAMRLPVTEFTGYSTCADEGKVVALIRDGEE
AlaS E.c.  437 REAS-F-ADYNAMI--   -S-S        --K--DHLELN---T--FV--KA

AlaS T.f.  480 VAFLEAGDMGVVILDRTPFYGESGGQAGDRGELQSDDALFAVTDTQKPMG
AlaS E.c.  480 -DAIN--QEA--V--Q----A-----V--K---KGANFS---E----Y -

AlaS T.f.  530 HLHVHLGRMESGRLRVGDMVVASIDEVARRATAAH HSATHLLHAALRNI
AlaS E.c.  529 QAIG-I-KLAA-S-K---A-Q-DV-- ----RIRLN------M-----QV

AlaS T.f.  579 LGSHVQQKGSLVNPERLRFDFSQPEPVTAAQLREIERVVNAAIRNNVGAE
AlaS E.c.  578 --T--SH------DKV------HN-AMKPEEI-AV-DL--TL--R-LPI-

AlaS T.f.  629 TRILPVAEAQALGAMALFGEKYGDE VRVVRMGDFSMELCGGTHVEALGD
AlaS E.c.  628 -N-MDLEA-K-K---------- --R---LS-----T-------ASRT--

AlaS T.f.  678 IGVFKILSESGVAAGIRRIEAVT?VALEAIQR DEERLQAAAGLLKVAPA
AlaS E.c.  677 --L-R-I----T---V-------GEGAI-TVHA-SD--SEV-H---GDSN
```

```
AlaS T.f.  727 ELDQRLAQTLERLRQLEKELEKVKRDEAARAGADLAAEAEDVGGVPVLIR
AlaS E.c.  727 N-ADKVRSV---T-------QQL-EQA--QES-N-SSK-I--N--KL-VS

AlaS T.f.  777 RLEGMDGKALRDALDRLRSQLPDGVIVLAG VEREKVALIAGVGKGLTGR
AlaS E.c   777 E-S-VEP-M--TMV-D-KN--GSTI----TV--G --S-----S-DV-D-

AlaS T.f.  826 VHAGELVNMVAQPLGGKGGGRPELAQAGAGNPAALDAALDAARDWVRGKL
AlaS T.f.  826 -K----IG----QV--------DM----GTDA---P---ASVKG--SA--

AlaS T.f.  876 G
AlaS E.c.  876 Q
```

Figure 2: Amino-acid sequence deduced from the nucleotide sequence of *T. ferrooxidans alaS* gene (*T.f.*) and alignment with the AlaS tRNA synthetases from *E. coli* (*E.c.*; 21), *R. meliloti* (*R.m.*; 22) and *R. leguminosarum* biovar *viciae* (*R.l.*; 22). Only the 221 amino-terminal residues of *R. leguminosarum* and *R. leguminosarum* biovar *viciae* are known. Dashes figure residues which are conserved in all the sequences. Characteristic motifs are in bold face. The numbers on the right of the figure indicate the distance from the N-termini.

Total RNA was isolated from *T. ferrooxidans* cells grown on ferrous iron medium, and *alaS* specific mRNA was analysed by Northern blotting with a non radioactive probe labelled with DIG-dUTP. Only one band, corresponding to a transcript of approximately 2.8 kb, was detected (data not shown). This result strongly suggests that the cloned *alaS* gene is actually transcribed in *T. ferrooxidans* and that this gene is monocistronic as indicated by the size of the transcript.

As expected from the high level of identity between the AlaS proteins, the *T. ferrooxidans alaS* gene can complement *E. coli alaS* mutant. However, this complementation is only observed when *alaS* is under the control of a promoter carried by the vector suggesting that its own promoter is not recognized by the *E. coli* RNA polymerase, or that it is not been included with cloned sequence.

Since the screening of a genomic bank with degenerated oligonucleotide probes corresponding to rusticyanin lead to negative results, we turned to a different approach to clone the *rus* gene. The polymerase chain reaction (PCR) allows amplification of a DNA fragment encompassing two convergent oliogonucleotides. Degenerated oligonucleotide probes deduced from the aminoacid sequence of rusticyanin were designed. All possible pairs of convergent probes were then tested for amplification (Figure 3). In one case, a single DNA fragment with the expected size was obtained. After cloning and sequencing of this fragment, the deduced amino-acid sequence was found to match that of rusticyanin.

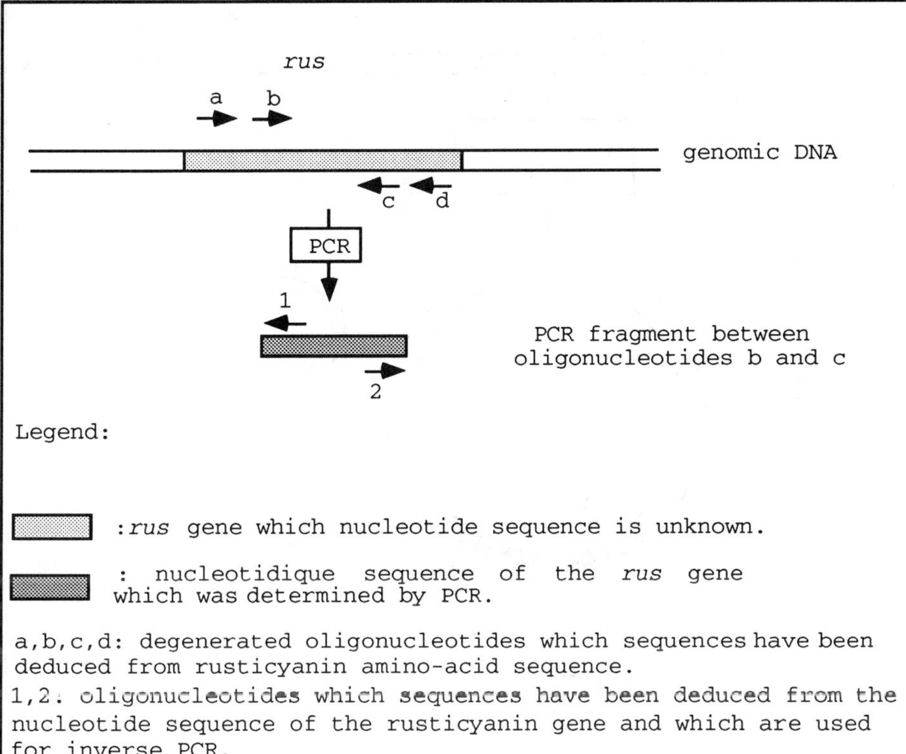

Figure 3: Polymerase chain reaction.

A Southern blotting experiment performed with this fragment as a probe, revealed the presence of only one copy of the *rus* gene per chromosome [data not shown].

To determine the sequence of the entire *rus* gene and of the flanking regions, the inverse PCR technique was then used. From this *rus* nucleotide sequence we then designed two divergent oligonucleotide probes. The *T. ferrooxidans* chromosomal DNA was digested with different restriction enzymes yielding small fragments and submitted to ligation to get small circles. This mixture was used for PCR amplification using the divergent probes. If both probes hybridise to the same DNA fragment, they will then be convergent on the corresponding circularised DNA structure. The new amplified fragement will correspond to the regions flanking the probes on the *rus* gene (Figure 4).

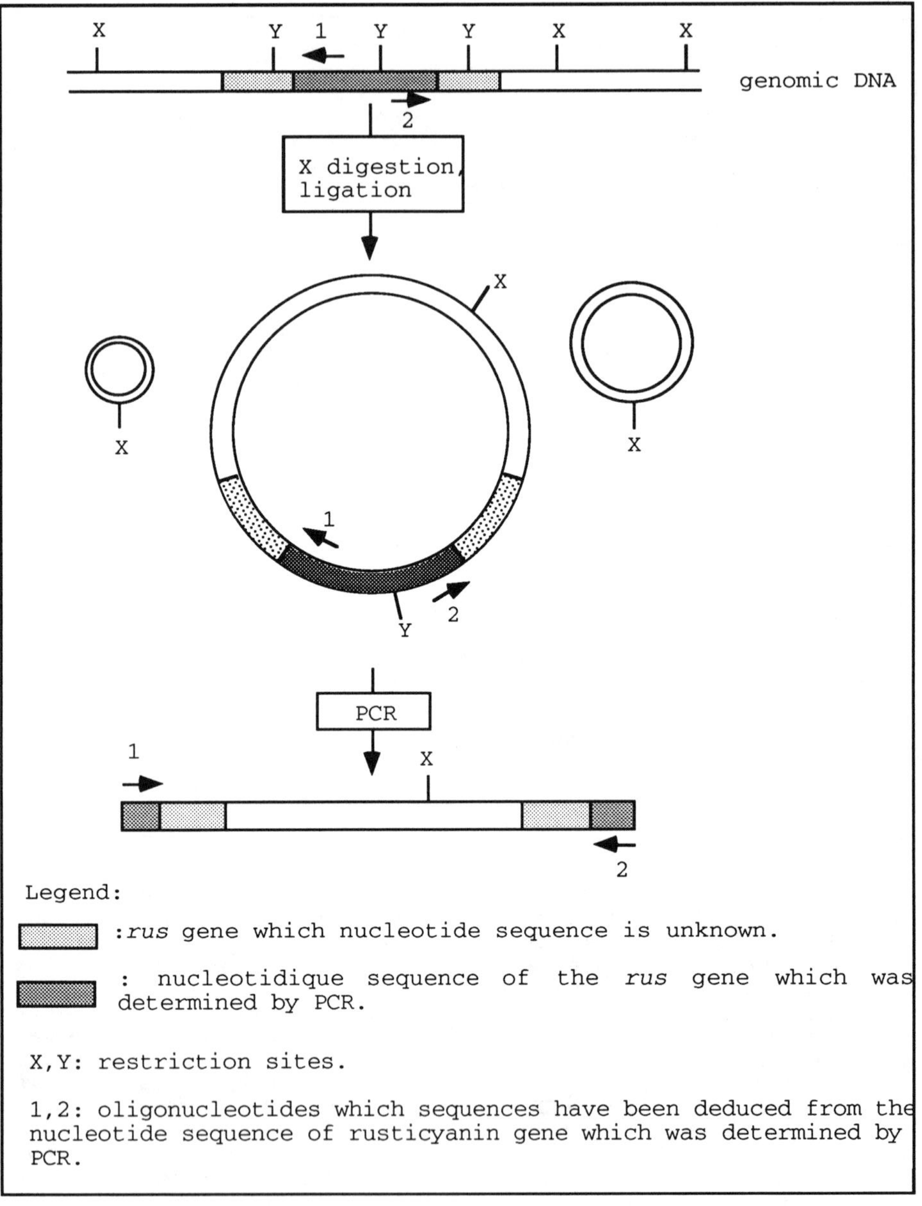

Figure 4: Inverse P.C.R.

This fragment can be cloned and sequenced leading to new oligonucleotide sequences which can be used once again in reverse PCR experiments. Using this approach, all the *rus* gene sequence was determined together with about 270 nucleotides downstream and 1430 upstream from the coding sequence (data not shown). As expected, the amino-acid sequence deduced from the nucleotide sequence is highly homologous to the other known rusticyanin sequences (7, 8, 9, 11; Figure 5).

```
(A)   1   GTLDSTWKEATLPQVKAMLEKDTGKVSGDTVTYSGKTVHVVAAAVLPGFP
(B)   1   ----T---------------------------------------------
(C)   1   ----T---------------------------------------------
(D)   1   -A--GS-------------Q-----A------------------------
(E)   1   -A---S-------E-----Q-----A---------- -------------

(A)  51   FPSFEVHDKKNPTLEIPAGATVDVTFINTNKGFGHSFDITKKGPPYAVMP
(B)  51   --------------------------------------------------
(C)  51   -----N--------D-------------------------Q-T--F----
(D)  51   --------------D------------------------------F----
(E)  50   --------------D------------------------------F----

(A) 101   VIDPIVAGTGFSPVPKDGKFGYTDFTWHPTAGTYYYVCQIPGHAATGMFG
(B) 101   --------------------------------------------------
(C) 101   -----------------------N--------------------------
(D) 101   N-K-------------------SE---N----------------------
(E) 100   N-K-------------------SE--------------------------

(A) 151    KIIVK
(B) 151    --V--
(C) 151   G-----
(D) 151    -----
(E) 150    --
```

Figure 5: Amino-acid sequence deduced from the nucleotide sequence of *T. ferrooxidans* ATCC33020 (A) *rus* gene encoding rusticyanin and comparison with the other known rusticyanin sequences: Ronk et *al.*(B; 7, 11), Ingledew et *al.* (C; 11), Nunzi et *al.* (D; 9) and Yano et *al.* (E; 8, 11). Dashes represent residues which are conserved in all the sequences. The numbers on the right of the figure indicate the distance from the amino termini.

Total RNA from *T. ferrooxidans* cells grown on ferrous iron or thiosulfate was isolated and *rus* specific mRNA analysed by Northern blotting with a fragment of DNA internal to *rus* gene as a probe. Preliminary results show that *rus* gene is monocistronic and that *rus* transcript from *T. ferrooxidans* grown on thiosulfate is larger than the one observed with *T. ferrooxidans* grown on ferrous iron (data not shown).

Permeabilization of *T. ferrooxidans* membrane.

To our knowledge, there are only two reports of successful genetic transfer into *T. ferrooxidans*. Peng et. al. (23) were able to conjugate broad-host range IncP-type plasmids between *E. coli* and *T. ferrooxidans* while Kusano and co-workers succeeded in introducing plasmid DNA in *T. ferrooxidans* after electropermeabilization of the cells (24), a technique in which an electric field generates small transient "holes" in membranes. However, in both cases, the strains of *T. ferrooxidans* which were used were not made available to other investigators.

Electropermeabilization is currently used to introduce DNA into cells which are resistant to routine methods. During the transient permeabilized state, both influx of foreign material and efflux of cellular material takes place. This last property was used to determine if *T. ferrooxidans* strain ATCC33020 is electropermeabilizable. After the application of electric pulses, efflux of ATP was monitored fluorometrically by using the luciferin/luciferase assay (17): the ATP-dependent oxidation of luciferin by luciferase produces light, the output of which is directly proportional to the concentration of ATP which, itself, reflects the percentage of electropermeabilized cells (17). By varying pulse time and elcetric field intensity and by monitoring light emission, the optimal conditions were determined. Figure 6 reports ATP efflux from ATCC33020 submitted to an electric field of increasing intensity during 12 ms and 20 ms.

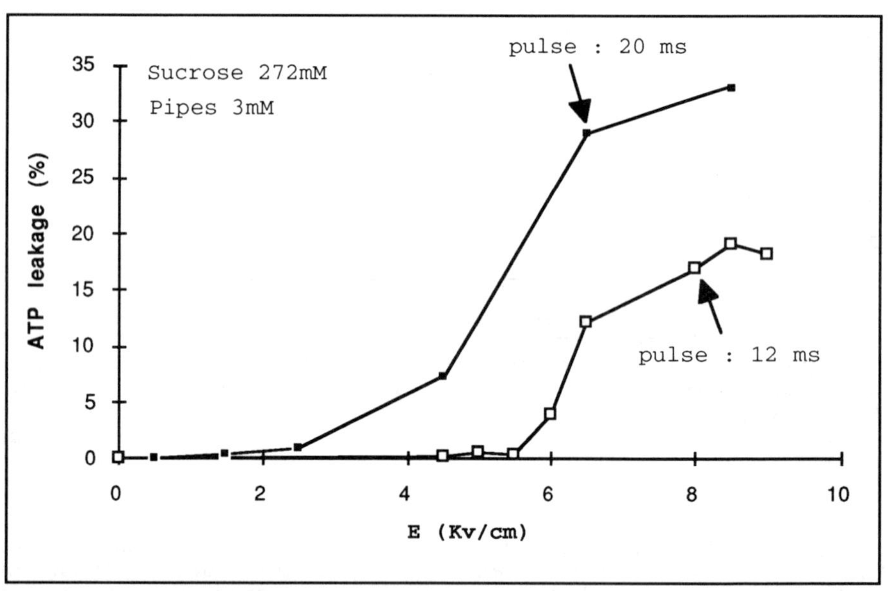

Figure 6: ATP leakage of ATCC33020 Thiobacillus ferrooxidans strain as function of pulse time and field intensity.

The optimal ratio of electropermeabilized to dead cells corresponds to one third of the plateau value obtained when percentage of ATP leakage is plotted against field intensity. In 272mM Sucrose and 3mM PIPES optimal pulse time and field intensity are respecitively 20 ms and 6.250 Kv/cm. In preliminary experiments, these conditions were used to introduce plasmids carrying the pTF-FC2 replicon as well as the chloramphenicol and mercury resistance markers (12), into *T. ferrooxidans*. Unfortunately, due to the very slow growth of the bacterium, to the absence of a reliable solid medium and to the lack of stability of mercury solutions, even the control (non-permeabilized cells) started to grow after a few days.

DISCUSSION

T. ferrooxidans rus gene

The amino-acid sequence corresponding to the mature rusticyanin deduced from the *rus* nucleotide sequence is highly homologous to all other known rusticyanins (7, 8, 9, 11) and particularly to that determined by Ronk et al. (7), with which only two conservative differences are observed: Ser to Thr and Val to Ile (Figure 5). Being a periplasmic protein, a signal sequence should be present at the amino terminus of unmature rusticyanin. Given the pH value of 3 for the periplasmic space fo *T. ferrooxidans*, one could wonder if the signal sequence would be similar to that of the Gram negative bacteria which have a periplasmic pH of 7 with a methionine followed by two basic residues (lysine) at the amino terminus, a core of hydrophobic amino acids, which permit the enzyme to cross the hydrophobic lipid cytoplasmic membrane, and a classical leader peptidase clevage site, Ala-X-Als (Figure 7). This site is in agreement with the known sequence of the mature rusicyanin.

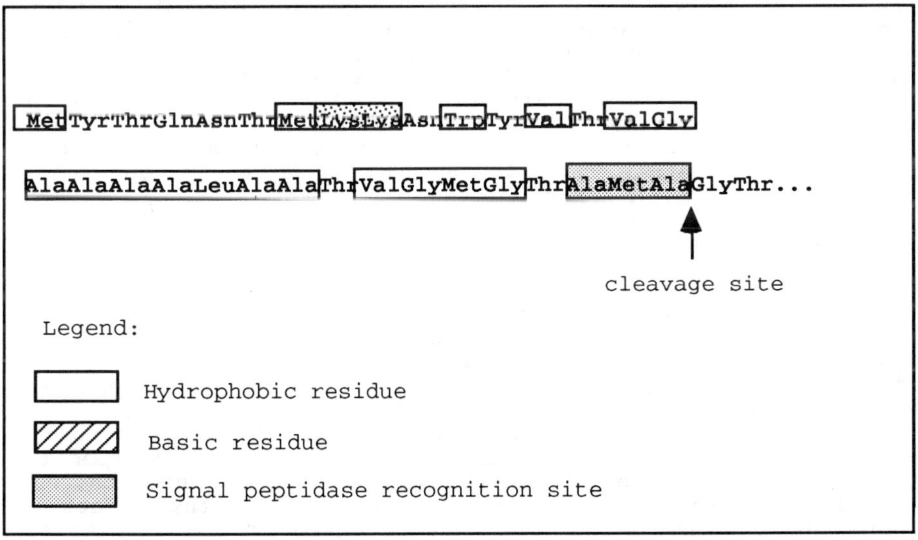

MetTyrThrGlnAsnThrMetLysLysAsnTrpTyrValThrValGly

AlaAlaAlaAlaLeuAlaAlaThrValGlyMetGlyThrAlaMetAlaGlyThr...

cleavage site

Legend:

Hydrophobic residue

Basic residue

Signal peptidase recognition site

Figure 7: Putative signal sequence of rusticyanin.

Whether this signal sequence is recognized by the *E. coli* leader signal peptidase will be tested by cloning all the *rus* gene in this organism and determining the subcellular localisation of the polypeptide. A putative correctly positioned canoniacal ribosome binding site is found upstream of the first methionine (data not shown).

Rusticyanin is a type I copper protein. One could wonder whether, in *E. coli*, copper would be incorporated into the aporusticyanin and, if such is the case, whether this would take place in the cytoplasm or in the periplasm. We are currently trying to answer these questions by cloning the *rus* gene with and without the nucleotides encoding the signal sequence. The presence of copper will then be detected spectrophotometrically.

When *T. ferrooxidans* is grown in the presence of iron, the concetration of some proteins, like rusticyanin, is much higher than when the cells are grown in the presence of reduced sulfur compound. In contrast, other proteins are synthesized preferentially when the cells are grown in the presence of reduced sulfur compounds (5, 18, 19, 26). One explanation is that *T. ferrooxidans* is able to synthesize proteins that are specifically involved in the oxidation of iron or sulfur according to the growth conditions. Pulgar et al. (27) have shown that regulation of ruticyanin occurs at the transcriptional level. Since the synthesis of several proteins is coinduced with that of rusticyanin, one could wonder whether the *rus* gene is transcribed alone, or belongs to a transcriptional unit corresponding to genes encoding proteins involved in iron oxidation. According to the nucleotide sequences upstream and downstrean of *rus* gene, this gene is monocristronic. Furthermore, the size of rusticyanin transcript from *T. ferrooxidans* grown on iron is incompatible with the idea that it is part of polycistronic mRNA. It has to be pointed out that the size of the transcript differs according to the growth medium: it is larger in cells grown on reduced sulfur compound than on iron. This was already noticed by Pulgar et al (27). We will determine the 5′ and 3′ ends of *rus* transcripts in cells grown on reduced sulfur compound or iron. This will help in identifying whether the difference in size is due to a difference in the transcription initiation site, in the termination site, or in both. The control region(s) of *rus* will then be studied.

T. ferrooxidans alaS gene

Probably because of the degeneracy of the oligonucleotides used when screening *T. ferrooxidans* genomic bank for *rus* gene, the *recA-alaS* region of *T. ferrooxidans* was inadvertendly cloned. The sequence of *alaS* gene and the intergenic region between *alaS* and *recA* was eventually determined.

An aminoacyl tRNA synthetase attaches its amino acid to the corresponding transfer RNA isoacceptor in a highly specific reaction. In spite of their common catalytic function, the proteins of this family are characterized by their lack of homology. Although the amino acid sequences are dissimilar, conserved sequence elements and structures were detected which have allowed the classification of synthetases in two families (28). Alanyl-tRNA-synthetase belongs to class II synthetases with an active site domain characterized by three degenerate sequence motifs (28) present in *T. ferrooxidans* AlaS protein (in bold face in Figure 2). They are located at the amino terminal half of the molecule in the consensus order and fit perfectly with the idiographic representation of the active domain of *E. coli* alanyl-tRNA synthetase proposed by Ribas de Pouplana et al. (29). The identity, even between conserved domains, is usually below 30%. The *T. ferrooxidans* AlaS protein primary structure shares 60% identity and 73.4% similarity with the AlaS protein from *E. coli*. The identity is even higher (63%) with the AlaS amino terminal parts of *Rhizobium meliloti* and *R. leguminosarum* biovar *viciae*. Such a high degree of homology of tRNA synthetases is unusual between organisms living in different biotops and belonging to three different phylogenetic groups of the "purple" bacteria. The homology is found not only in the conserved motifs but also in other parts of the sequence.

The homology between recA-alaS sequences of the four organisms is only observed in the coding region. We have found that the genetic complementation of *E. coli recA* or *alaS* mutants by *T. ferrooxidans recA* or *alaS* gene, occurs only when the genes are expressed from a vector-born promoter. Northern experiments have shown that *alaS* is trascribed in *T. ferrooxidans* as a monocistronic unit. From these results, it can be concluded that the *alaS* and *recA* promoters are not recognized by *E. coli* RNA polymerase or that they are not present in the chimical clones. Except *iro*, encoding the Fe(II) oxidase (2), all the *T. ferrooxidans* genes cloned in *E. coli* can be expressed from their own promoter. Our data suggest that *T. ferrooxidans*, possesses a form of RNA polymerase holoenzyme no present in *E. coli*. This attractive possibility will be investigated further.

Electropermeabilization of *T. ferrooxidans*

We have demonstrated that, when subjected to an electric field, ATCC33020 cells can release ATP in the medium. This result strongly suggests that this strain can be electropermeabilized. We have determined the optimal conditions of field intensity and pulse time. A shuttle plasmid has been constructed which contains chloramphenicol and mercury resistance markers and the replicon of pTF-FC2 plasmid, isolated from a *T. ferrooxidans* strain (12). Using the conditions determined for electropermeabilization, we have tried to introduce this plasmid in ATCC33020 strain by selecting mercury resistant clones. In our hands, however, the mercury solution is unstable and, even the non electropermeabilized cells start growing after a few days. Chloramphenicol resistance will be tried and other selectionable markers will be looked for.

Conclusion.

We have cloned and sequenced the *rus* gene and its putative control region. The gene seems to be monocistronic. The availability of the *rus* nucleotide sequence makes it possible to precisely define its promoter(s) and constitutes a very promising opportunity in better understanding the regulation of the genes involved in the iron oxidation by *T. ferrooxidans*.

The *recA* and the *alaS* promoters are not recognized by the *E. coli* RNA polymerase holoenzyme. It is then a likely possibility that these promoters require a special sigma subunit, not present in *E. coli* or only expressed in special conditions in the organism. The structural gene for this sigma subunit will be searched in a genomic bank of *T. ferrooxidans* by looking for the expression of a reporter gene cloned downstream the *recA* or the *alaS* promoter.

Reverse genetics in *T. ferrooxidans* strain ATCC33020 will only be possible once an easily reproducible technique of DNA transfer is available. We have shown that a transient permeabilization of the mebrane of this bacterium is possible. The electropermeabilization technique then appears as a promising approach provided a reliable selectionable marker is found. As soon as this requirement will be met, we will construct a mutant bearing a null allele of the *rus* gene and study the physiological properties of a strain devoid of rusticyanin.

ACKNOWLEDGEMENTS

The authors are grateful to Dr. Rawlings for his gift of a ATCC33020 recombinant cosmid library and of pTF200 plasmid. We acknowledge B. Bachman for the *alaS* mutant. We are indebted to J. Teissier and his group for their contribution in the electropermeabilization studies. The authors thank "le Centre de Séquencage du Laboratoire de Chimie Bactérienne" for its contribution in the sequence works. We also thank Vincent Méjean for helful advice on PCR experiments. This work was supported by "l'Agence de l'Environment et de la Maítrise de l'Energie" (A.D.E.M.E.), "le Bureau de Recherce Géologique et Miniére" (B.R.G.M.) and "la COGEMA".

REFERENCES

1. Y. Fukumori, T. Yano, A. Sato and T. Yamanaka, "Fe(II)-oxidizing enzyme purified from *Thiobacillus ferrooxidans*", FEMS Microb. Lett., 50 (1988), 169-172.

2. T. Kusano et al., "Molecular cloning of the gene encoding *Thiobacillus ferroxidans* Fe(II) oxidase, J. Biol. Chem., 267 (1992), 11242-11247.

3. T. Yamanaka et al., "The electron trasport system coupled to the oxidation of Fe^{2+} in *Thiobacillus ferrooxidans*", Biohydrometallurgical technologies, ed. A.E. Torma, M.L. Apel and C.L. Brierley, (Warrendale, PA: The Minerals, Metals and Material Society, 1993) 453-461.

4. T. Sugio et al., "Molybdenum oxidation by *Thiobacillus ferrooxidans*", Appl. Environ. Microbiol., 58 (1992), 1768-1771.

5. Cobley, J.G. and Haddock, B.A., "The respiratory chain of *Thiobacillus ferrooxidans*: the reduction of cytochromes by Fe^{2+} and the preliminary characterization of rusticyanin, a blue copper protein", FEBS Lett., 60 (1975), 29-33.

6. W.J. Ingledew, "Ferrous iron oxidation by *Thiobacillus ferrooxidans*", Biotechnol. Bioeng. Symp., 16 (1986) 23-33.

7. M. Ronk et al., "Amino acid sequence of the blue copper protein rusticyanin from *Thiobacillus ferrooxidans*", Biochemistry, 30 (1991), 9435-9442.

8. T. Yano, Y. Fukumori and T. Yamanaka, "The amino acid sequence of rusticyanin isolated from *Thiobacillus ferrooxidans*", FEBS Lett., 288 (1991), 159-162.

9. F. Nunzi et al., "Amino-acid sequence of rusticyanin from *Thiobacillus ferroxidans* and its comparison with other blue copper proteins", Biochim. Biophys. Acta, 1162 (1993), 28-34.

10. A. Sato et al., "*Thiobacillus ferrooxidans* cytochrome c-552: purification and some of its molecular features", Biochim. Biophys. Acta, 976 (1989) 129-134.

11. R.C. Blake et al., "Enzymes of aerobic respiration on iron", FEMS Lett., 11 (1993), 9-18.

12. N. Guiliani et al., "Perspectives in the gentics of *Thiobacillus ferroxidans*, Biohydrometallurgical tecnologies, ed. A.E. Torma, M.L. Apel and C.L. Brierley, (Warrendale, PA: The Minerals, Metals and Material Society, 1993), 645-658.

13. J.H. Miller, A short course in bacterial genetics: a laboratory manual and handbook for *Escherichia coli* and related bacteria, Cold Spring Harbor, New York: Cold Spring Harbor Laboratory Press, 1992.

14. F.M. Ausubel et al., Current protocols in molecular biology, New-York: Greene publishing associates/ Wiley-interscience (1992).

15. C.T. Chun, S.L. Niemela and R.H. Miller, "One step prepartion of competent *Escherichia coli*. Transformation and storage of bacterial cells in the same solution. Proc. Natl. Acad. Sci., 86 (1989), 2172-2175.

16. J. Devereux, P. Haeberli and O. Smithies, "A comprehensive set of sequence analysis programs for the VAX", Nucl. Acids Res., 12 (1984), 387-395.

17. S. Sixou et al., "Optimized conditions for electrotransformation of bacteria are related to the extend of electropermeabilization", Biochim. Biophys. Acta, 1088 (1991) 135-138.

18. Kulpa, C.F., Roskey, M.T. and Mjoli, N.: Construction of genomic libraries and induction or iron oxidation in *Thiobacillus ferrooxidans*. Biotech. Appl. Biochem. 8 (1986) 330-341.

19. E. Jedlicki et al., "Rusticyanin: initial studies on the regulation of its synthesis and gene isolation", Biotech. Appl. Biochem., 8 (1986) 342-350.

20. R.S. Ramesar et al., "Nucleotide sequence and expression of a cloned *Thiobacillus ferrooxidans recA* gene in *Escherichia coli*", Gene, 78 (1989), 1-8.

21. S.D. Putney et al., "Primary structure of a large aminoacyl-tRNA synthetase", Science, 213 (1981), 1497-1501.

22. W. Selbitschka et al., "Characterization of *recA* genes and *recA* mutants of *Rhizobium meliloti* and *Rhizobium leguminosarum* biovar *viciae*", Mol. Gen. Genet., 229 (1991), 86-95.

23. J.B. Peng, W.M. Yan and X.Z. Bao, "Plasmid and transposon transfer to *Thiobacillus ferrooxidans*", J. Bacteriol., 176 (1994), 2892-2897.

24. Kusano, T. et al. "Electrotransformation of *Thiobacillus ferrooxidans* with plasmids containing a *mer* determinant", J. Bacteriol. 174 (1992) 6617-6623.

25. A.P. Pugsley, "The complete general secretory pathway in Gram-negative bacteria", Microbiol. Rev., 57 (1993) 50-108.

26. G. Osorio et al., "Changes in global gene expression of *Thiobacillus ferrooxidans* when grown on elementary sulphur", Biohydrometallurgical technologies, ed. A.E. Torma, M.L. Apel and C.L. Brierley, (Warrendale, PA: The Minerals, Metals and Material Society, 1993), 565-575.

27. V. Pulgar et al., "Expression of rusticyanin gene is regulated by growth condition in *Thiobacillus ferrooxidans*", Biohydrometallurgical technologies, ed. A.E. Torma, M.L. Apel and C.L. Brierley, (Warrendale, PA: The Minerals, Metals and Material Society, 1993), 541-548.

28. D. Moras, "Structural and functional relationships between aminoacyl-tRNA synthetases", T.I.B.S. 17 (1992), 159-164.

29. L. Ribas de Pouplana et al, "Idiographic representation of conserved domain of a class II tRNA synthetase of unknown structure", Protein Science, 2 (1993), 2259-2262.

DIFFERENTIAL GENE EXPRESSION OF *Thiobacillus ferrooxidans*

UNDER DIFFERENT ENVIRONMENTAL CONDITIONS.

Carlos A. Jerez*, Patricia Varela, Gonzalo Osorio, Michael Seeger, Ana M. Amaro, and Héctor Toledo.

Departamento de Bioquímica, Facultad de Medicina, Universidad de Chile, Casilla 70086, Santiago, Chile. Fax: (56 2) 73 555 80.

Abstract

We have studied the global changes in gene expression of *Thiobacillus ferrooxidans* when the cells are subjected to phosphate starvation, to changes of growth between elementary sulfur and ferrous iron, and to a switch from planktonic to solid-attached living (sulfide ore or elementary sulfur). Several proteins that changed their levels of synthesis under these conditions have been identified by microsequencing of their N-terminal ends. The information generated will allow us to use reverse genetics to isolate some of their genes, making possible the study of their regulation. In addition, the knowledge of several of the proteins induced under the environmental changes studied, will provide a means of generating probes to monitor the *in situ* physiological state of the cells during bioleaching in industrial operations.

*: corresponding author.

Mineral Bioprocessing II
Edited by David S. Holmes and Ross W. Smith
The Minerals, Metals & Materials Society, 1995

Introduction.

Microorganisms are capable of detecting a wide variety of signals from the environment. One possible response to these changes, is to move to a more favorable environment, if the cell possesses the capacity to migrate. This can be achieved by a fast and transient change in motility. In a paralel and/or alternative way, the cell can adapt to the new environment changing certain enzymatic activities and altering the expression of specific or groups of genes (1).

Starvation of carbon, phosphorus and nitrogen constitute a stress for microorganisms such as *Escherichia coli* (7). Therefore, the lack of phosphorus may be a stress condition which can affect the bioleaching of minerals (8). During this process, the majority of microorganisms are normally found attached to the ores, but many are also found under free-living or planktonic conditions (2, 3). In this habitat, the microorganisms are subjected to different kinds of environmental alterations such as temperature or pH variations, the lack of essential nutrients or changes in the availability of oxidizable substrates, all of which may affect the activity of the bacteria, and therefore, the efficiency of the bioleaching process. In response to these stressing changes, bacteria are known to alter the genetic expression of several cellular polypeptides, inducing in some cases stress proteins and others as a defense or remediating mechanism to overcome the adverse condition (1). Consequently, knowing these adaptive responses of extreme acidophiles at the molecular level is not only interesting from the fundamental point of view, but will help us also to monitor the *in situ* state of these microorganisms, making possible to improve their activity in industrial processes.

We have previously reported the existence of a heat shock (4, 5) and pH-stress (6) responses in *T. ferrooxidans*. We also analyzed the changes in total protein synthesis of *T. ferrooxidans* when it was subjected to phosphate starvation (see below).

The genes involved in the oxidation of ferrous iron and sulfur make *T. ferrooxidans* and other bioleaching bacteria industrially important. Therefore, there is great interest in studying the structure, expression and regulation of these genes (9). Several proteins involved in ferrous ion oxidation by *T. ferrooxidans* are known, including rusticyanin, cytochromes a and c and other proteins (10-12). On the other hand, the proteins involved in sulfur oxidation have not been studied in detail and their cellular localization and exact roles are still unknown. We have analized the changes in global gene expression of *T. ferrooxidans* when the bacteria was grown on ferrous iron or elementary sulfur as energy sources (13).

In the present work, we review some of our data on the proteins whose synthesis varies during the global molecular responses of *T. ferrooxidans* to phosphate starvation, to switching from planktonic living to ore-attached living and to a shift from ferrous iron to elementary sulfur as energy sources.

Experimental procedures.

Bacterial Strain and Growth Condition.

T. ferrooxidans strain ATCC 19589 was used in these studies. Growth on ferrous iron was done in modified 9K medium as before (6, 14). Growth under phosphate-limiting conditions was done in the same medium, except that the phosphate salt was omitted (17). Growth on elementary sulfur was done employing sulfur prills (3, 15). Stationary phase cells (10^{10}) in 0.5 ml of the corresponding medium, were labeled in the presence of 0.1 mCi of ^{35}S-methionine (sp. act. = 1087 Ci/mmol) for 2 h. For labeling with $Na_2{}^{14}CO_3$, 1×10^{11} stationary phase *T. ferrooxidans* cells grown in sulfur were suspended in 500 μl of the medium (with 4-5 sulfur prills) and were added of 12 μCi of the radioisotope (sp. act. = 51.2 mCi/mmol), followed by incubation for 12 h at 30°C. The cells present in the liquid supernatant were considered to be free-living. To recover the sulfur-attached cells from the sulfur prills, they were first washed gently with the salts of the medium, to remove the loosely bound cells. The attached cells were then released from the prills by washing them three times by strong vortexing in the presence of the salts of the medium. Both kinds of radioactively-labeled cells were recovered by centrifugation, and were processed for SDS-PAGE as before (6). Growth in mineral, labeling and processing of free-living and ore-attached cells for standard two-dimensional (2-D) PAGE was done as described before (14).

2-D NEPHGE, SDS-PAGE and Autoradiography.

Standard 2-D PAGE (pH 5 to 7 in the first dimension)(16) or 2-D non-equilibrium pH polyacrylamide gel electrophoresis (2-D NEPHGE) (pH 3 to 10 in the first dimension)(16) was performed as described before for *T. ferrooxidans* (6, 14, 17). SDS-PAGE consisted of 7.5-15% or 11.5-20% polyacrylamide gradients.

The radioactively-labeled proteins from the cells or their subcellular fractions were separated by one-dimensional SDS-PAGE or by 2-D NEPHGE followed by staining with Coomassie Blue and autoradiography (5, 6, 13, 14, 17). The relative intensity of protein spots was determined by using a Hewlett-Packard Scan Jet Plus device and the GelPerfect image analysis program (kindly developed by S. Bozzo, Universidad de Chile).

Microsequencing of Proteins Extracted from Dried 2-D Gels.

Proteins of interest were recovered from Coomassie Blue-stained and heat-dried 2-D gels by excising the protein spots. After rehydration and concentration of the spots by SDS-PAGE, the proteins were electroblotted onto a polyvinylidene difluoride membrane (18) and were subjected to microsequencing by the Center for the Synthesis and Analysis of Biomolecules at the University of Chile, Santiago, Chile.

Results and Discussion.

Changes in Gene Expression Under Phosphate Starvation.

For most of our studies, we have employed 2-D NEPHGE system to separate the total cellular proteins from *T. ferrooxidans* as shown in Fig. 1. Under these conditions, both acidic and very basic polypeptides, such as rusticyanin, which has an isoelectric point of 9.2, can bee seen in the gel. On the contrary, when the standard O' Farrel 2-D PAGE is employed, only the proteins with isoelectric points between 5 and 7 are ressolved (14, 16).

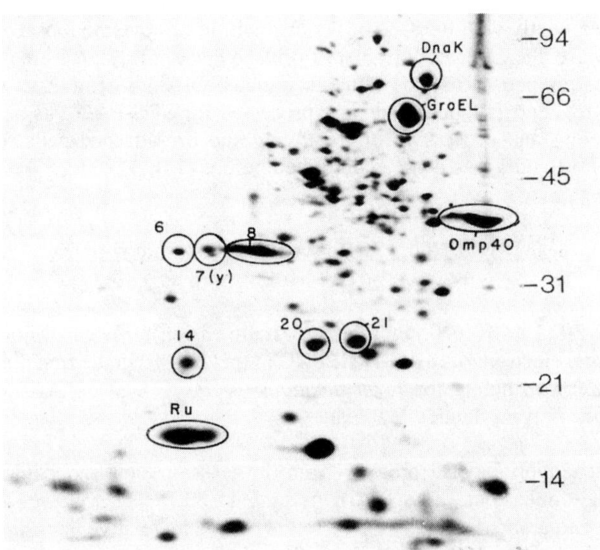

Fig. 1. NEPHGE of *T. ferrooxidans* total proteins synthesized in ferrous iron-grown cells. Total proteins from cells grown in ferrous iron in the presence of ^{35}S-methionine were separated by 2-D NEPHGE with a pH gradient between 3.0 (right side of the gel) and 10.0 (left side of the gel) and were detected by autoradiography. Some of the main identified polypeptides whose levels of synthesis varies under different environmental conditions are indicated by circles or ovals. Ru is rusticyanin. The spot numbering system employed is the same used for the original description of the proteins in the references. Numbers to the right of the gel indicate the molecular mass standards in kilodaltons.

We have found that when *T. ferrooxidans* is grown in ferrous iron under phosphate limitation, at least 25 proteins change their extent of synthesis (17-19). A few outer membrane proteins greatly changed upon phosphate starvation (17, 20). One of this proteins is Omp40, an outer membrane protein whose N-terminal end sequence showed 38.5% identity with the outer membrane channel-forming protein NosA from *Pseudomonas stutzeri* (20), whose expression is also regulated environmentally. In addition, we detected the induction of an acidic periplasmic phosphatase present in *T. ferrooxidans* (possibly spot 20 in Fig. 1)(18). Interestingly, the phosphate-starved *T. ferrooxidans* cells showed a higher capacity of attachment to a sulfide ore, perhaps due to an increase in the cell surface hydrophobicity (21). The induction or derepression of this kind of proteins suggest the appearence of an environmental phosphate scavenging system in *T. ferrooxidans* which may be similar to the *E. coli* emergency system known as the *pho* stimulon and whose expression is regulated by the levels of external phosphate (1). We are currently identifying some of the other phosphate starvation-derepressed proteins, and have found that a few of them also change their degree of phosphorylation (unpublished data).

Differential Gene Expression of *T. ferrooxidans* Grown in Sulfur or in Ferrous Iron.

To identify some of the gene products involved in the oxidations of elementary sulfur and ferrous iron, we have initiated the study of the changes in global gene expression in *T. ferrooxidans*, when the bacteria is grown on either of these substrates. A few polypeptides were exclusively synthesized when the cells were grown in ferrous iron, and others were apparently synthesized only when *T. ferrooxidans* was grown in sulfur (13). The levels of rusticyanin are greatly induced when cells are grown in the presence of increasing ferrous iron concentrations, as it has also been described by others (22, 23). In addition, the levels of protein spot 14 (see Fig. 1) were also increased under these conditions. This 23 kDa protein may correspond to a cytochrome of the c type (unpublished results). As expected, both rusticyanin and spot 14 decreased their levels of synthesis when the cells were grown in sulfur. It is clear then, that the expression of these and other proteins are regulated by the levels of ferrous iron to be oxidized. Nevertheless, it will be interesting to study the behaviour of these proteins when both oxidizable substrates are available to the bacteria.

We also observed changes in protein spots 6, 7 and 8 (Fig. 1) when comparing cells grown in sulfur with those grown on ferrous iron (13). To identify some of these proteins, they were subjected to microsequencing of their N-terminal ends. Spots 7 and 8 had identical sequences, suggesting a possible post-translational modification in one of them. In addition, the 50% of hydrophobic residues present in both polypeptides suggests their possible interaction with the membranes (13).

Changes in Protein Synthesis in Planktonic Versus Attached *T. ferrooxidans*.

T. ferrooxidans has been described as forming biofilms on iron or mild steel pipe (24) and the bacteria normally adheres to ores in its habitat (2). The physiological activity of cells attached to surfaces may be different from that of comparable free-living cells (25). Therefore, one could expect that the bacteria change their gene expression when passing from the free-living or planktonic form to the adhered form.

To analyze the behaviour of solid-attached *T. ferrooxidans*, we labeled cells growing in sulfur prills and a sulfide mineral in the presence of $Na_2{}^{14}CO_3$ and analyzed both planktonic and solid-attached cells by SDS-PAGE and 2-D PAGE as shown in Fig. 2.

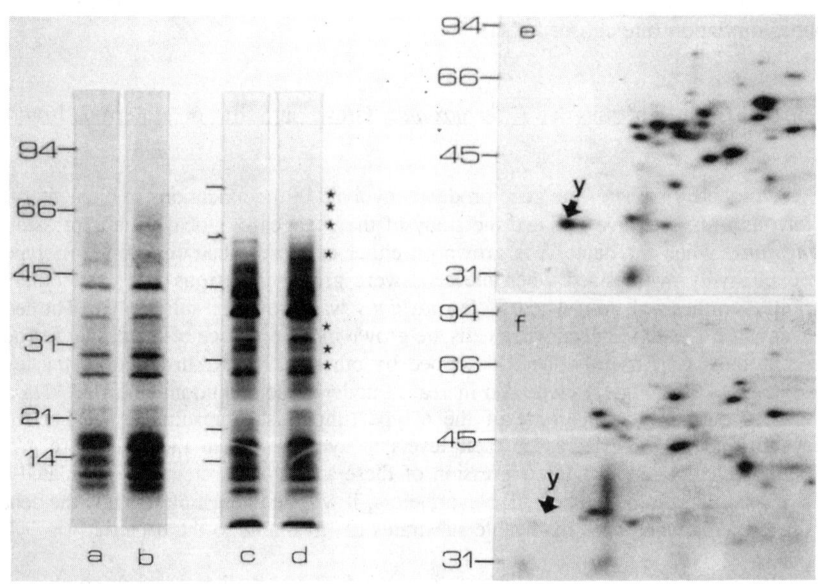

Fig. 2. Proteins synthesized by free-living and sulfur attached *T. ferrooxidans* cells. Cells grown in the presence of elementary sulfur prills (a, b, c, d) or a sulfide mineral (e, f) were labeled in the presence of $Na_2{}^{14}CO_3$ and the planktonic cells (a, c, e) and the attached to the solid ones (b, d, f) were analyzed by SDS-PAGE (a to d) or by standard 2-D PAGE (e, f). The protein spots were detected by Coomassie Blue staining (a, b) or by autoradiography (c, d, e, f). The asterisks refer to the main proteins changing between planktonic and attached cells. The arrow indicates the position of spot y (14). Numbers to the left of the gels represent the molecular weight standards in kilodaltons.

Several proteins change their levels of synthesis in the cells attached to the solids. However, these variations are rather difficult to see in Coomassie Blue-stained gels (fig. 2, a and b), and only when radioactively-labeled cells are used, are clearly seen. In sulfur attached cells (Fig. 2, lane d), at least seven polypeptides change between planktonic and attached cells (see asterisks in Fig. 2). These results are in general agreement with our previous results in which *T. ferrooxidans* attached to a sulfide mineral changed its global pattern of proteins synthesized (Fig. 2, e, f)(14). Several of the main spots changing in both solid substrates have similar sizes. The protein spot labeled as y (e and f, Fig. 2)(14) is particularly interesting, since its synthesis is greatly reduced in the cells attached to the sulfide ore. According to its relative migration in the standard two-dimensional gel (pH 5 to 7 in the first dimension), it could correspond to protein spots 6 or 7 separated by NEPHGE (pH 3 to 10 in the first dimension)(see Fig. 1). However, several other polypeptides also vary their levels, perhaps reflecting the bacterial adaptation to oxidize the different solid substrates. Therefore, further studies will be obviously necessary before ascribing a particular role if any, to some of these proteins in cell attachment.

Figure 3 shows some of the proteins from *T. ferrooxidans* in which we have determined their N-terminal end partial sequences.

Omp40	1 **A D T S N A D T G P V V F G Y A Q I T G A Q Q F G T** 26	(20)
Spot 20	1 **A V L V G K A A P D F V A P A V M P D N** 20	(18)
Spot 21	1 **S E Y S V R E E L K P S G L D G I S D A** 20	(18)
DnaK	1 **A K V I G I D L G T T N S ? V A V M E G** 20	(5)
GroEL	1 **P A K Q V A F A E H A R E K M L R G V N V L A** 23	(5)
Ru	1 **G T L D T T W K E A T L P Q V K A M L E K D T** 23	(13)
Spot 6	1 **A D M G W N G K A E A P R Y Q E Q V F P P W** 22	(13)
Spot 7	1 **A P T I S L L E T G S T L L Y P L F N L A V** 22	(13)

Fig. 3. N-terminal end sequences of several *T. ferrooxidans* proteins isolated from 2-D gels. The sequences indicated were obtained as described in Materials and Methods. Numbers in parenthesis are the references from which the sequences were taken. Usually, only around 100-200 pmoles of each protein is required to generate the corresponding sequence.

A summary of the main proteins we have identified is presented in Table I.

Table I.

Some proteins from *T. ferrooxidans* and their levels of synthesis during different environmental conditions

protein	condition	level of synthesis	general properties	reference
Omp40	-Pi	decreases	40 kDa, outer membrane protein, possibly an ionic channel	(20, 17)
Spot 20	-Pi	increases	26 kDa, possibly a periplasmic phosphatase with a pH optimum of 3.8	(18)
Spot 21	-Pi	increases	26 kDa, phosphorylated *in vivo*	(18)
DnaK	-Pi	unchanged	heat shock or stress protein	(5, 17)
GroEL	-Pi	unchanged	heat shock or stress protein	(5, 17)
Rusticyanin	Fe S	increases decreases	17 kDa blue-copper protein involved in iron oxidation	(13)
Spot 6	Fe S	decreases increases	unknown function	(13)
Spots 7 & 8	Fe S	increase decrease	unknown function	(13)
Spot 14	Fe S	increases decreases	23 kDa protein, possibly a c type cytochrome	(13)
Spot y	ore atthachment	decreases	unknown function, possibly related to spots 6 or 7	(14)

"Condition" refers to growth of *T. ferrooxidans* in ferrous iron (Fe), sulfur (S) or in ferrous iron under limiting phosphate conditions (-Pi).

Conclusions.

i. Modification of environmental conditions such as phosphate starvation, switching from planktonic to attached living or changing from one oxidizable substrate to another, alter *T. ferrooxidans* global gene expression.

ii. Some of the observed changes under phosphate limitation could reflect the appearence of an environmental phosphate scavenging system in *T. ferrooxidans*.

iii. Depending on the oxidizable substrate, *T. ferrooxidans* changes the levels of synthesis of diverse proteins, including cytochromes.

iv. The protein sequences we are generating will allow us to employ reverse genetics for the isolation in *T. ferrooxidans* of some of the genes involved in the global response analyzed, making possible the study of their regulation.

v. It is expected that by measuring the relative levels of expression of some of the proteins that change with specific environmental conditions, it might be possible to estimate *in situ* the physiological state of these biomining acidophiles, and therefore to assess some of the rate-limiting steps in the bioleaching process.

Aknowledgements.

Supported by FONDECYT 91-1010 and 194/0379, SAREC, ICI, Universidad de Chile and UNIDO 91/049.

References.

1. J. B. Stock, A. J. Ninfa, and A. M. Stock, "Protein Phosphorylation and Regulation of Adaptive Responses in Bacteria", Microbiol. Revs., 53, (1989), 450-490.

2. C. L. Brierley, "Bacterial leaching," Crit. Rev. Microbiol., 6, (1978), 207-262.

3. R. T. Espejo and P. Romero, "Growth of *Thiobacillus ferrooxidans* on Elementary Sulfur", Appl. Environ. Microbiol., 53, (1987), 1907-1912.

4. C. A. Jerez, "The Heat Shock Response in Meso and Thermoacidophilic Chemolithotrophic Bacteria", FEMS Microbiol. Lett., 56, (1988), 289-294.

5. P. Varela, and C. A. Jerez, "Identification and Characterization of GroEL and DnaK homologues in *Thiobacillus ferrooxidans*", FEMS Microbiol. Lett., 98, (1992), 149-154.

6. A. M. Amaro, D. Chamorro, M. Seeger, R. Arredondo, I. Peirano and C. A. Jerez, "Effect of External pH Perturbations on In vivo Protein Synthesis by the Acidophilic Bacterium *Thiobacillus ferrooxidans*," J. Bacteriol., 173, (1991), 910-915.

7. A. Matin, "The Molecular Basis of Carbon-Starvation-Induced General Resistance in *Escherichia coli*", Mol. Microbiol., 5, (1991), 3-10.

8. O. Tuovinen, "Biological Fundamentals of Mineral Leaching Processes," *In* H.L. Ehrlich and C.L. Brierley (ed.), Microbial Mineral Recovery (McGraw-Hill Book Co., New York, 1990), 55-77.

9. D. Rawlings, and T. Kusano, "Molecular Genetics of *Thiobacillus ferrooxidans*", Microbiol. Rev., 58, (1994), 39-55.

10. W. J. Ingledew, "*Thiobacillus ferrooxidans*. The Bioenergetics of an Acidophilic Chemolithotroph," Biochim. Biophys. Acta., 683, (1982), 89-117.

11. R. Mansch and W. Sand, "Acid-stable Cytochromes in Ferrous ion Oxidizing Cell-free Preparations from *Thiobacillus ferrooxidans*," FEMS Microbiol. Lett. 92, (1992), 83-88.

12. R. C. Blake, II, and E. A. Shute, "Respiratory Enzymes of *Thiobacillus ferrooxidans*. Kinetic Properties of an Acid-Stable Iron:Rusticyanin Oxidoreductase", Biochemistry, 33, (1994), 9220-9228.

13. G. Osorio, P. Varela, R. Arredondo, M. Seeger, A. M. Amaro, and C. A. Jerez, "Changes in Global Gene Expression of *Thiobacillus ferrooxidans* when Grown on Elemental Sulfur", p. 565-575. In A. E. Torma, M. L. Apel, and C. L. Brierley (ed.), Biohydrometallurgical technologies, vol 2. TMS Press, Warrendale, Pa.

14. D. Chamorro, R. Arredondo, I. Peirano, and C. A. Jerez, "The Programme of Proteins Synthesized by *Thiobacillus ferrooxidans* Under Different Environmental Conditions: Analysis by Two-dimensional Gels", *in* P. R. Norris and D. P. Kelly (ed.), Biohydrometallurgy: proceedings of the International Symposium. Warwick, 1987, Great Britain (Science and Technology Letters, London, Great Britain, 1988), 135-143.

15. R. Arredondo, A. García, and C. A. Jerez, "Partial Removal of Lipopolysaccharide from *Thiobacillus ferrooxidans* Affects its Adhesion to Solids", Appl. Environ. Microbiol., 60, (1994), 2846-2851.

16. P. Z. O'Farrell, H. M. Goodman and P. H. O'Farrell, "High Resolution Two-Dimensional Electrophoresis of Basic as well as Acidic Proteins", Cell, 12, (1977), 1133-1142.

17. M. Seeger, and C. A. Jerez, "Phosphate Limitation Affects Global Gene Expression in *Thiobacillus ferrooxidans*", Geomicrobiol. J., 10, (1992), 227-237.

18. M. Seeger, and C. A. Jerez, "Phosphate-Starvation Induced Changes in *Thiobacillus ferrooxidans*", FEMS Microbiol. Lett.,108, (1993), 35-42.

19. M. Seeger, and C. A. Jerez, "Response of *Thiobacillus ferrooxidans* to Phosphate Limitation", FEMS Microbiol. Revs., 11, (1993), 37-42.

20. C. A. Jerez, M. Seeger and A. M. Amaro, "Phosphate Starvation Affects the Synthesis of Outer Membrane Proteins in *Thiobacillus ferrooxidans*", FEMS Microbiol. Lett., 98, (1992), 29-34.

21. A. M. Amaro, M. Seeger, R. Arredondo, M. Moreno, and C. A. Jerez, "The Growth Conditions Affect *Thiobacillus ferrooxidans* Attachment to Solids", p. 577-585. In A. E. Torma, M. L. Apel, and C. L. Brierley (ed.), Biohydrometallurgical technologies, vol 2. TMS Press, Warrendale, Pa.

22. R. T. Espejo, B. Escobar, E. Jedlicki, P. Uribe and R. Badilla-Ohlbaum, "Oxidation of Ferrous Iron and Elemental Sulfur by *Thiobacillus ferrooxidans*," Appl. Environ. Microbiol., 54, (1988), 1694-1699.

23. N. Mjoli and C. F. Kulpa, "Identification of an Unique Outer Membrane Protein Required for Iron Oxidation in *Thiobacillus ferrooxidans*," *in* P.R. Norris and D.P. Kelly (ed.), Biohydrometallurgy: proceedings of the International Symposium. Warwick, 1987, Great Britain (Science and Technology Letters, London, Great Britain, 1988), 89-102.

24. W. G. Characklis, K. Marshall, and G. A. McFeters, "The Microbial Cell", in W. G. Characklis and K. C. Marshall (ed.), Biofilms. John Wiley & Sons, (1990), 131-159.

25. M. Fletcher, "Measurement of Glucose Utilization by *Pseudomonas fluorescens* that are Free-Living and that are Attached to Surfaces", Appl. Environ. Microbiol., 52, (1986), 672-676.

COMPARATIVE GENOMIC ORGANIZATION OF Thiobacilli USING PULSED FIELD GEL ELECTROPHORESIS TECHNIQUES: TAXONOMIC, GENETIC, ECOLOGICAL AND TECHNOLOGICAL APPLICATIONS.

I. Marín, D. Moreira, J.P. Abad and R. Amils.

Centro de Biología Molecular, Universidad Autónoma de Madrid, Cantoblanco, Madrid 28049, Spain.

ABSTRACT

In spite the fundamental as well as biotechnological potential of acidophilic chemolithotrophic microorganisms and the efforts of many groups, very little progress has been made in developing the tools with which to generate the genetic information related with their unique mode of obtaining energy. The introduction of pulsed field gel electrophoresis techniques (PFG) in microbiology has facilitated the generation of genomic information from microorganisms for which conventional techniques are not applicable or difficult to implement. In this communication we present the genomic organization and macrorestriction patterns for different members of the thiobacilli group, using PFG techniques. Their possible use for genetic, taxonomic, ecological and technical purposes related with biohydrometallurgic processes are discussed.

Mineral Bioprocessing II
Edited by David S. Holmes and Ross W. Smith
The Minerals, Metals & Materials Society, 1995

INTRODUCTION

Biohydrometallurgy, with its broad applications in metal extraction, coal desulfurization and metal recovery from diluted solutions, constitutes a biotechnological area with enormous industrial potential, mainly due to its low environmental impact and to its possible use in bioremediation processes. Different studies estimate the value of cooper recovery using these techniques since the beginning of its industrial application at 1 billion dollars (Holmes and Debus, 1991), and the potential of biomining until the end of this century (Brierley, 1984) at 5 billion dollars. Since the discovery that strict chemolithotrophic microorganisms were mainly responsible for the production of acidic waters and the associated high level of corrosion of metallic structures in mines (Colmer and Hinckle, 1947), an increasing interest in the study of these microorganisms has risen, even though their unusual growth conditions such as low pH, opaque mineral substrates, slow growth rates, etc., make progress in the field difficult. The biotechnological potential of these microorganisms accelerated the basic research required for the improvement of the different industrial processes and the subsequent opening of new areas of application.

Biohydrometalurgic processes offer several important advantages over pyrometalurgic techniques, some related with the economy of the processes: energy, possibility to treat minerals with low metal content, low cost of operation, and others with the restrictive environmental standards that metallurgic processes must meet (Decker, 86; Ballester et al., 1988, Amils et al., 1994).

Although bioleaching has been used for the past thirty years, the microorganisms responsible for this process are still not well characterized. Growth by oxidation of sulfidic ores is known only for a few microorganisms, for example the mesophiles, *Thiobacillus ferrooxidans, thiooxidans* and *Leptospirillum ferrooxidans*. These microorganisms have an obligate chemolithoautotrophic metabolism, growing optimally at pH values of around 2. For thermophiles, the best characterized systems belong to the sulfolobales group of archaebacteria. Several moderate thermophilic bacteria have been isolated from acidic mine drainages (Norris, 1990; Gómez et al., 1993) but most of them have not yet been taxonomically characterized.

Although the importance of a complex microbial ecology in the industrial bioleaching processes performed under non sterile conditions is recognized, most of the basic knowledge in the field has been concentrated on *T. ferrooxidans* (Rawlings and Kusano, 1994).

Classical genetic approaches have not proven to be very useful in characterising these microorganisms. Their growth requirements make it difficult to detect and select mutants that might lead to the identification and isolation of the genes involved in their unique mechanism of energy transduction. Due to the difficulty inherent in the study of chemolithotrophic microorganisms, our group has begun to explore the use of pulsed field gel electrophoresis (PFG) to obtain genomic information for extremophilies in general (Abad and Amils, 1991), and acidophilic chemolithotrophs in particular. In this report a general view of the state of the art in the application of PFG techniques to study of thiobacilli and its possible applications in the bioleaching field are presented.

MATERIALS AND METHODS

Bacterial strains and culture media. The following bacterial strains have been obtained from culture collections: *T. intermedius* ATCC 15466, *T. acidophilus* ATTC 27807, *T. ferrooxidans* ATTC 23270, *T. thiooxidans* ATTC8085, *T. thioparus* ATTC 8158, *T. perometabolis* ATTC 23370, *T. organoparus* ATTC 27977, *T. denitrificans* ATTC 23644, *Thiobacillus* sp. ATCC 2779, *T. cuprinus* DSM 5495. Also natural isolates from the Tinto river RT-1 and RT-2, characterized by our group, have been used in this study. Cells were cultured in different media using the specifications for the strain or media 9K supplemented with different inorganic substrates: elemental sulfur, mineral concentrates of Río Tinto.

Total intact DNA preparations. The basic method described by Smith and Cantor (1987), with the modifications introduced by Marín (1989), for the preparation of intact DNA from chemolithotrophic microorganisms have been used throughout the work.

Pulsed field electrphoretic conditions. DNA samples were analyzed by PFG using inhomogeneous fields in a Pulsaphor apparatus (Pharmacia-LKB). Gels were made up of 1% agarose (SeaKem LE Agarose,

FMC) in modified TBE buffer (100 mM Tris, 100 mM boric acid and 0.2 mM EDTA, final pH 8-8.4) and run at 15°C in the same buffer at different resolution windows: running time, voltage and pulse time. The size of the DNA fragments separated by PFG was estimated by comparison of their mobilities with those used as size markers: *Schizosaccharomyces pombe* 972 h$^-$ and *Saccharomyces cerevisiae* YP80 chromosomes (Smith et al., 1988), lambda phage DNA concatamers (48,5 Kb monomer), and restriction fragments of lambda phage DNA digested with EcoR I and Hindd III.

DNA restriction digestions. Restriction enzymes were purchased from New England Biolabs or Boehringer-Mannheim. Prior to restriction endonuclease digestion, plugs were thoroughly washed in a buffer containing 10 mM Tris-HCl, 1 mM EDTA, pH 8. For digestion, DNA samples in agarose were incubated for at least 6 hours at the appropriate temperature with 20 units of enzyme/ug of DNA in a final volume of 200 ul. After incubation the solution was then pipetted off and 250 ul of ESP solution (0.5 M EDTA pH 9.5, 1% sodium lauroyl sarcosine; 1 mg/ml of proteinase K) added and incubated for 2 hours at 50°C before samples were loaded on the PFG gel or stored at 4°C.

RESULTS

Due to the difficulties found with the conventional genetic techniques, our group decided to study the possibility of applying pulsed field gel electrophoresis techniques (PFG) to obtain genomic and genetic information for the thiobacilli group and related microorganisms. This technique permits the resolution of megabase pieces of DNA, allowing the determination of the number, size and topology of the different genomic elements: chromosomes and extrachromosomal elements. The obtention of macrorestriction patterns for different rare cutters allows the generation of the correspondent physical maps on which different genes can be placed, using homologous and heterologous probes, resulting in the construction of indirect genetic maps.

Obtaining Intact DNA. One of the limitations in the application of PFG techniques is the absolute requirement to obtain intact DNA. Partially degraded DNA could be due to problems in the preparation of DNA such as contamination, presence of endogenous nucleases, etc., or to mechanical damage produced to the DNA prior to the immobilization of the cells. Working with microorganisms which grow in the presence of important concentrations of metallic cations, and knowing the damage that heavy metals can produce to naked DNA, several experiments were carried out to design a protocol allowing intact DNA from strict chemolithotrophs to be prepared (detail practical conditions are described in Marín, 1989).

Number of genomic elements. The possibility of preparing intact DNA for different chemolithotrophic thiobacilli, in different growth conditions, allowed their genomic organization to be compared in terms of number and topology. Figure 1 shows the electrophoretic run at a long pulse time of total intact DNA from different thiobacilli species. As can be seen most of the DNA remains in the wells, which strongly suggests that it corresponds to circular chromosomes. Some of the samples show a band below the compression band which corresponds to linear forms of the chromosome generated by mechanical stress produced during the electrophoresis. It can be established that these bands correspond to the linear form of the chromosomes because the size, determined by comparison with their relative mobility to DNA markers, agrees with the size obtained by addition of the macroresticition bands (see below). In these conditions some of the extrachromosomal elements can be resolved, although shorter pulse times allow better characterization of these important genomic elements.

Extrachromosomal elements. PFG permits an adequate characterization of extrachromosomal elements such as number, topology and size. Linear elements run faster than supercoiled forms. Also linear forms do not show changes in their relative mobility when compared to linear markers at different pulse times. The correspondence between linear and supercoiled DNAs can be established by hybridization using one of the forms extracted from the gel. Also hybridization experiments can be used to study the homology between different extrachromosomal elements from different related microorganisms. Table I shows preliminary data on the number and size of different extrachromosomal elements present in different thiobacilli species. As can be seen there is a lack of correlation between the number and estimated size of different megaplasmids detected in different species. Most of the species studied show only one megaplasmid of different size with small homology when compared to the rest. *T. thiooxidans* is the only memeber of the *Thiobacillus* genus that exhibits five extrachromosomal elements. In the electrophoretic conditions used in this work small plasmids, less than 10Kb, are lost in the front of the gel. Several plasmids of small size have been reported and characterized for several *T. ferrooxidans* strains (Holmes et al., 1984; Rawlings et al., 1993) which have probably been underestimated in our characterization.

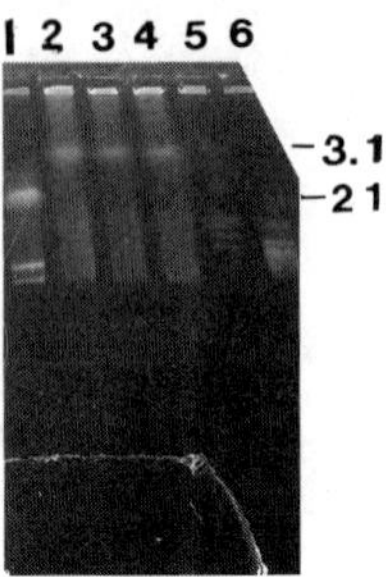

Figure 1: PFG of intact DNA from different acidophilic thiobacilli. Electrophoretic conditions: seven days run, 4500 seconds pulse time and 3 v/cm. Samples: 1: size markers; 2: *T. ferrooxidans*; 3: RT-1; 4: RT-2; 5: chromosomes of *H. wingei* and 9: chromosomes of *S. cerevisiae*

Species	Plasmids	Estimate size
- *T. thiooxidans*	pTT1	30 Kb
	pTT2	45 Kb
	pTT3	50 kB
	pTT4	100 Kb
	pTT5	250 Kb
- *T. thioparus*	pTTh1	275 Kb
- *T. perometabolis*	pTP1	150 Kb
- *T. organoparus*	pTO1	50 Kb
	pTO2	70 Kb
	pTO3	85 Kb
- *T. denitrificans* 23644	pTD1	50 Kb
	pTD2	120 Kb
- *T. denitrificans* 25259	pTD3	60 Kb
- *T. cuprinus*	pTC1	50 Kb
- *T. sp* 27793	pTSp1	150 Kb

Table I: Extrachromosomal elements of different thiobacilli species detected by PFG.

Special mention should be given to a linear plamid detected in *T. cuprinus* and related strains which is induced under chemolithotrophic conditions. Figure 2 shows the presence of a strong DNA signal in the 50 Kb range when the cells are grown in the presence of sulfur or in the presence of minerals from Río Tinto. This signal is not seen when the cells are grown heterotrophically. These results strongly suggest that this megaplasmid is induced in the presence of inorganic substrates probably because it carrys genetic information required for its chemolithotrophic growth. Further characterization of the genetic information stored in this plasmid is required before understanding its role in the bioleaching processes mediated by this microorganism.

Macrorestriction patterns. A number of different restriction enzymes have been used to obtain macrorestriction patterns for different thiobacilli species. The patterns originated with enzymes that produce an adequate number of restriction bands, which can be resolved by PFG, have different uses:

- i) comparison with patterns obtained with related microorganisms to evaluate similarity (karyotyping).
- ii) identification of genes using homologous and heterologous probes to facilitate their isolation and characterization (a comparative pattern of hybridization with different gene probes can increase the accuracy to the karyotyping obtained in i.

- iii) obtention of physical maps of the different genomic elements: chromosomes and extrachromosomal elements.
- iv) facilitate the obtention of genetic maps by combination of ii and iii.

Figure 3 shows the macrorestriction patterns for different thiobacilli species obtained with Xba I. The patterns, as observed for other microorganisms, (López-García et al.,1993) are very characteristic for each species. Similar results have been obtained with other restriction enzymes. Interestingly enough, the two *T. denitrificans* used in this work show quite distinct patterns of restriction, suggesting that their adscription to the same species might have to be reconsidered. Further analysis with other restriction enzymes and other techniques like rRNA sequencing, ribotyping or FRLP analysis is required to further clarify this situation. The strain *Thiobacillus sp.* ATTC 27793 has been deposited as a strain related with *T. perometabolis*. The restriction patterns exhibited by both systems are so different that this assumption becomes very unlikely. The restriction patterns for *T. thiooxidans* and *T. ferrooxidans* show a high degree of homology, which agrees with the results obtained with 16S rRNA sequences that places both species in the same cluster, although they exhibit interesting metabolic differences (Lane et al., 1992). The two isolates from the Tinto river, RT-1 and

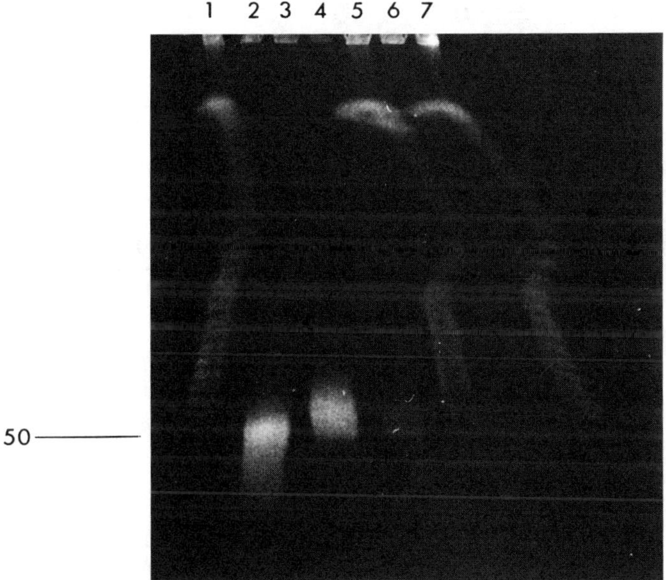

Figure 2: PFG of DNA from *T. cuprinus* cells grown in different conditions. Electrophoretic conditions: 48 hours, 100 seconds pulse time, 10 volts/cm. Samples: 1 and 7: lambda phage DNA concatamers; 2: *T. cuprinus* grown in mineral; 3 and 4: *T cuprinus* grown in different concentratios of S; 5 and 6: *T. cuprinus* grown in yeast.

RT-2, which phenotipicaly correspond to *T. ferrooxidans* strains, show some differences with the type collection strain, which raises some doubts about its taxonomic adscription to this species.

Obviously this type of approach requires further analysis once the macrorestriction database of this taxonomically ill defined group of microorganisms has been increased. Furthermore, other analytical techniques, like those mentioned above, will be required to clarify the situation. It would be useful if this group of microorganisms, which is extremely important due to its phylogenetic implications and its biotechnological applications and which is difficult to study, mainly due to its unusual growth conditions and the lack of appropiate techniques, could be adequately defined at a taxonomic level, to avoid unnecessary complications produced by the utilization of only one criteria. The macrorestriction patterns obtained so far agree with the separation of the thiobacilli species into at least two different clusters, the strict chemolithotrophs and the facultative heterotrophs, suggested by rRNA sequence data comparison (Lane et al., 1985; Lane et al., 1992).

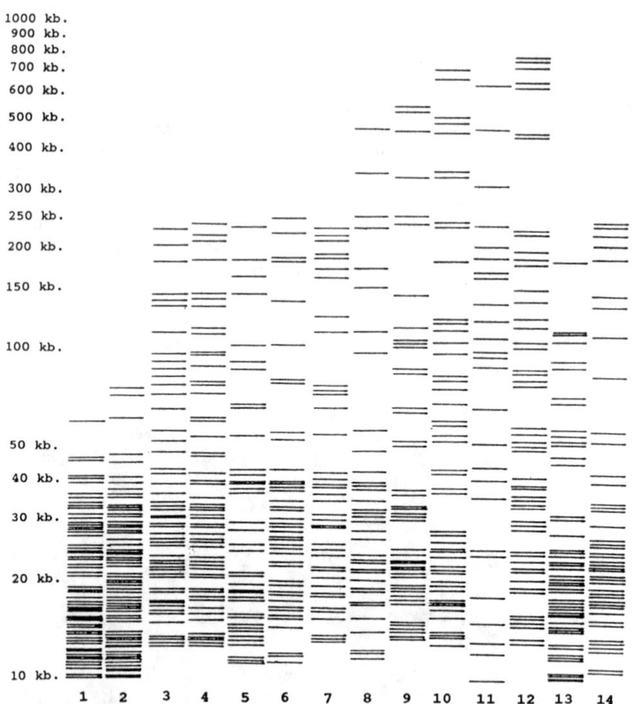

Figure 3: Comparative macrorestriction patterns of different thiobacilli species for Xba I. Different electrophoretic conditions were used to resolve the different restriction fragments produced by Xba I. Lane 1: *T. perometabolis*; 2: *T. intermedius*; 3: *T. thiooxidans*; 4: *T. Ferrooxidans*; 5: *T. denitrificans* ATCC 23644; 6: *T. denitrificans* ATTC 25259; 7: *T.acidophilus*; 8: *T. thioparus*; 9: *T. tepidarius*; 10: *T. thioparus*; 11: *T. cuprinus*; 12: *T. sp.* ATCC 27793; 13: isolate RT-1; 14: isolate RT-2.

Physical and genetic maps of genomic elements. One of the most important characteristics of PFG analysis is that the data generated in each of the different steps of the analysis can be used to obtain additional complementary information, helping to build up the complete genomic information of the microorganism being studied. The preparation of intact DNA allows the identification and isolation of the different genomic elements, and to study their stability under diverse growth conditions. Their macrorestriction patterns can be of use for karyotyping, taxonomic adscription of new isolates and patent protection, but most importantly, they are the first step for the obtention of the correspondent physical maps, which will eventually generate genetic maps by location of genes through the use of appropiate gene probes (López-García et al., 1994). Using the top-down approach (López-García et al., 1992) and adequate restriction enzymes the physical maps of *T. cuprinus* and *T. ferrooxidans* chromosomes are being refined. The location of different genes on the physical maps will facilitate the obtention of low resolution genetic maps for the chromosomes of both microorganisms.

A serious limitation on the generation of reverse genetic maps for thiobacilli species is the lack of gene probes for the identification of the genetic information commited with their unusual physiological and metabolic properties, the most important being those related with bioleaching. An interesting approach to overcome this problem is to use differential gene expression analysis in order to identify gene products induced at different conditions, e.g., different mineral substrates, presence of high concentrations of heavy metals, starvation for important elements, etc. This pioneering approach introduced into the bioleaching field by the Jerez group (Jerez et al., this volume) is being used to identify gene products which are differentialy expressed under different growth conditions. In figure 4 the differential expression of *T. cuprinus* proteins is analyzed in

2D gel electrophoresis after switching from heterotrophic to chemolithotrophic growth conditions. The microsequence of the differential purified proteins extracted from the gel should allow to synthesize the correspondent oligonucleotide. This gene probe could be used for locating the gene in the macrorestriction patterns. The cloning of the correspondent fragment will facilitate the obtention of the gene sequence and the screening for similar genes using the macrorestriction patterns of related microorganisms. This approach should facilitate the genetic analysis of these microorganisms and its comparison with others related phylogenetically, thus increasing the comprehension of the evolution of this unusual mode of energy obtention.

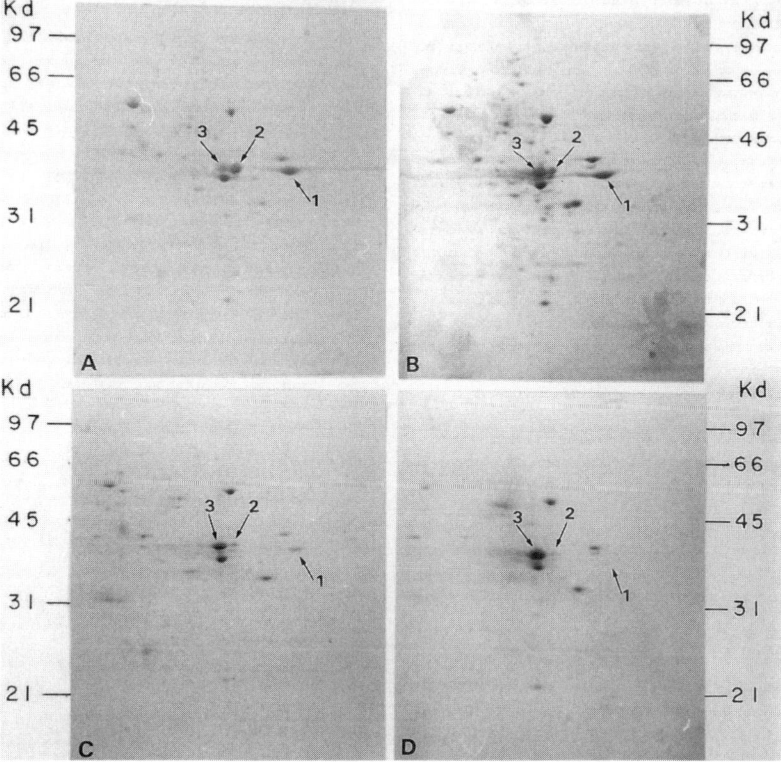

Figure 4: 2D-electrophoretic analysis of total proteins extracted from *T. cuprinus* cells grown heterotrophically or chemolithotrophically in the presence of sulfidic minerals (from Marín and Amaro, 1993, reproduced with permission of the editor).

Genomic organization of rRNA genes and sequence analysis. As part of the comparative genomic analysis of thiobacilli we have been interested in their rRNA genes for different reasons: to gain comparative information on the number of operons involved in the synthesis of ribosomes in this group of microorganisms, to obtain comparative hybridization patterns for these extremely conserved genes (ribotyping), to use them as references to compare the relative positions of other genes on the genetic maps (López- García et al., submitted), and to obtain rRNA sequences suitable for phylogenetic analysis. To date we have analyzed the rRNA operon of *T. cuprinus*, from which the complete sequence of the 23S rRNA gen has been obtained (figure 5).

```
   1 GATCAAGCGA CTAAGTGCAT ACGGTGGATG CCTTGGCGAG TACAGGCGAC
  51 AAGGACGTGG AAGCCTGCGT TAAGCTGCGG GGAGCTGTGC AATCAAGCTT
 101 TGATCCGTAA ATGTCCGAAT GGGGAAACCC ACCCGTAAGG TATCTTGCTG
 151 AATACATAGG CAAAGAAGGC GACCGAGTGA ACTGAAACAT CTAAGTAACT
 201 CGAGGAAAAG AAATCCAACC GAGATTCCGA AAGTAGTGGC GAGCGAAATC
 251 GACCAGCCTG GTACGTTTTA GCGTAGATAT CAGGGGAATC TCTTGGAAAA
 301 GAGAGCCACA GTGGGTGATA GCCCCGTACC TAAAAATATT TGCGTGGAAC
 351 TAGGCGTACG ACAAGTAGGG CGGGACACGT GTAATCCTGT CTGAACATGG
 401 GGGGACCATC CTCCAAGGCT AAATACTCGT AGTCGACCAT AGTGAACAAG
 451 TACCGTGAGG GAAAAGGTGA AAAGAACCCC GGGAGGGGAG TGAAATAGAT
 501 CCTGAAACCG TATGCATACA AAAAAGTCGG AGCCTCGCAA GGGGTGACGG
 551 CGTACCTTTT GTATAATGGG TCAGCGACTT ACATTCAGTG GCAAGCTTAA
 601 CCGAATAGGG AAGGCGTAGG GAAACCGAGT CCGAACAGGG CGTTCAGTCG
 651 CTGGGTGTAG ACCCGAAACC GGATGATCTA TCCATGGCCA GGATGAAGGC
 701 TGGGTAACAC CAGCTGGAGG TCCGGATCCA CTAGTGTTGC AAAATTAGCG
 751 GATGAGCTGT GGATAGGGGT GAAAGGCTAA ACAAATCCGG AAATAGCTGG
 801 TTCTCCTCGA AAACTATTTA GGTAGTGCCT CGTGTATCAC CTTCGGGGGT
 851 AGAGCACTGT TTTGGCTAGG GGGTCATGGC GAGTTACCAA ACCAAGGCAA
 901 ACTCCGAATA CCGAAGAGTG TCAGCACGGG AGACAGACAC CGGGTCGTAA
 951 CGCTCGGACA CAAGAGGGAA ACAACCCAGA CCGCCAGCTA AGGTCCCTAA
1001 ATATAGCTAA GTGGGAAACG AAGTGGGAAG GCTTTGACAG TCAGGAAGTT
1051 GGCTTAGAAG CAGCCATCCT TAAAGAAAGC GTAATAGCTC ACTGATCGAG
1101 TCGTCCTGCG CGGAAGATGT AACGGGGCTA AGCTATATAC CGAAGCTGCG
1151 GATGTGCGTA ACGTGGTAGA GGAGCGTTCC GTTCCGTAAG CCTGCGAGGG
1201 TGTTCTGTAA GGAATGCTGG AGGTATCGGA AGTTCGAATC GTGACATGAG
1251 TAGCGTTAAA GGGGGTGAAA GCTCCCCTCG CCGTAAGTCC AAGGTTTCCT
1301 ACGCAACGTT CATCGGCGTA GGGTAGTCGG CCCCTAAGGC GAGGCAGAAA
1351 TGCGTAGCTG ATGGGAAACA GGTTAATATT CACTGTACCA CATGTCTAGT
1401 GCGATGTGGG GACGGAGAAA GTTAGGTTAT CCAGGTGTTG GATGTCCTGG
1451 TTCAAGCAGG TAGGCGTGCC CCTTAGGCAA ATCCGGGGGG CTAAGCTGAG
1501 ATGTGATGAC GAGCGGGGCT TGCCGCTGAA GTAACTGATA CTCTGCTTCC
1551 AAGAAAAGCC ACTAAGCTTC ACTAGACGTG ACCGTACCGC AAACCGACAC
1601 TGGTGCGCGA GATGAGTATT CTCAGGCGCT TGAGAGAACT CGGGAGAAGG
1651 AACTCGGCAA ATTTGTACCG TAACTTCGGG ATAAGGTACG CCCCTAGTAT
1701 GTGACGCCGT ACAGGCTGAG CAGAATGGGG CCGACGTGAA AAGGTGGCTG
1751 CGACTGTTTA TTAAAAACAC AGCACTCTGC AAACACGAAA GTGGACGTAT
1801 AGGGTGTGAC GCCTGCCCGG TGCCGGAAGG TTAAGTGATG GGGTGCAAGC
1851 TCTTGATCGA AGCCCGGTAA ACGGCGGCCG TAACTATAAC GGTCCTAAGG
1901 TAGCGAAATT CCTTGTCGGG TAAGTTCCGA CCTGCACGAA TGGCGTAACG
1951 ATGGCCACAC TGTCTCCTCC CGAGACTCAG CGAAGTTGAA ATGTTTGTGA
2001 TGATGCAATC TCCCCGCGGA AAGACGGAAA GACCCCATGA ACCTTTACTG
2051 TAGCTTTGCA TTGGACTTTG AACAGATCTG TGTAGGATAG GTGGGAGGCT
2101 TTGAAGCCGG GATGCTAGTT CCGGTGGAGC CAACCTTGAA ATACCACCCT
2151 GGTGTGTTTG AGGTTCTAAC CTTGGTCCTT GAATCAGGAC TGGGGACAGT
2201 GCATGGTGGG CAGTTTGACT GGGGCGGTCT CCTCCTAAAG CGTAACGGAG
2251 GAGTTCGAAG GTACGCTAGG TACGGTCGGA AATCGTGCTG ATAGTGCATG
2301 CATAGCGTGC TTACTGCGAG ACTGACACGT GACGAATGCG AAACGGACAT
2351 AGTGATCCGA GAACAGTATG AAGGGCCATC GCTCAACGGA TAAAAGGTAC
2401 TCTGGGATAA CAGGCTGAAG TACCGCCCAA GAGTTCATAT CGACGGCGGT
2451 GTTTTGGCAC CTCGATGTCG GCTCATCTCA TCCTGGGGCT GTAGCCGGTC
2501 CCAAGGGTAT CCTGTTCGCC ATTTAAAGAG GTACGTGAGC TGGGTTTAAA
2551 ACGTCGTGAG ACAGTTTTGT CCCTATCTTC CGTGGGCGCT GGAAGCTTGA
2601 GAGGGGCTGC TCCTAGTACG AGAGGACCGG AGTGGACGCA CCCCTGGTGT
2651 ACCGGTTGTC ACGCCAGTGC ATCGCCGGGT AGCTATGTGC GGAAGAGATA
2701 ACCGCTGAAA GCATCTAAGC GGGAAACTCG CCTCAAGATT AGGCTTCCCT
2751 GGAGACTTGA TCTCCCTAAA AGGGTCGTTG AAGACGACAA CGTTGATAGG
2801 CTGGTGTGGA AGCGCAGTAA TGCGTTGAGC TAACAGTACT AATTGCCCTA
2851 GCTTGATC
```

Figure 5: *T. cuprinus* 23S rRNA gene sequence.

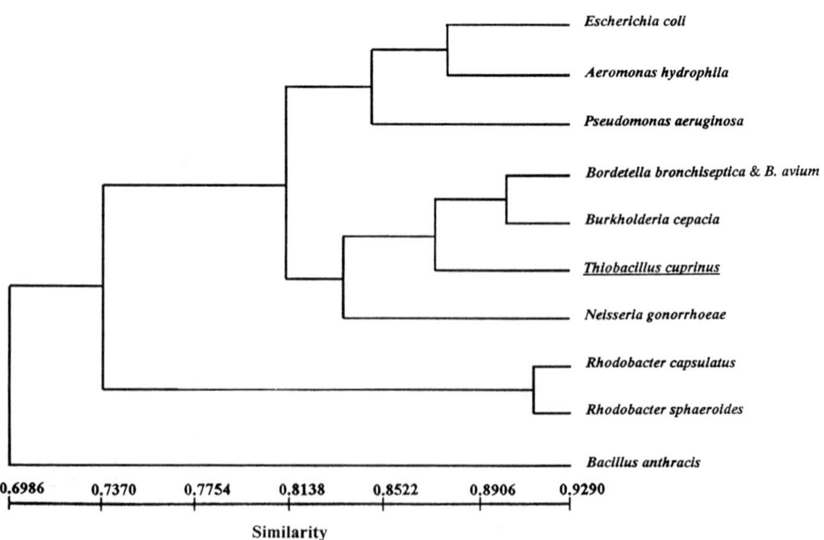

Figure 6: Sequence homology dendogram for 23S rRNA sequences. From Moreira et al, 1994, reproduced with permision of the editor.

Using the LSU rRNA sequences of the EMBL databank the 23S rRNA from *T. cuprinus* shows the highest homology with the 23S rRNA from *Burkholderia cepacia* (88.7%) (figure 6). This result confirms the affiliation of *T. cuprinus* to the B1 subgroup of the purple bacteria (Moreira et al., in press). A previous phylogenetic study based on 5S rRNA sequences showed that *B. cepacia* is related with members of the thiobacilli group sharing some metabolic characteristics with *T. cuprinus*: *T. intermedius* and *T. perometabolis* (Lane et al., 1985). This data suggest a phylogenetic clustering of the facultative heterotrophic thiobacilli, which agrees quite well with the macrorestriction patterns shown earlier, and their possible relationship with some *Pseudomonadaceae*. More rRNA sequences are required to further clarify this hypotesis, mainly due to the high phylogenetic dispersion of the thiobacilli group. Determination of complete sequences of the different thiobacilli rRNA genes will increase our knowledge of the phylogenetic adscription of this outstanding group of bacteria and will facilitate its comparison with other microorganisms with similar metabolic capabilities.

CONCLUSIONS

PFG techniques have been implemented for the study of acidophilic thiobacilli. The use of different electrophoretic conditions allowed to ascertain the presence of circular chromosomes and a variable number of extrachromosomal elements with little homology between them. Different restriction enzymes have been used to generate comparative macrorestriction patterns, which are of use for karyotyping the different strains, for the taxonomic adscription of new isolates, and for the generation of physical maps. The use of heterologous and homologous gene probes should allow the construction of genetic maps. The combination of this information together with complementary information obtained with other techniques should allow to better understand the role of these microorganisms in nature and to facilitate the screening for related microorganisms with improved technological applications.

AKNOWLEDGEMENTS

This work has been supported by the following grants: CICYT (BIO94-0733-C03-01) and EEC (BIO2-CT93-0274). Also institutional grants from the Fundación Areces and the Fondo de Investigaciones Sanitarias to the CBM are knowledge.

REFERENCES

ABAD J.P., AMILS R. "Genomic organization studies in halobacteriaw using pulse field gel electrophoresis" In "General and Applied Aspects of Halophilic Microorganisms". F. Rodriguez Valera ed. (New York, Plenum Press, 1991) 295-303.

AMILS R., MARÍN I., LOPEZ A.I., MOREIRA D., IRAZABAL N., GARRIDO X. "Environmental Applications of Biohydrometallurgy" (Paper presented in Environmental Biotechnology Workshop, Biotec 94, Algarve, Portugal, 2 october 1994), BAm C05.

BALLESTER A., GONZALEZ F., BLAZQUEZ M.L. "Biolixiviación de menas naturales. Posibilidades actuales de utilización". Rev. Metal. Madrid. 24 (1988) 91-102.

BOTELLA J.A., LOPEZ GARCIA P., ABAD J.P., SMITH C., AMILS R. "Aligment of genes and *Swa I* restriction sites to the *BamH I* genomic map of *Haloferax mediterranei*". FEMS Microbiol., 117 (1994) 53-60.

BRIERLEY C.L. "Microbial mining: Technology status and commercial opportunities". World Biotech. Rep., 1 (1984) 599-609.

COLMER A.R., HINCKLE M.E. "The role of microorganisms in acid mine drainage: a preliminary report". Science. 106 (1947) 253-256.

DECKER R:F: "Biotechnology/Materials: The growing Interface". Metall. Trans., 17A (1986) 5-14.

GOMEZ E., LOPEZ A.I., MARIN I., AMILS R. "Isolation and characterization of novel microorganisms from Río Tinto". In "Biohydrometallurgic Technologies, Vol II". A.E. Torna, M.L. Apel, C.L. Brierley eds. (Warrendale, TMS, 1993) 479-486.

HOLMES D.S., LOBOS J.H., BOPP L.H., WELCH G.C. "Cloning of a *Thiobacillus ferrooxidans* plasmid in *Escherichia coli*". J. Bacteriol., 157 (1984) 324-326.

HOLMES D.S., DEBUS K.A. "Oportunities for biological metal recovery". In "Engineering Foundation Conference, Mineral Bioprocessing".R.W. Smith & M.Misra eds.(Reno, 1991) 57-78.

KUSANO T., SUGAWARA K., INOUE C., TAKESHIMA T., NUMATA M., SHIRATORI T. "Electrotransformation of *Thiobacillus ferrooxidans* with plasmids containing a *mer* determinant as the selective marker by electroporation". J. Bacteriol., 174 (1992) 6617-6623.

LANE D.J., STHAL D.A., OLSEN G.J., HELLER D.J., PACE N.R. "Phylogenetic analysis of the genera *Thiobacillus* and *Thiomicrospira* by 5S rRNA sequences". J. Bacteriol., 163 (1985) 75-81.

LANE D.J., HARRISON A.P., STHAL D. Jr., PACE B., GIOVANONI S.J., OLSEN G.J., PACE N.R. "Evolutionary relationships among sulfur- and iron-oxidizing eubacteria". J. Bacteriol., 174 (1992) 269-278.

LOPEZ GARCIA P., ABAD J.P., Smith C., AMILS R. "Genomic organization of the halophilic archaeon *Haloferax mediterranei*: Physical map of the chromosome". Nucl. Acids Res., 20 (1992) 2459-2464.

LOPEZ GARCIA P., ABAD J.P., AMILS R. "Genomic analysis of different species of *Haloferax mediterranei* using pulse field gel electrophoresis" Syst. Appl. Microbiol., 16 (1993) 310-321.

MARIN I. "Asignación filogenética y estudio del genoma de Hö5, un nuevo microorganismo lixiviador" (Ph.D. thesis, Universidad Autónoma de Madrid, 1989), 35-45.

MARIN I., AMARO A.M. "Study of proteins specifically involved in the chemolithotrophic metabolism of *Thiobacillus cuprinus*" In "Biohydrometallurgic technologies, Vol II". A.E. Torma, M.L. Apel, C.L. Brierley eds. (Warrendale, TMS, 1993) 473-478.

MOREIRA D., AMILS R., MARIN I. "Complete primary structure of the 23S rRNA coding gene from *Thiobacillus cuprinus* and its similarity with that of *Burkholderia cepacia*". Syst. Appl. Microbiol., (1994) in press.

NORRIS P.R. "Acidophilic bacteria and their activity in mineral sulfide oxidation". In "Microbial mineral recovery". H.L. Ehrlich and C.L. Brierley eds. (New York, McGraw-Hill Book Co.,1990) 3-27.

RAWLING D.E., DORRINGTON R.A., ROHRER J., CLENNEL A.M. "A molecular analysis of the replication and mobilization regions of a broad-hoste-range plasmid isolated from *Thiobacillus ferrooxidans*" FEMS Microbiol. Rev., 11 (1993) 3-7.

RAWLING E.D., KUSANO T. "Molecular genetics of *Thiobacillus ferrooxidans*" Microbiol. Rev. 58 (1994) 39-55.

SMITH C.L., CANTOR C.R. "Purification, specific fragmentation and separation of large DNA molecules". In "Methods in Enzymology: Recombinant DNA". R. Wu ed.(New York, Academic Press, 1987) 449-467.

SMITH C.L., KLCO S.R., CANTOR C.R. "Pulsed field gel electrophoresis and the technology of large DNA molecules" In "Genomic analysis: A practical approach". K. Davies, ed. (Oxford, IRL Press, 1988) 41-72.

III.

BIOLEACHING

PILOT SCALE MICROBIAL LEACHING OF GOLD AND SILVER FROM AN OXIDE ORE IN ELSHITZA MINE, BULGARIA

S.N. Groudev[1], I.M. Ivanov[1], I.I. Spasova[1]
and V.I. Groudeva[2]

[1] Research and Training Centre of Mineral Biotechnology, University of Mining and Geology, Sofia 1156, Bulgaria

[2] Department of Microbiology, Faculty of Biology, University of Sofia, Sofia 1421, Bulgaria

Abstract

A heap containing 415 tons of a low-grade ore with a gold content of 2.2 g/ton was leached by solutions containing both thiosulphate and amino acids of microbial origin as gold-complexing agents. 36.3% of the gold was leached within 32 days. The degree of extraction depended strongly on the size of the ore particles. About 68% of the gold was extracted from the finest ore fraction (less than 4 mm in size) but only 9% was extracted from the coarsest ore fraction (more than 25 mm in size). Silver was leached together with gold. When the pregnant solutions were treated by cementation with metallic zinc high-grade gold-silver concentrates were obtained. The barren solutions from the cementation unit were supplemented with make up water and reagents to the desired levels and were recycled to the heap.

Mineral Bioprocessing II
Edited by David S. Holmes and Ross W. Smith
The Minerals, Metals & Materials Society, 1995

Introduction

World-wide present practice for extracting precious metals finely disseminated in oxide ores is leaching by means of cyanide solutions. However, the cyanides are highly toxic and can cause substantial environmental problems. Moreover, cyanides are costly reagents which, in some cases, makes their use economically disadvantageous.

Some non-toxic chemical reagents such as thiosulphate and thiourea form stable complexes with gold and silver and leach efficiently these precious metals from oxide ores. However, because of different reasons these reagents are not used in industrial scale operations.

It is known that different heterotrophic bacteria are able to leach gold from oxide ores (1-8). The dissolution of gold is connected with the microbial production of gold-oxidising (peroxides) and gold-complexing (amino acids, peptides, proteins and nucleic acids) agents. The oxidising agents turn the native gold into an ionic state. The gold ions are then complexed by the different microbial metabolites. Different gold complexes are formed depending on the nature of the complexing metabolites but the gold - amino acid complexes are the most stable and always dominate quantitatively in the most active bacterial cultures. The formation of such complexes involves nitrogen of the amino groups but other functional groups may also participate in the formation of complexes between gold and amino acids. Silver is also solubilized in this way.

The efficacy of the process is dependent on the nature and levels of oxidising and complexing agents in the culture medium. The culture medium contains a suitable organic source of carbon and energy for the bacteria. Peptone or different sugars are usually used as such sources. The best results are obtained by some *Bacillus* strains producing peroxide compounds and one or more of the amino acids (aspartic acid, histidine, serine, cysteine, methionine) which form the most stable complexes with the ionic gold under slightly alkaline conditions (pH from 8 to 10). These strains also possess high catalase, superoxide dismutase and oxidase enzymatic activities. The different oxidase enzymes degrade the organic substrates in the culture medium and form some radicals (mainly O_2^- and OH^-) which are very strong oxidising agents but, at the same time, are very toxic for the bacteria. The enzymes catalase and superoxide dismutase prevent the toxic action of these radicals and of the peroxide by involving them into reactions as a result of which molecular oxygen is formed.

The process above-mentioned is more economically attractive when some waste organic compounds are used as a source of carbon and energy (3,7). In some cases it is necessary to maintain the pH of the culture solution at the desired level, i.e. between 8 and 10, by addition of some alkalising agents. In acidic media the gold complexes with the amino acids are degraded, the gold reduced to metallic state and precipitated.

In some cases active gold-leaching microbial strains are isolated from deposits of gold ores. The microflora of such deposits are quite varied but the most active gold-leaching strains are related to different species of the genera *Bacillus*, *Pseudomonas*, and *Bacterium*. It must be noted, however, that very active strains can be isolated also from other biotopes which contain no gold. There is no evidence as yet that microbial strains possessing a specific adaptation to gold-bearing mineral raw materials or capable of growing autotrophically on gold exist. In any case, the rates of gold solubilisation and concentrations obtained by using microbial strains growing in the presence of the gold-bearing ores are much lower than those of current industrial gold retrieval systems. Furthermore, such type of leaching is possible only under aseptic conditions. Under non-aseptic conditions the leach systems are contaminated by different heterotrophic microorganisms which utilize the initial organic substrate as well as the amino acids produced by the gold-solubilising bacteria.

The most important factor limiting the bacterial activity is the relatively low production of gold-oxidising metabolites. The gold solubilisation is markedly accelerated when a chemical gold oxidant is added to the microbial cultures. However, the most efficient gold oxidants (potassium permanganate, potassium persulphate, different peroxides, etc) are toxic for the bacteria. For that reason, these oxidants are added to culture fluids previously formed by different amino acid-producing bacteria, i.e. the bacterial growth and secretion of amino acids are separated from the leaching of the gold-bearing ores. However, the most attractive, from an economical point of view, is the utilisation of protein hydrolysates as sources of amino acids. Such hydrolysates are obtained from different rich-in-protein waste products, including microbial biomass (9, 10). Alkaline solutions containing gold-complexing amino acids obtained after hydrolysis of biomass of the yeast *Saccharomyces lactis* as well as the oxidant potassium permanganate leach efficiently gold and silver from oxide ores (11). The microbial strain used as a source of amino acids is an efficient producer of the enzyme superoxide dismutase. The protein hydrolysate is obtained from the microbial biomass as a by-product after the separation of the enzyme-bearing fraction. The hydrolysate contains 36.5% protein and includes mainly such amino acids (aspartic acid, histidine, serine, cysteine, methionine) which form the most stable complexes with the ionic gold under slightly alkaline conditions (pH from 8 to 10).

Very efficient is the use of leach solutions containing both the above-mentioned protein hydrolysate and thiosulphate as gold-complexing agents as well as copper and sulphite ions (12,13). In the presence of amino acids the leaching rates and the final extractions of gold are similar to those obtained by solutions containing only thiosulphate as a gold-complexing agent, but the optimum concentrations and the overall consumption of thiosulphate are lower than those in the absence of amino acids. In this system the native gold is oxidized and turned into an ionic state by the molecular oxygen dissolved in the leach solution. The gold ions are then complexed by the thiosulphate ions and the amino acids. The copper ions act as a catalyst in both oxidation and complexation. The sulphite ions are necessary to inhibit the decomposition of the thiosulphate and to prevent the precipitation of already dissolved gold. Silver is solubilised together with gold.

This leaching method is similar to cyanidation with respect to the rates of gold and silver leaching as well as to the final extractions of these precious metals. The technological application is also similar. The pregnant solutions containing gold and silver dissolved as complexes with amino acids and thiosulphate are treated efficiently by the well-know methods for recovering these metals from solutions (sorption with activated carbon or biomass, ion-exchange and cementation with zinc). However, the reagents used in this combined chemico-biological method are not toxic and the method as a whole is more economically attractive than cyanidation.

A large number of different gold-bearing oxide ores from deposits in Bulgaria and other countries were efficiently leached by the above-mentioned method under laboratory conditions. The excellent results obtained under such conditions with ore from the well-known Bulgarian gold deposit Petelovo (14) gave the possibility of applying the microbial leaching of this ore under pilot scale conditions. The main results from these pilot scale experiments are shown in this paper.

Materials and Methods

The gold-bearing ore from the Petelovo deposit consists of secondary, highly modificated quartzites, with a high content of iron oxides and alumina. The gold is finely disseminated in the ore and reaches relatively higher concentrations in the iron oxides. The size of the gold particles is less than 2 microns.

In 1993 a heap containing 415 tons of the above-mentioned ore was formed on a treatment site near the Elshitza mine in Central Bulgaria (at a distance of about 25 km

Table I. Chemical Analysis of an Oxide Ore Sample from the Petelovo Deposit

Component	Content	Componenet	Content
SiO_2	69.17 %	Ag	10.5 g/Mg
Fe_2O_3	15.28 %	Au total	2.2 g/Mg
Al_2O_3	6.45 %	Gold phases (in % from	
TiO_2	0.42 %	the total gold content):	
Cu	0.16 %	- free gold	18.10 %
Zn	0.10 %	- gold encapsulated in	
S total	0.88 %	iron oxides (recoverable	
S sulphatic	0.78 %	by cyanidation)	71.10 %
CaO	1.25 %	- gold finely disseminated	
BaO	0.35 %	in sulphide and silicate	
MgO	0.20 %	minerals	10.80 %
Na_2O	0.15 %		
Loss on burning	3.23 %	- Total	100.00 %

from the Petelovo deposit). The heap was formed on a slightly steep ground covered by a corrosion-resistant cement to facilitate the collection and to prevent the seepage of solutions. The heap had the shape of a truncated pyramid and was 16 m long, 1.70 m high, 6 m wide at the top end and 10 m wide at the bottom end.

Data about the chemical composition and the particle size of the ore are shown in Tables I and II, respectively.

The ore was initially treated by alkaline solutions (with a pH in the range of 9 - 10 obtained by addition of ammonium hydroxide) which were pumped to the top of the heap and after percolation through the ore mass and readjustment of the pH were recycled to the heap. When the pH of the heap effluents was stabilized in the range of 8.5 - 9.5, the leaching of the precious metals started with leach solutions containing protein hydrolysate from *Saccharomyces lactis* biomass - from 0.5 to 2.0 g/l, thiosulphate ions (added as ammonium thiosulphate) - from 5 to 20 g/l, copper ions (added as $CuSO_4 \cdot 5H_2O$) - from 0.5 to 1.0 g/l, sulphite ions (added as $(NH_4)_2SO_3$) - from 0.5 to 2.0 g/l. The pH of the solutions was maintained in the range of 8.5 - 9.5 by addition of ammonium hydroxide.

The leach solutions were pumped to the top of the heap at a rate from 50 to 200 l/ton ore for 24 h. The solutions were introduced onto the heap by means of spraying and flooding. The spraying was carried out by means of plastic sprinkler heads, while the flooding was carried out by perforated plastic pipes put on the surface of the heap.

The solutions percolated through the heap and dissolved gold and silver. The heap effluents were collected in a collection pond and then were pumped to the cementation unit where the dissolved precious metals were precipitated from the solutions being treated by means of cementation reactions with metallic zinc. The cementation unit consisted of four cementators which worked simultaneously but independently from each other under continuous-flow conditions.

The depleted solutions from the cementation unit were collected in a regeneration pond where make up water and reagents were added to the desired levels. The leach solutions adjusted in this way were recycled to the heap.

The dissolved metals were determined by ICP and AA spectroscopy. The amino acid concentrations were determined by an amino acid analyzer. The total concentration of soluble sulphur compounds was measured by ICP and AA. The thiosulphate ions were determined titrimetrically with iodine. The sulphate ions were determined gravimetrically by using $BaCl_2$. The identification of the soluble sulphur compounds was carried out by ascending paper chromatography, using a tertiary mixture of n-butanol : acetone : water (2 : 2 : 1.2) (7).

After the end of the leaching the ore was washed with water and a sampling was carried out to determine the residual contents of gold and silver in the different classes of ore particles.

Results and Discussion

The leaching of precious metals from the heap was very efficient (Figure I). 36.36% of the gold and 27.50% of the silver were leached within 32 days. The degree of extraction highly depended on the size of the ore (Table II). The maximum concentrations of dissolved gold and silver in the pregnant solutions were as high as 0.8 and 1.2 mg/l, respectively.

The optimum concentrations of thiosulphate ions, microbial protein hydrolysate and copper ions in the leach solutions were found to be 10 g/l, 0.5 - 1.0 g/l and 0.5 g/l, respectively. Sulphite ion concentrations in the range of 0.5 - 1.0 g/l were sufficient to

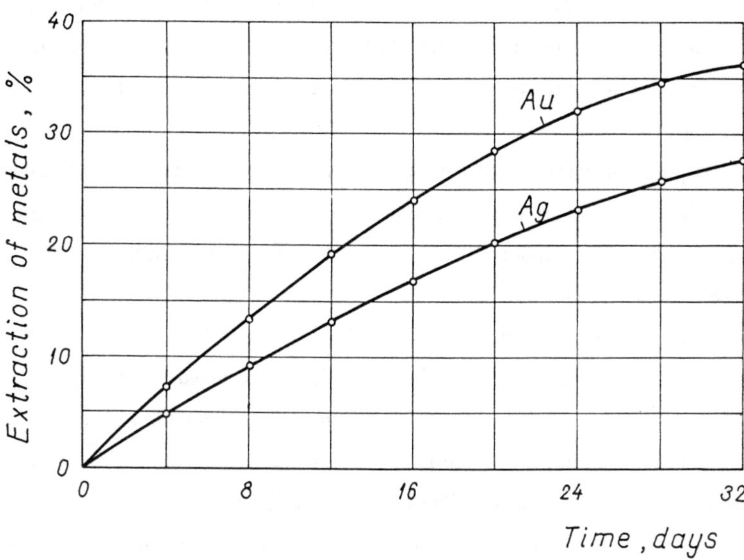

Figure I. Microbial Leaching of Gold and Silver from a Heap Consisting of an Oxide Ore from the Petelovo Deposit

Table II. Particle Size Analysis of the Ore Before and After the Leaching of the Heap

Ore size, mm	Yield, %		Gold content, g/Mg		Gold extraction, %
	I	II	I	II	
+25	10.70	10.44	1.5	1.4	8.9
-25 +10	55.82	56.52	1.9	1.6	14.7
-10 +4	14.03	13.73	2.5	1.1	56.9
-4	19.45	19.31	3.2	1.0	68.4
Total	100.00	100.00	2.2	1.4	36.36

I - Before leaching; II - After leaching.

inhibit the decomposition of thiosulphate and to reduce its consumption during the leaching process. The pH of solutions during their percolation through the ore heap decreased by 0.1 - 0.5 pH units, and the redox potential (Eh) increased from the initial 10 - 50 mV to about 60 - 100 mV.

The optimum irrigation rate was 100 l/ton ore for 24 h. It was found that spraying gave better leaching presumably because it provided more uniform distribution of the solutions as well as more dissolved oxygen. However, the evaporation rates were higher than those observed at flooding and amounted to about 10%.

It was found that different chemolithotrophic bacteria and heterotrophic microorganisms grew well in the ore heap and in the recycled solutions. Most of the chemolithotrophs were related to the species *Thiobacillus thioparus* and they grew at the expense of the thiosulphate. The number of these bacteria in the pregnant solutions in some cases exceeded 10^6 cells/ml. The heterotrophs grew mainly at the expense of the amino acids in the leach solutions. Most of these microorganisms were related to the genera *Bacillus* and *Pseudomonas* and their total number in the pregnant solution sometimes exceeded 10^5 cells/ml. Some moulds and yeasts were also present. A significant portion of the microorganisms were attached to the ore being leached.

It was possible to eliminate almost completely these undesirable microorganisms and to prevent the renewed outside contamination of the leach system by supplementing the leach solutions with an appropriate disinfectant, sodium merthiolate, or by passing these solutions through toxic plastic fibres. These fibres were put on the surface of the heap and the solutions passed through them before penetrating into the ore mass. The inhibition of the undesirable microorganisms reduced considerably the consumption of reagents during the leaching and cementation. This consumption amounted to 2.4 kg ammonium thiosulphate, 0.3 kg protein hydrolysate, 1.0 kg copper sulphate and 0.4 kg ammonium sulphite per ton of ore.

The degree of extraction of the precious metals from the pregnant solutions by cementation was higher than 95 %. The product from the cementation unit was mixed gold-silver concentrate which was then processed by the conventional procedure for obtaining pure gold and silver.

The results from these pilot scale experiments confirmed the conception that the combined chemico-biological leaching is a technologically efficient, technically feasible, economically attractive and environmentally safe method for recovering precious metals from oxide ores.

Acknowledgements

The authors acknowledge the financial support given by the ROMB Company and the excellent assistance of B. Penin, I. Pishinkov and A. Karagjozov from the Elshitza mine.

References

1. Pares, Y., 1968. Mise en solution de l'or par voie bacterienne. Etude biologique du phenomene. Problemes d'application pratique, Paper presented at the 8th International Mineral Processing Congress, Leningrad, 11 pp.

2. Korobushkina, E.D., Karavaiko, G.I. and Korobushkin, I.M., 1983. Biogeochemistry of gold, In: Environmental Beogeochemistry (R. Hallberg, ed.), Ecological Bulletin (Stockholm), 35, 325-333.

3. Mineev, G.G., 1989. Biometallurgy of Gold, 160 pp., Metallugiya, Moscow (in Russian).

4. Goudev, S.N. and Goudeva, V.I., 1989. Microbiological solubilization of gold, In: Precious and Rare Metal Technologies, Process Metallurgy 5 (A.E. Torma and I.H. Gundiler, eds.), pp. 255 262, Elsevier, Amsterdam.

5. Groudev, S.N. and Groudeva, V.I., 1989. Microbial leaching of gold from ores and mineral wastes by means of heterotrophic bacteria, In: Biohydrometallurgy 1989 (J. Salley, R.G.L. McCready and P.L. Wichlacz, eds.), pp. 261-269, CANMET SP89-10.

6. Goudev, S.N. and Goudeva, V.I., 1990. Small-scale feasibility study of the biological leaching of gold from ores, In: Minerals, Materials and Industry, pp. 221-227, the Institution of Mining and Metallurgy, London.

7. Goudev, S.N., 1990. Microbiological transformations of mineral raw materials, Doctor of Biological Sciences Thesis, 538 pp., University of Mining and Geology, Sofia, Bulgaria (in Bulgarian).

8. Goudev, S.N. and Goudeva, V.I., 1993. Biohydrometallugy of gold: present-day status and future prospects, Preprints of the XVIII International Mineral Processing Congress, Sydney, 23-28 May 1993, pp. 1385-1387.

9. Polkin, S.I., Adamov, E.V. and Panin, V.V., 1982. Technology of Bacterial Leaching of Non-Ferrous and Rare Metals, 288 pp., Nedra, Moscow (in Russian).

10. Karavaiko, G.I. and Verikova, L.M., 1988. Role of microorganisms in solubilization of native gold, In: Biogeotechnology of Metals, Manual (G.I. Karavaiko, G. Rossi, A.D. Agate, S.N. Goudev and Z.A. Avakyan, eds.), pp. 323-328, Centre for International Projects GKNT, Moscow.

11. Groudev, S.N., Genchev, F.N., Groudeva, V.I., Burzev, G.I., Spasova, I.I., Nenova, S.T., Doycheva, A.S., Savov, V.A., Junev, N.I. and Davidov, E.R., 1992. Method for extraction of precious metals from ores, Russian Patent reg. No 5036072 (in Russian).

12. Goudev, S.N., Goudeva, V.I. and Spasova, II., 1993. Extraction of gold and silver from oxide ores by means of a combined biological and chemical leaching, In: Biohydrometallurgical Technologies, vol. 1, Bioleaching Processes (A.E. Torma, J.E. Wey and V.I. Lakshmanan, eds.), pp. 417-425, The Minerals, Metals & Materials Society, Warrendale, Pennsylvania.

13. Groudev, S.N., Ivanov, I.M., Spasova, I.I., Goudeva, V.I., Doycheva, A.S., Genchev, F.N., Karagjozov, A.K., Pishinkov, I.L., Penin, B.L., Savov, V.A. and Davidov, E.R., 1994. Method for extraction of precious metals from oxide ores, Bulgarian Patent reg. No 98365 (in Bulgarian).

14. Groudev, S.N., Groudeva, V.I., Spasova, I.I., Genchev, F.N. and Ivanov, I.M., 1993. Biological leaching of gold from oxide ores, Annual of the University of Mining and Geology, Sofia, 39, 1, 133-135.

GOLD RECOVERY FROM PYRRHOTITE BY BIOLEACHING AND CYANIDATION:

A PRELIMINARY STUDY USING STATISTICAL METHODS.

S. Ubaldini, F. Vegliò*, L. Toro*, C. Abbruzzese

CNR, Institute of Mineral Processing
via Bolognola 7, 00138 Roma, Italy
*Dep. Chemical Engineering and Materials, University of L'Aquila
67040 Monteluco di Roio, L'Aquila, Italy

Abstract

Gold is usually recovered by cyanidation, however the efficiency of this technology, in the case of refractory ores, is low. The aim of the present work has been to study the main factors that can influence the biooxidative pretreatment by factorial experiments, in order to increase the gold recovery in a conventional cyanidation process. The biological treatment with Thiobacillus ferrooxidans has been investigated at bench scale on a gold-containing pyrrhotite (10 g/t Au) ore coming from the Atoroma mine in Bolivia.

Mineral Bioprocessing II
Edited by David S. Holmes and Ross W. Smith
The Minerals, Metals & Materials Society, 1995

Introduction

Gold is often very finely disseminated inside the sulphide crystals (pyrrhotite, pyrite, arsenopyrite) and an oxidative pretreatment is necessary before the conventional extractive process such as roasting, nitric acid oxidation, pressure leaching, and bacterial oxidation to remove the refractory nature of the ore to liberate gold particles [1-4].
Roasting, consisting in the oxidation of the sulphides to sulphur dioxide, is sometimes used, but it is energy intensive and involves a costly gas scrubbing system to prevent atmospheric pollution by the evolved arsenious and sulphur oxide gases [2,3,5-7]. Chemical oxidation used to recover gold from arsenopyrite ores (using pressure oxidation, when sulphide is oxidized to sulphate and oxidation with nitric acid), is also expensive requiring high temperatures, high pressures and corrosion resistant materials [1,8].
Biohydrometallurgical pretreatment has been developed to an industrial process on a small scale, and involves oxidation of sulphides to sulphuric acid in solution; it is very interesting if compared with the environmental effects and costs of the conventional pretreatment processes [2,9,10].
Gold bearing pyrrhotite behaves as a refractory ore to the cyanidation, adversely affecting gold dissolution during cyanide leaching. As a matter of fact, the leachant cannot physically reach the entrapped gold particles due to extremely low liberation size [11]. In addition, pyrrhotite decomposition generates ferrocyanide which removes free cyanide from cyanide solution, resulting the consumption of large amounts of both cyanide and oxygen [12,13].
The alternative pretreatment process for the refractory auriferous pyrrhotite may be achieved by _Thiobacillus ferrooxidans_ [14]. The biological treatment is carried out after the crushing and grinding stages and before cyanidation in the process flowsheet. Microorganisms catalyse the oxidation of the sulphide matrix, permitting the subsequent dissolution of the gold by cyanide leaching [15,16].
The biooxidation of pyrrhotite [17,18] involves several steps; in the initial chemical phase H_2S is generated:

$$FeS + 2H^+ \longrightarrow Fe^{2+} + H_2S \quad (1)$$

Afterwards, at pH value of 2-2.5, the microorganisms activity increases. This phase is very fast in comparison with the chemical stage. Bacteria oxidize pyrrhotite according to the following reactions:

$$2FeS + 4.5O_2 + 3H^+ \longrightarrow 2Fe^{3+} + SO_4{}^{2-} + HSO^{4-} + H_2O \quad (2)$$

$$2FeS + 1.5O_2 + 6H^+ \longrightarrow 2Fe^{3+} + 2S^\circ + 3H_2O \quad (3)$$

Moreover, sulphur can be producted consequently from the H_2S oxidation:

$$H_2S + 2Fe^{3+} \longrightarrow 2Fe^{2+} + 2H^+ + S^\circ \quad (4)$$

sulphur is then oxidized by <u>Thiobacillus ferrooxidans</u> according to the reaction (5):

$$2S^{\circ} + 3O_2 + 2H_2O \longrightarrow 3H^+ + H_2SO_4 + SO_4^{2-} \quad (5)$$

Important reserves of gold associated to pyrrhotite occur in the world [9,10] and this justifies the efforts to develop new treatment technology [2,7].

The aim of the present work has been the study of a biooxidative pretreatment of a refractory gold-containing pyrrhotite (10 g/t Au) coming from the Atoroma mine (Bolivia) by <u>Thiobacillus ferrooxidans</u>, in order to increase the gold extraction in a subsequent conventional cyanidation process.

After 24 hours leach time 17% Au yield has been obtained by direct cyanidation of the -74 µm ground ore, with an high NaCN concentration. When grinding the ore at -43 µm and after roasting, cyanide leaching has allowed to solubilize 26% Au and 30% Au, respectively. After bioleaching pretreatment, gold recoveries were higher than those obtained from cyanidation after fine milling and roasting: after 5 hours leaching gold dissolution reached about 67% Au and finally after 24 hours 83% Au. Reagent consumption was higher than in the direct leaching, in particular lime, to neutralize excess acid of the bioleached pulp.

The investigation has been performed out using statistical methods of experimental design such as Full Factorial Design (FFD) to study the influence of several process parameters on the biooxidative process. The investigated factors have been: pulp density, pH, inoculum and bioleaching time.

The plan of experimental runs using a factorial planning 2^4 and cheking the date by Yate's method were carried out, in order to determine the influence of the main factors of the bioleaching process on the gold recovery [3,11,19].

Experimental

A sample of pyrrhotite ore from Atoroma Mine in Bolivia was used in this study. It is a typical refractory gold ore, the fine particles of the precious metal being encapsulated in a quartzitic mass.

The samples for the experimental tests were prepared by comminution to -74 µm.

Qualitative analysis was achieved by X-ray fluorescence spectrometry; while AAS and ICP were used for the quantitative determinations, revealing 10 g/t Au. The main minerals, identified by X-ray diffractometry, were pyrrhotite (FeS) and quartz (SiO_2).

Bioleaching experiments

The microorganisms used in this study were strain of <u>Thiobacillus ferrooxidans</u> [6]. The selection was evaluated by the solubilization capacity of the bacterial cultures versus time. The biomass content of the pulp was determined by the Kjeldhal technique and the bacterial activity was measured through the oxygen consumption by the Warburg method. The cultures were grown in a 9K medium without ferrous sulphate [20].

The batch tests were carried out in 300 ml glass flasks continuously shaken on a rotary incubator (200 revolution per minute). The age of the inoculum utilised were 2 and 7 days. The temperature was fixed at 30°C. Solution samples for chemical analysis were taken at various time intervals. Fe was analyzed by AAS.

At the end of the bioleaching experiments, the solid residue was filtered, dried, weighed and analysed for gold before being submitted to cyanide leaching. A preliminary study was conducted on the basis of a two-levels factorial design. The factors and the levels considered in the biological tests are shown in Table I.

Table I. 2^4 full factorial experiment: factors and levels investigated

FACTORS		LEVELS –	LEVELS +
A	pulp density (%)	5	10
B	pH	2.0	2.5
C	inoculum	NO	YES
D	time of treatment (days)	2	7

The surface responce of the factorial experiment has been the total iron extracted (mg/l) and the iron extraction yield (%): both are reported in Table II.

Table II. Experimental results: total iron solubilized (mg/l) and iron extraction yield (%)

Treatments	Fe (mg/l)	Fe (%)
1	1390	9.44
a	2148	7.29
b	1054	7.16
ab	1809	6.14
c	4203	28.53
ac	7060	23.97
bc	3934	26.71
abc	6027	20.46
d	1375	9.33
ad	2215	7.52
bd	1119	7.60
abd	1951	6.62
cd	6935	47.08
acd	13229	44.91
bcd	6659	45.21
abcd	12388	42.05

Data analysis

In the analysis of the factorial design on two levels, a systematic check of Yates's Method has been used for the interpretation of the experimental results [3,11,19].
These were then worked out by ANOVA analysis.

Cyanidation

The residues from the bioleaching tests were treated by standardised cyanidation leaching for gold extraction.
The cyanide leaching was carried out in 500 ml hemispherical glass reactor, at 25% solids density for 24 hours with dilute aqueous cyanide solutions (0.2-18.3 Kg/t $Ca(OH)_2$, pH 11.0, and 2.2-21.2 Kg/t NaCN) at atmospheric pressure and room temperature (20°C). The ore/cyanide solution ratio was about 1:4. The slurry was mechanically stirred by paddle-blades rotating at 250 rev/min. Samples of solution were periodically taken periodically for determinations of gold and iron.
At the end of the tests, the pulp was filtered and the pregnant solution was submitted to gold analysis by AAS after solvent extraction with Aliquat in DIBK.

Results and discussion

Bioleaching of pyrrhotite

Table III reports the data worked out by ANOVA analysis limited to the significant parameters and indicates main effects and their interactions on iron extraction yield.

Table III. ANOVA: total iron solubilized (mg/l) and iron extraction yield as surface responce (%) *

FACTORS	EFFECTS ppm	%
pulp density (%)	+2520.2	- 2.76
pH	- 451.8	- 2.00
inoculum	+5921.8	+27.22
time of treatment (days)	+2280.8	+10.08
pulp density (%) x inoculum	+1723.4	- 1.27
pulp density (%) x time of treatment	+ 903.9	n.s.
inoculum x time of treatment	+2216.1	+ 9.82
pulp density (%) x inoculum x time	+ 864.2	n.s.

* S.D.= 194 ppm with 48 d.f.

As it can be seen, the total iron solubilized (mg/l) and the iron extraction yield (%) are higher in the inoculated samples. The time of treatment affects positively the process.
The effect of content of pulp influences positively the total iron dissolution, but it is negative for the iron extraction yield; moreover, the iron dissolution decreases when increasing pH.
All second and third-order interactions (excluded those with pH) are significants and positives as a consequence of the previous observations.
These considerations hold well within the experimental region investigated.
On the basis of these results, further tests were performed in order to evaluate the effect of pulp density on iron dissolution in a more wide experimental range. In particular 4 differents pulp density levels (5%-10%-15% and 20%) versus time (from 1 to 7 days) were studied.
In this phase of the work it was possible to ascertain qualitatively the biooxidation of FeS by SEM (Scanning Electron Microscopy) analysis (the results are not been reported). The extraction yields of iron in solution for each pulp density versus time are shown in Fig. 1.

Fig. 1 - Effect of the pulp density on the iron extraction yield versus time at pH = 2.00

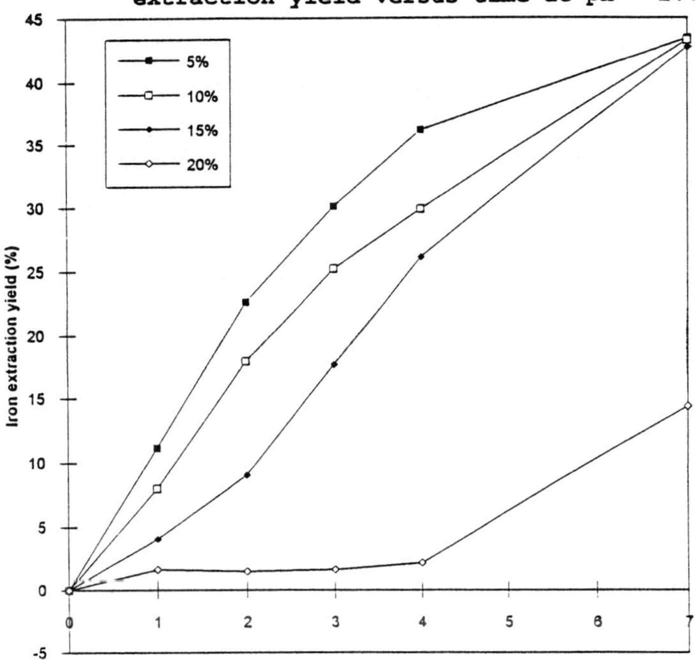

From the analysis of these results it was observed that:

1) the concentration of metal ions in solution has a sigmoid trend typical of that of the microbial growth (in the cases of 5%-10% and 15% pulp content);

2) in the tests at 5%-10% and 15% pulp content, similar iron extraction yield were obtained (40%-45% Fe);

3) an initial low iron extraction yield in the case of 20% pulp density was due to a lag phase (until to the fourth day) of the activity of the microorganisms. Successively the iron extraction increases until about 15% Fe.

<u>Cyanidation of the biooxidation residues</u>

The biooxidation residues obtained at 7 days of pretreatment time were blended homogeneously in order to obtain a representative feed samples for the cyanidation.
For each type of experiment, cyanidation tests have been carried out in order to check the effect of the biooxidative process on gold extraction and the consumption of reagents.
Preliminary studies (the main results were reported in Table IV) had already shown that without bioleaching pretreatment low gold recovery was obtained by cyanide leaching at the same particle size (-74 μm): 17.3% Au without roasting against 20.1% Au after roasting.
Moreover 25.6% and 29.8% gold extraction were achieved without any preliminary bioleaching but using a fine particle size fraction (-43 μm), after 24 hours of cyanidation, without and with roasting, respectively.

Table IV. Influence of the different pretreatments on the gold recovery

Particle size µm	Gold recovery %	NaCN Kg/t	$Ca(OH)_2$ Kg/t	Pre-treatment
<589	8.3	4.5	9.0	no-pretreated
<589	10.2	4.0	8.0	roasting
<208	10.7	4.0	9.5	no-pretread
<208	16.7	3.0	3.5	roasting
<104	15.6	14.9	10.4	no-pretreated
<104	17.1	7.9	2.5	roasting
< 74	17.3	15.8	10.7	no-pretreated
< 74	20.1	2.3	0.2	roasting
< 74	83.3	21.2	18.3	bioleaching
< 43	25.6	14.4	8.3	no-pretreated
< 43	29.8	2.2	0.2	roasting

Fig.2 - Effect of the pulp density of the bioleaching process on gold recovery

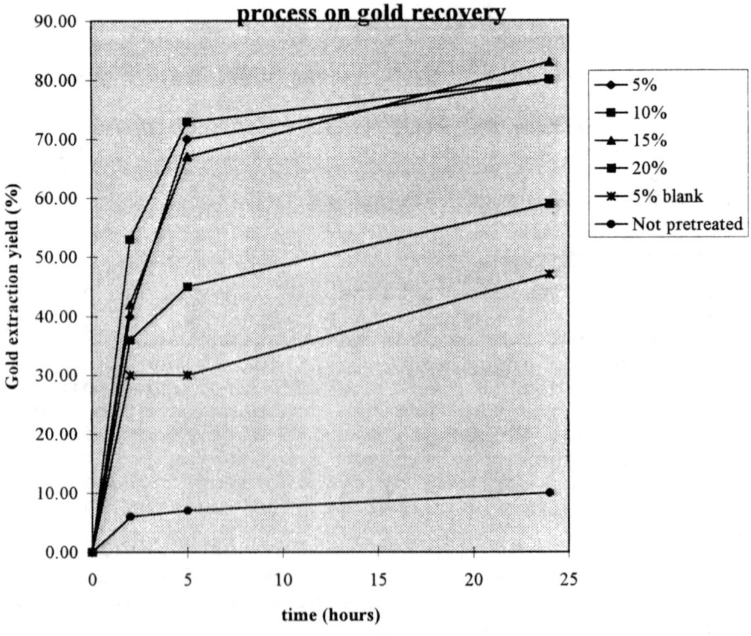

The analysis of the results illustrated in Table V, permits to verify the effect of the bacterial oxidation on gold recovery. It was observed that after bacterial oxidation gold recoveries were higher than those obtained from direct cyanidation. Even after 2 hours of cyanidation the recovery was 66.7% Au for the residues after 7 days bioleaching while 83.3% Au was achieved after 24 hours cyanidation.
As shown in Fig. 2, higher gold recovery was obtained utilizing the biooxidated residues at 15% of pulp content, according to the curves reported in Fig. 1.

Table V. Gold recovery by cyanidation, without or after bioleaching (15% pulp density, pH=2 and 7 days time)

Time	Without bioleaching (%)		After bioleaching (%)
	43 µm	74 µm	74 µm
2 h	15.6	11.3	41.7
5 h	21.0	15.7	66.7
24 h	25.6	17.3	83.3

Lime and cyanide consumption during the cyanidation of the bioleached samples was compared with the consumption of the directly cyanided samples.
It was possible to observe that lime consumption was less than 1 Kg/t for the untreated samples and higher for the pretreated (about 18 Kg/t).
Also cyanide consumption increases consequently to the biotreatment; this may be attributable to the concurrent consumption of cyanide by iron as precipitates in the pretreated samples. Besides pyrrhotite decomposition generates ferrocyanide which removes free cyanide from cyanide solution.
The economics of the leaching process depends mainly on the consumption of the reagents. The preliminary bacterial assisted attack allows to increase gold recovery in the subsequent conventional cyanidation. The additional consumption of chemicals is attributable to lime and cyanide. These reagents are not too expensive, but they permit to increase gold extraction from 17.3% Au to 83.3% Au.

Conclusions

The gold recovery obtained after bioleaching and cyanidation and that one obtained by grinding and cyanidation and by roasting and roasting were compared, in order to evaluate the effect of the different pretreatments. The experimental results have shown the higher efficiency of the biooxidative pretreatment on pyrrhotite for gold extraction.
In fact, after the direct cyanidation carried out on the -589 µm fraction of pyrrhotite ore, only 8.3% Au recovery was

obtained, even after 24 hours leaching with a high NaCN concentration (20 g/l).
After has the pyrrhotite to -74 µm and to -43 µm, cyanide leaching solubilized 17.3% Au and 25.6% Au respectively.
After bench scale bacterial oxidation experiments, at particle size of 74 µm, gold recoveries were higher than those obtained from direct cyanidation and after roasting at the different particle size. An interesting gold recovery has been achieved for the samples bioleached for 7 days with 15% of pulp density and at pH = 2: after 5 hours cyanidation gold dissolution reached 66.7% Au and after 24 hours 83.3% Au.
The consumption of reagents, lime and cyanide, was higher than that of the direct leaching process, particularly lime to neutralize excess acid produced in the bioleached pulp.
The experimental results have outlined the technical feasibility of the biooxidative pretreatment for gold recovery from auriferous pyrrhotite.

Acknowledgements

The authors are grateful to Messers Roberto Massidda and Marcello Centofanti for their helpful collaboration in the experimental work and Dr Paolo Plescia for X-ray analysis.

References

1. C. Abbruzzese, S. Ubaldini, F. Vegliò, and L. Toro, "Preparatory Bioleaching to the Conventional Cyanidation of Arsenical Gold Ores", Minerals Engineering, 7 (1) (1994), 49-60.

2. S. Ubaldini and C. Abbruzzese, "La biolisciviazione: Applicazione alla valorizzazione di minerali contenenti metalli preziosi", Industria Mineraria, 2 (1991), 25-32.

3. F. Vegliò, S. Ubaldini, C. Abbruzzese, and L. Toro, "Gold Recovery from Arsenopyrite Ore by Bacterial Leaching and Cyanidation", 6th European Congress on Biotechnology, Firenze: ECB6, (1993), 34-35.

4. C. Gasparrini, "The mineralogy of gold and its significance in metal extraction", CIM Bulletin, 851 (76) (1983), 144-153.

5. S. Ubaldini, P. Fornari, and C. Abbruzzese, "Smaltimento e bioriciclaggio dei rifiuti urbani ed industriali", Industria Mineraria, 6 (1992), 8-13.

6. B.A. Paponetti, S. Ubaldini, C. Abbruzzese, and L. Toro, "Biometallurgy for the recovery of gold from arsenopyrite ores", Mineral Bioprocessing, ed. Ross W. Smith and Manorajan Misra (TMS, 1991), 179-188.

7. B.A. Paponetti, F. Vegliò, and C. Abbruzzese, "Biotecnologie nel trattamento di minerali e di scarti industriali tossici", Biologi Italiani, 7 (1990), 14.

8. I. Nagy, P. Mrkusic, and H.W. McCullock, "Chemical treatment of refractory gold ores", Nat. Inst. Metall., Johannesburg, Rep. 38 (1966).

9. P.C. Van Aswegen, A.K. Haines, and H.J. Marais, "Design and operation of a commercial bacterial oxidation plant at Fairview", Proc. Randol. Gold Symp., Perth (1988), 144-147.

10. P.C. Van Aswegen et al., "Developments and innovations in bacterial oxidation of refractory ores" (SME Annual Meet., Denver, Colo., February 1991).

11. F. Vegliò, C. Abbruzzese, S. Ubaldini, and L. Toro, "Biooxidative pretreatment to enhance the gold recovery by cyanidation from an arsenopyrite refractory ore: a preliminary study using statistical methods", Proc. First Conference on Chemical and Process Engineering, Firenze: AIDIC, (1993), 589-593.

12. L. Lorenzen, and J.S.J. Van Deventer, "Electrochemical interactions between gold and its associated minerals during cyanidation", Hydrometallurgy, 30 (1-3) (1992), 177-94.

13. K.A. Natarajan, "Electrochemical aspects of bioleaching multisulfide minerals", Miner. Metall. Process., 5 (2) (1988), 61-5.

14. L.K. Yakhontova, and L.G. Nesterovich, "Role of bacteria in a supergene process in ore deposits", Mineral. Zh., 4 (1) (1982), 3 9.

15. T.M. Bhatti et al., "Mineral products of pyrrhotite oxidation by Thiobacillus ferrooxidans", Appl. Environ. Microbiol., 59 (6) (1993), 1984-90.

16. L. Ahonen, and O.H. Tuovinen, "Catalytic effects of silver in the microbiological leaching of finely ground chalcopyrite-containing ore materials in shake flasks", Hydrometallurgy, 24 (2) (1990), 219-36.

17. L. Ahonen, P. Hiltunen, and O.H. Tuovinen, "The role of pyrrhotite and pyrite in the bacterial leaching of chalcopyrite ores", Process Metall., 4 (No Fundam. Appl. Biohydrometall.) (1986), 13-22.

18. J. Pinka, "Bacterial oxidation of pyrite and pyrrhotite" Erzmetall, 44 (11) (1991), 571-3.

19. F. Vegliò et al., "Factorial Experiments in the Analysis of the Arsenopyrite Bioleaching using Thiobacillus ferrooxidans", Proc. Ninth International Symposium, Division of Bhiochemical Technology, Crystal City: American Chemical Society (1992).

20. L. Toro et al., "Biochemical and chemical events in copper solubilization from a chalcopyrite concentrate by Thiobacillus ferrooxidans in batch culture", International Journal of Mineral Processing, 26 (1989), 153-162.

PREVIOUS STUDIES OF CHARACTERIZATION AND BIOLEACHING

OF A NOVEL THERMOPHILIC MICROORGANISM

E. Gómez (*), I. Marín (**), A. Ballester (*), F. González (*)
R. Amils (**), and M. L. Blázquez (*)

(*) Departmento de Ciencia de Materiales
Facultad de Ciencias Químicas. Universidad Complutense
28040 Madrid (Spain)
(**) Centro de Biología Molecular. CSIC-UAM
Cantoblanco, Madrid (Spain)

ABSTRACT

A new thermophilic and acidophilic microorganism growing at 65-68°C and pH 1.8-2.0 has been isolated from Rio Tinto mines (Huelva, Spain). Due to similarities with bacteria of the *Sulfolobus* genus, its DNA was studied by pulsed field gel electrophoresis in addition to its protein pattern. These studies show that the newly isolated microorganism might be a non-described species. Its earlier metallurgical characterization has suggested interesting properties for possible use in bioleaching processes. Therefore, different tests to determine the influence of variables such as temperature, stirring rate, ore pulp density, metal toxicity and leaching ability were carried out. Results showed a high copper recovery from chalcopyrite concentrates (more than 85%) in optimal conditions. The new isolate displays a great resistance to metallic cations. Finally, bioleaching studies in mechanically stirred reactors, were also performed.

Mineral Bioprocessing II
Edited by David S. Holmes and Ross W. Smith
The Minerals, Metals & Materials Society, 1995

Introduction

Since the discovery of *Thiobacillus ferooxidans* and the determination of its role in bioleaching (Colmer and Hinkle, 1947), other microorganisms capable of participating in this kind of process have been isolated particularly from acid mine waters. However, initially the majority of the new microorganisms were mesophyllic bacteria like *T. ferrooxidans*, these included, *T. thiooxidans, Leptospirilum ferrooxidans*, and other species of *Thiobacillus* genus.

Industrial processes employing these bacteria have problems because their kinetics are slow. Success in isolating several thermophilic species capable of growth with mineral sulphides as their energy source has provided a new means of resolving this important problem in bioleaching and, in fact, several researchers have recommended their use to improve the leaching kinetics of a variety of ores and concentrates (Brierley, 1984, Ballester, 1988). In addition, the increase in the process temperature favours the chemical reaction rate.

The thermophilic bacteria used most often in bioleaching are those of the *Sulfolobus* genus (Brock et al., 1972). Recently however, a new thermophilic microorganism has been isolated in a collaborative project between two Spanish groups (Autonoma and Complutense Universities of Madrid). The samples were collected from the sulphide mines of Rio Tinto (Huelva, Spain). This microorganism is similar to other bacteria of *Sulfolobus* genus and presents, as its most interesting property, a significant capacity for extracting metal from chalcopyrite and complex sulphide concentrates. Data concerning its characterization and capacity to be used in bioleaching processes are presented for the first time in the present paper.

Experimental Methods

Bacteria Culture

In all the experiments a pure culture of a new thermophilic bacteria, named SLRT, which was isolated from mine samples in Rio Tinto, was used. It was grown in iron free 9K medium (Silverman and Lundgren, 1959): 3.00 g/L $(NH_4)_2SO_4$; 0.1 g/L KCl; 0.5 g/L K_2HPO_4; 0.5 g/L $MgSO_4.7H_2O$. A chalcopyrite concentrate was used as energy source.

Temperature, pulp density and stirring speed were fixed in accordance with the experiment to be carried out. The bacteria growth was controlled by direct recount of the cells using an optical microscope with phase contrast.

Proteins Pattern

This was carried out according to the method describe by Laemmi (1970).

Restriction Pattern

Intact DNA preparation was carried out according to Smith and Cantor's method (1987). For the enzymatic digestion of DNA the Smith et al. (1988) method was used.

Leaching Capacity

Different bioleaching experiments were carried out in shaken flasks on a chalcopyrite concentrate from Rio Tinto mines (Huelva, Spain). Its chemical composition was 27.6% Cu, 29.7% Fe, 34.2% S, 0.15% Zn and 0.03% Pb. Its mineralogical composition, determined by X-ray diffraction, was chalcopyrite, pyrite and silica as the main phases.

The flasks were stirred in an orbital shaker. Before taking samples, the stirring was stopped to settle the solids at the bottom of the flasks. Then, distilled water was added to compensate for the evaporation loss and a 1 mL sample was taken to determine the metal content by atomic adsorption spectroscopy.

Toxicity Tests

The bacteria's resistance to silver, cobalt, zinc, copper, mercury, arsenic, molybdenum, iron, bismuth and cadmium was determined by adding their corresponding soluble salts to the culture medium. These salts were $AgNO_3$; $CoSO_4$; $ZnSO_4$; $7H_2O$; $CuSO_4$; $5H_2O$; $HgSO_4$; AsO_2; $Na_2MoSO_4.2H_2O$; $FeSO_4.H_2O$; $BiCl_3$, respectively. In these experiments, the energy source was also a chalcopyrite concentrate.

Study of the Bioleaching Variables

The testes were carried out in shaking flasks with an orbital stirring. The nutrient medium (90mL) was an iron free 9K medium supplemented with a chalcopyrite concentrate as energy source. The different variables tested were: Temperature 63, 65 and 68°C; Stirring speed: 100, 150 and 200 rpm; Pulp density: 1, 3,5 and 7% (w/v).

Leaching in Mechanically Stirred Reactors

The tests were carried out in a 500 mL cylindrical glass reactor provided with a condenser to avoid the evaporation loss. The effective volume of reactor was 300 mL.

Sterilization

The nutrient media were sterilized under steam at 120°C for 20 min. The glass material and the ores were sterilized in the absence of water at 140°C for 12 h. The solutions of salts were sterilized by titration using a Millipore filter (0.2 μm).

Results and Discussion

Restriction Pattern

The pattern of DNA restriction fragments is generally characteristic for each microorganism. The comparison between the patterns of different microorganisms carried out with the same restriction enzyme often permits a filogenetic relationship to be established. When the new microorganism, termed SLRT, is compared with other members of the Sulfolobales family distinct restriction patterns are obtained for different rare cutters [data not shown].

Figure 1: SDS-PAGE of SLRT (a) and *S. solfataricus* (b). Sizes are indicate in kiloDaltons

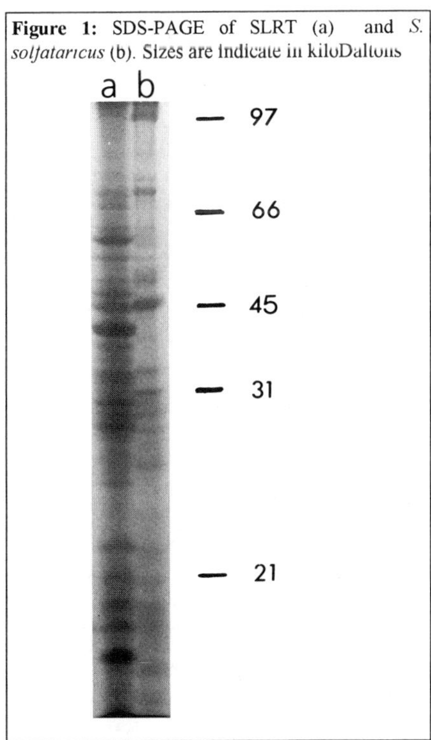

Protein Pattern

The 2-dimensional pattern of proteins frequently permits one to distinguish among different microorganisms. In this case, the total proteins contained in SLRT and *S. solfataricus* were compared (Figure 1). SDS-page shows a large difference in the total protein of the two microorganisms.

Bioleaching Experiments

Effect of different variables

Stirring speed. Normally, bioleaching tests performed in shaking flasks have a better extraction efficiency at a fast stirring speed because under these conditions most of the solids are in constant movement and in contact with the solution. However, when thermophilic microorganisms are used, stirring cannot be increased excessively due to the sensitivity of the membranes of these bacteria due to attrition. When the stirring speed increases the contacts between ore particles and microorganisms increase as well. It is possible that their death rate, caused by attrition, may be higher than their growth rate, resulting in a decrease in the number of cells and, therefore, also in the efficiency of bioleaching.

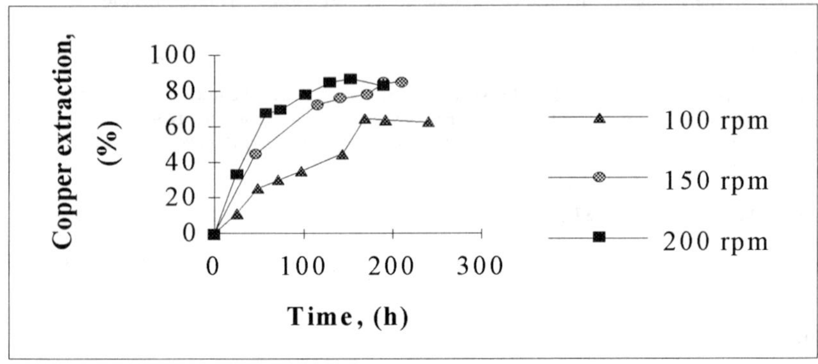

Figure 2: Copper extraction in tests carried out with different stirring speeds

The extraction of copper from a chalcopyrite concentrate was measured using SCRT and varying time as is shown in figure 2. The final extraction efficiencies were similar in tests at 150 and 200 rpm, and were superior to those obtained at 100 rpm. However, the dissolution rate was slightly faster at 200 rpm. Cell growth in the bioleaching was greater at stirring speeds of over 100 rpm (Figure 3) thus increasing the amount of copper leached.

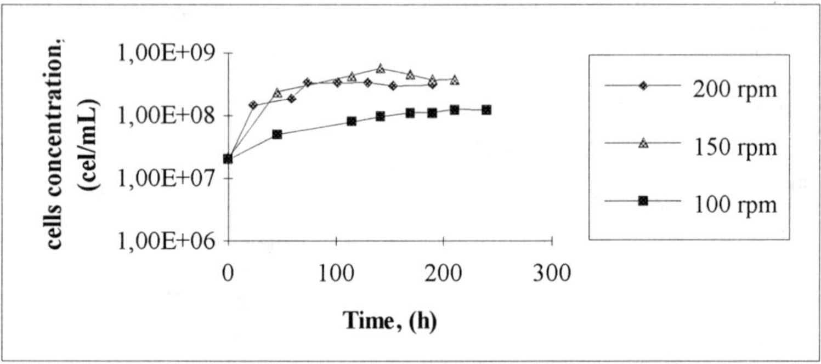

Figure 3: Cells growth in experiments with different stirring speeds.

The difference in the extraction efficiency between the tests at 150 and 200 rpm was marginal even though the number of bacteria at 150 rpm was somewhat higher due possibly to the fact theat attrition phenomena at 150 rpm had a less adverse influence on the cell growth. However, at 200 rpm the solids remained in suspension more easily and therefore the kinetics of the system were improved.

Temperature. Many thermophilic microorganisms can grow over a wide range of temperatures. However, when they are used in biotechnological applications it is necessary to incubate them under optimum conditions in order to achieve maximum efficiency. In addition, other factors must be considered in bioleaching. For instance, a higher temperature increases the leaching rate from the kinetic point of view. Therefore, it is quite necessary to take these two aspects into account in the control of temperature, so that the bacteria can grow favourably and the chemical reactions can proceed with the best kinetics.

Having considered the fact that SLRT can grow most favourably at temperatures between 65 and 68°C, different tests were carried out at 63, 65, and 68°C. The pulp density and the stirring speed were 1 % and 100 rpm, respectively.

After 10 days, the differences between the results of these three tests reached a minimum and their culture growth was also similar (between 3.5 and $4.0x10^8$ cells/mL). Therefore, the extraction efficiencies should also have been quite similar. In reality, with the copper extraction (Figure 4) these three cultures started their growth and metal dissolution at the same time; however, the test at 65°C had a small advantage because the stock cultures were maintained at this temperature. Nevertheless, at the end of the experiments the extraction efficiency was highest at the highest temperature (68°C) since the optimum temperature of SLRT's growth is close to 68°C.

Figure 4: Copper extraction in experiments with different incubation temperatures.

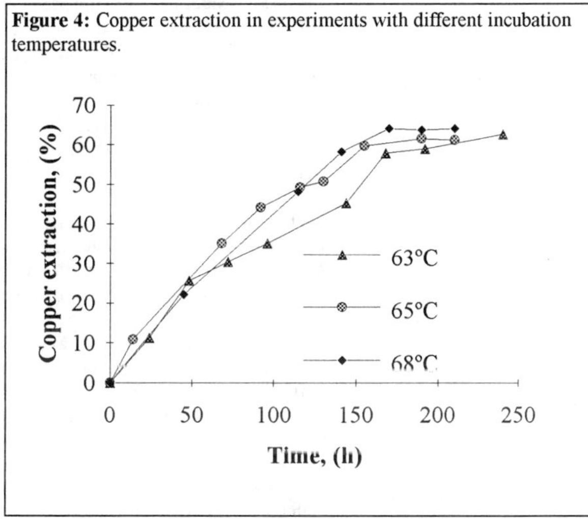

It can be seen from these results that it seems suitable to run the bioleaching experiments either at the optimum temperature of microorganisms's growth or only slightly higher because SLRT's maximum and optimum growth temperatures are close to each other. So, during the bioleaching the working temperature should not be too high because although the chemical reaction rate could be improved the biological influence could be very unfavourable.

Pulp density. When the experiment is carried out in shaking flasks, the stirring speed must be increased at a higher pulp density in order to maintain the solids in suspension. However, under these conditions attrition may have an adverse effect. In the present work the stirring speed was 200 rpm which took into account the fact that at that speed the leaching was effective, the microorganisms grow efficiently, and the solids in the solution can be suspended sufficiently.

Four different pulp densities were tested 1, 3, 5, and 7 (w/v). Each experiment was inoculated from a stock culture grown at 1% pulp density and, therefore, the microorganisms needed an initial period of adaptation under the new conditions particularly for the higher pulp densities.

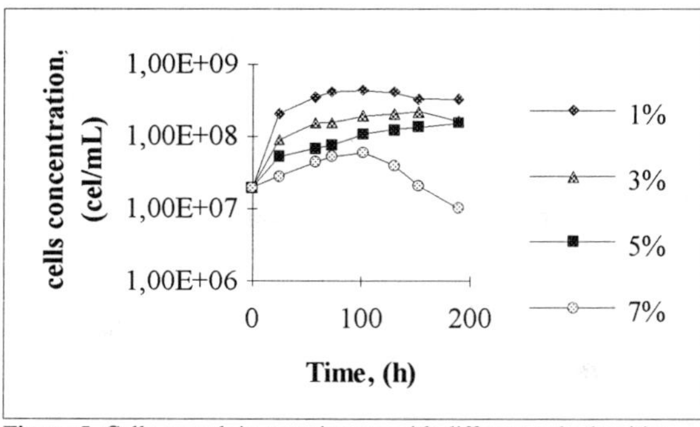

Figure 5. Cells growth in experiments with different pulp densities

In figure 5 the cell concentration was plotted as a function of time at different pulp densities. The highest concentration was achieved at 1% pulp density. At other densities the bacteria growth was not ideal because the number of cells did not increase fast enough. For the tests with 7% of solids, the number of bacteria occasionally started to decrease, particularly when strong stirring was applied. A higher pulp density and a faster stirring speed can lead to serious attrition of the culture.

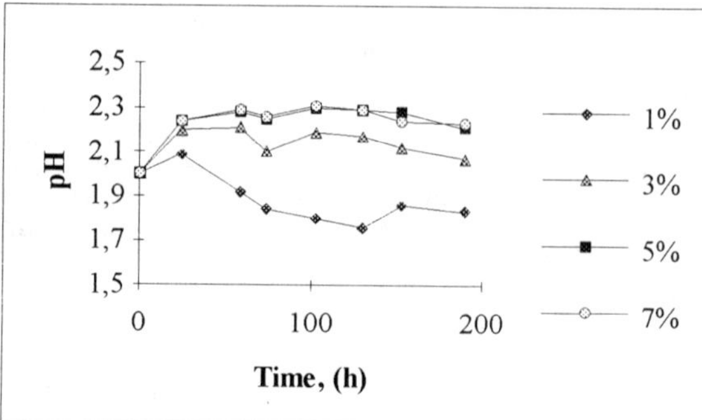

Figure 6. pH changes in experiments with different pulp densities.

In figure 6 changes in pH are plotted as a function of time at different pulp densities. The pH changed in a manner similar to that of figure 5. The experiments with 1% and 3% pulp density showed behaviour that can be considered normal in the bioleaching tests because the initial pH increase was due to the reaction between the medium and the ore and the subsequent pH decrease was due to the bacterial oxidation of sulphides to form sulphuric acid. In other tests, the pH rose during the first hours but remained at the maximum until the end of experiment.This fact may be related to the difficulty of microorganisms to grow at higher pulp densities.

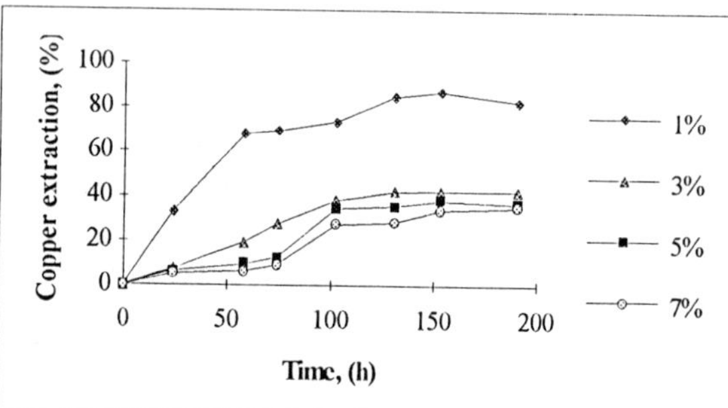

Figure 7. Copper extraction in experiments with different pulp densities.

The metal extraction efficiency confirmed what has been discussed above. Only the test at 1% pulp density reached a significantly efficient extraction (Figure 7). The other experiments extracted smaller amounts of copper. It should be pointed out here that in the presence of 5 and 7% solids, the copper ion concentrations in the solution were as low as in the case of the sterile test, which indicates a very slow growth of bacterial culture at high pulp densities. Nevertheless, after 5 days of incubation, the copper efficiency did increase somewhat but never to a satisfactory level.

The iron concentration in the solution was always very low (Figure 8) because most of the iron precipitated as jarosites. This precipitation is favoured in a medium very rich in salts such as the 9K medium at a pH above 2 during the leaching, particularly at a high temperature.

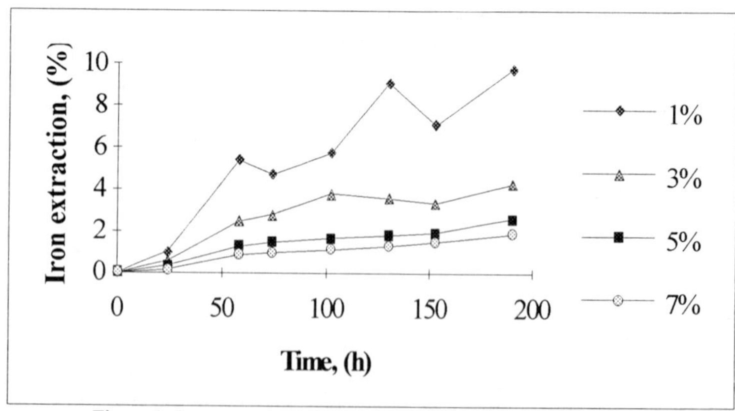

Figure 8: Iron extraction in experiments with different pulp densities.

Toxicity of Metallic Ions

The resistance of microorganisms to the different metallic cations studied is shown in Table I. A test without metallic cations was also carried out and its maximum number of cells was taken as the reference value. When there was a cell growth in the presence of a specific metallic ion but the number of cells was significantly smaller than the reference value, this cation concentration was considered inhibitory to the growth of the microorganism. If the number of bacteria was lower than or equal to the number of bacteria in the initial inoculum, this cation concentration was considered toxic. In Table I the last two cation concentrations (mM) are those under which one of the microorganisms had a similar growth to that in the reference tests and the other in which there was no growth.

Metal	Toxic concentr. (mM)	Inhibitory concentr. (mM)	Metal	Toxic concentr. (mM)	Inhibitory concentr. (mM)
Fe	121.7	60.9	Cd	13.6	3.63
Co	59.1	29.5	Ag	1.854	1.39
Cu	287.3	143.7	Hg	0.668	0.067
Zn	180.4	90.2	As	5.3	2.67
Mo	0.154	0.077	Bi	1.435	0.718

Table I. Toxicity of different metals on SLRT.

The results indicate that SLRT had a great resistance to the cations tested and especially to copper, iron, and zinc which are very common in the sulphide concentrated which are its energy source. These experiments are important given that some of these cations can be used as potential catalysts in the leaching processes. Some of them can form insoluble sulphides on the ore surface causing different chemical reactions which catalyze the solid dissolution. This effect has been verified by several researchers, such as Scott and Dyson, 1968 (Bi and Mo), Ballester, 1987, Ahonen, 1990a and 1990b (Ag), Mier, 1993 (Bi, Ag, Mo, Co, As). The resistance of SLRT to these cations indicates that it can be used in bioleaching processes even in the presence of these metallic cations.

In any case, it is important to point out that the concentrations shown in Table I were obtained without a prior adaptation of bacteria to the cations in the solution. So, with ions like Hg^{2+} and Mo^{6+}, which are very toxic, an adaptation process should be carried out to obtain microorganisms capable of growing at concentrations suitable for the bioleaching catalysis.

Bioleaching In Mechanically Stirred Reactors

From an industrial point of view, thermophilic bioleaching should be carried out in mechanically stirred reactors to improve the kinetics of the process. In addition, the results obtained with this type of reactors are more reliable when scaling up the process. Different tests were performed to evaluate the performance of this microorganism in this type of reactor.

The bacteria grew very well in mechanically stirred reactors from the first hours of the process. The growth of bacteria was very satisfactory once they had adapted to the medium.

In relation to the pH, it can be seen from Figure 9 that the pH of the solution increased up to 2.1 in the beginning of the incubation and then decreased to 1.5.

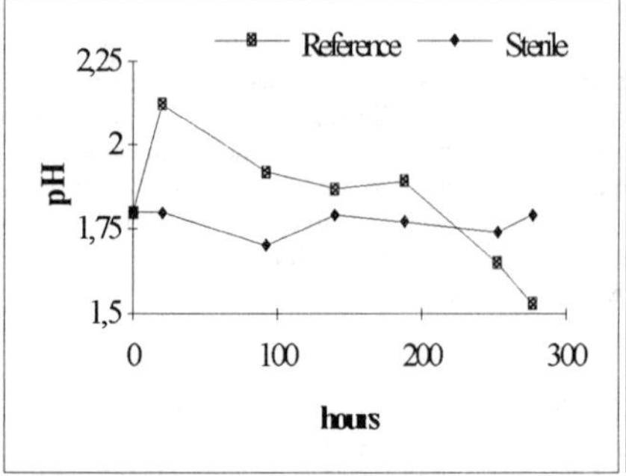

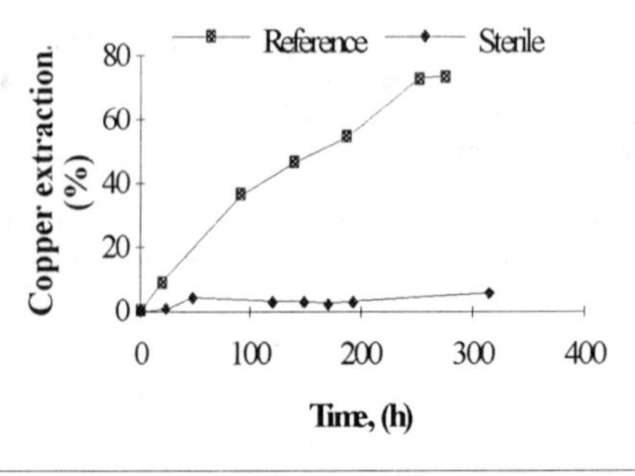

Figure 9. pH changes in experiments carried out in mechanically stirred reactors.

Figure 10. Copper extraction in experiments carried out in mechanically stirred reactors.

The copper recovery (Figure 10) was high, close to 80%. These results are similar to those obtained with shaking flasks. The iron concentration in the solution was very low due to the precipitation of jarosites.

Conclusions

The data from the physiological characterization of SLRT together with the preliminary results obtained for total proteins and DNA macrorestriction patterns indicates that this thermophilic microorganism is a novel archaebacteria related to the Sulfolobales group with interesting leaching properties and biotechnological potential. This microorganism can stand high stirring conditions (up to 200 rpm) without being affected by attrition, which is a common problem reported for other thermophiles. The optimal growth temperature is 68°C although small variations do not affect its growth efficiency drastically. Its tolerance to high concentrations of heavy metals is an important factor to be considered for its biotechnological applications. Pilot experiments in mechanically stirred reactors showed high efficiencies of copper extraction from recalcitrant minerals underlining the potential of this microorganism in bioleaching technologies.

Acknowledgments

This work has been supported by a joint grant from the CICYT (BIO 91-0223-C03)

References

AHONEN L., TUOVINEN O.H.: "Catalytic Effects of Silver in the Microbiological Leaching of Finely Ground Chalcopyrite-Containing Ore Materials in Shaked Flasks". Hydrometallurgy. 24, (1990) 219.

AHONEN L., TUOVINEN O.H.: "Silver Catalysis of the Bacterial Leaching of Chalcopyrite-Containing Ore Material in Column Reactors". Miner. Eng. 3 (5) (1990), 437.

BALLESTER A., GONZÁLEZ F., BLÁZQUEZ M.L.: "Biolixiviación de menas naturales. Posibilidades actuales de utilización". Rev. Metal. Madrid. 24 (1988), 91-102.

BALLESTER A.: "A Study of Silver-Catalysed Bioleaching of Chalcopyrite". In Separation Processes in Hydrometallurgy. Davies G. A. Ed. 9. (1990).

BRIERLEY C.L. "Microbiological mining: Technology status and commercial opportunities". World Biotech. Rep. 1 (1984), 599-609.

BROCK T.D., BROCK K.M., BELLY R.T., WEISS R.L. "*Sulfolobus;* a New Genus of Sulfur-Oxidizing Bacteria Living at Low PH and High Temperature". Arch. Mikrobiol. 84, (1972), 54-68.

COLMER A.R., HINCKLE M.E. "The role of microorganisms in acid mine drainage: a preliminary report." Science. 106 (1947), 253-256.

GROUDEV S.N., GENCHEV F.N., GAIDARJEV **S.S.** "Observation of the microflora in an Industrial Copper Dump Leaching Operation". In Metallurgical Applications of Bacterial Leaching and Related Microbiological Phenomena. Murr L.E., Torma A.E. and Brierlery J.A. (Eds.). Academic Press. New York. (1978) 253.

GROUDEVA V.I., GROUDEV S.N., VASSILEV D.V. "Microflora of two industrial copper dump leaching operations". In "Proceedings on the International Seminar: Dump and Underground Bacterial Leaching". G.I. Karavaiko, G. Rossi & Z.A. Avakyan eds. Moscu, 1990. 210-217.

LAEMMLI U.K. Nature, 227 (1970), 680-685.

MIER J.L. "Estudios de catálisis y toxicidad con iones metálicos en la lixiviación de la calcopirita por organismos termófilos". (Ph.D. Thesis. Complutense University, Madrid. 1993).

SCOTT T.R. and DYSON N.F.: "The Catalyzed Oxidation of Zinc Sulfide under Acid Pressure Leaching Conditions". Transactions of the Metallurgical Society of AIME. 242. (1968), 1815.

SILVERMAN M.P., LUNDGREN D.G.. "Studies on the Chemoautotrophic Iron Bacterium *Ferrobacillus ferrooxidans* I. An Improved Medium and a Harvesting Procedure for Securing High Cell Yields". J. Bacteriol. 77 (1959), 642.

SMITH C.L., CANTOR C.R. "Purification specific fragmentation and separation of large DNA molecules". In Methods in Enzimology, Recombinant DNA. R. Wu Ed. Academic Press. New York, (1987) 155: 449-467.

SMITH C.L., KELCO S.R., CANTOR C.R. "Pulsed field gel electrophoresis and the technology of large DNA molecules. In Genome Analysis: A Practical Approach". K. Davies Ed. IRL Press, Oxford, England (1988). 41-72.

OXYGEN MASS TRANSFER SCALE-UP IN THE BIOLEACHING OF MANGANIFEROUS MINERALS BY HETEROTROPHIC MICROORGANISMS.

F. Veglio', S. Ubaldini [1] , C. Abbruzzese [1], L. Toro

Dipartimento di Chimica, Ingegneria Chimica e Materiali, Universita' degli Studi de L'Aquila, 67040 Monteluco di Roio, L'Aquila, Italy.
[1] Istituto per il Trattamento dei Minerali (C.N.R.), via Bolognola 7, 00138 Roma, Italy.

ABSTRACT

In the present work, the bioleaching of italian manganiferous minerals (mainly constituted by pyrolusite and psilomelane) was carried out by a mixed culture isolated from naturals environments. The biological reduction occours in microaerobic conditions that are naturally formed in the shaken flask tests. In fact, this system is a very common and useful apparatus to study small-scale submerged fermentation systems, and, considering its lay-out, the oxygen supply may become limiting if the oxygen requirement is larger than the overall oxygen transfer capacity. Considering its extensive use, an empirical model with dimensionless parameters has been proposed in order to correlate the oxygen mass transfer coefficient in shaken flasks to predict the oxygen mass transfer in several experimental conditions. The comparison between the bioleaching tests in shaken flasks and in microfermenter indicate how the oxygen mass transfer is the main parameter for controlling the scaling-up of the bioreductive process.

Mineral Bioprocessing II
Edited by David S. Holmes and Ross W. Smith
The Minerals, Metals & Materials Society, 1995

Introduction

Mass transfer of oxygen, carbon dioxide and water vapor are very important in the microbiological cultivation of heterotrophic and autotrophic cultures [1]. Considering the heterotrophic microorganisms, the manganese dioxide bioleaching has been studied with interest by several workers [2-8].

The microbiological reduction of an italian manganiferous mineral of industrial interest is obtained by a savage heterotrophic mixed culture [6-8]. It is remarkable how this bio-reduction (from MnO_2 to Mn^{2+}) occours in aerobic conditions in the first phase of the process; in a second phase microaerobic condition takes place. Several hypotheses [6] were formulated in order to explain this behaviour and, however, it was concluded that the microaerobic conditions are necessary to obtain a microbiological reduction of the MnO_2 [2]. The preliminary tests permitted to establish that a direct mechanism is involved in the reductive process. For this reason it is very important to study the oxygen mass transfer capacity of the experimental systems utilized in the work: shaken flasks (V= 100 mL) and a lab-scale microfermenter (V= 5L).

Shaken flasks in a rotating incubator have been used in the first phase of the study. In general, these flask are usually provided with closures such as cotton plugs, filter paper, specific filters made of PVA etc. [9], to preclude airborne contaminants from penetrating through into the flasks. Therefore the shake flasks is a very common and useful device for the study of small-scale submerged fermentations. In fact, the use of shake flasks gives a number of advantages: they are very easy to manipulate, little material and manpower expense is required and it is possible to realize numerous of experiments. For these reasons, this experimental system is used for primary screening, testing of fermentation media, strain selection, growth evaluation [10] and testing of the main factors that influence the microbiological process. However, the oxygen supply may become limiting if the oxygen demand exceeds the oxygen transfer capacity through the shake flask closure or/and the gas-liquid interface.

The next phase of the development of the bioleaching process had to be carried out on a bigger scale: in this phase the kinetic of the process, technology problems and so on, had to be determined: for this reason a scale-up problem arose in this phase of the study [6]. In fact while in the shaken flasks the bioleaching of MnO_2 proceeds with high manganese extraction rate, the bioreductive process in the lab-scale microfermenter is a function of the areation conditions of the system [8,11].

Main goal of this study was to establish the main parameters that must be taken into consideration to perform a correct scale-up between shaken flask and lab-scale microfermenter tests. For this reason gassing-out tests to measure the oxygen mass transfer in both systems were carried out. Empirical models were employed to correlate oxygen mass transfer with the experimental conditions in both systems. A comparison between the two experimental systems utilized was carried out considering oxygen mass transfer coefficent and the manganese extraction yield. The experimental runs were performed considering full factorial experiments in order to determine wheter interaction among the investigated factors takes place.

Materials and Methods

Isolation, selection and maintenance of the microbial strains utilized within this work has been described elsewhere [6,7]. Batch cultures performed in 250 ml Erlenmeyer flasks were carried out with 100 ml of medium; incubation of flasks was generally done in an orbital rotatory shaker at 250 rpm at 30°C. Batch cultures performed in the lab-scale reactor were carried out at 30°C, volume 5 L and with several mixing and aeration conditions; 10% of inoculum of a 7 day old culture was used in both cases. The manganiferous ore has been collected by ITM (Istituto Trattamento Minerali - CNR Rome); Pyrolusite and Psilomelane are the main components of the mineral: the complete characterization of the mineral has been described elsewhere [6,7]. All the analytical tests were carried out as reported in a number of previous papers.

In the oxygen mass transfer measures, standard shaken flasks (volume = 300 mL) have been used in the experimental runs. The flasks were equipped with one sidearm in which oxygen probes could be set (see Fig.1). The oxygen concentration was measured by an oxygen probe with a small response time (AMEL mod.332) connected with a recorder device. The experimental runs have been performed using two types of flasks having different neck diameters (D_B=3-4 cm). The shaken flasks were placed in a New Brunswich shaker with an orbiting speed range of 0.66-6.67 Hz (40-400 rpm). The diameter of the orbit was 32 mm. A L.K.B. 1601 microfermenter was used in the experimental runs in lab-scale tests; this apparatus was used in microbiological tests and in the oxygen mass transfer evaluations.

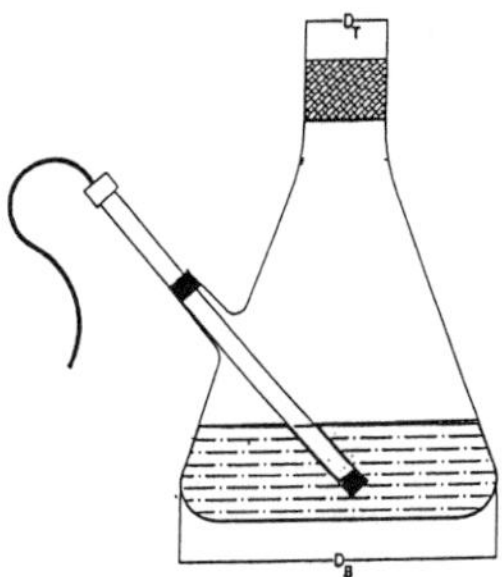

Figure 1 - Experimental system employed in the gassing-out tests in shake flaskes

The measurement of oxygen transfer coefficent in the liquid phase was determined by the well-known method of "dynamic gassing out" given by Bandyopadhyay et. al. [10]. In this method the oxygen in the shake flasks or in the lab-scale fermenter, is first driven out by nitrogen in order to remove the oxygen present in the liquid phase. The nitrogen gas is sparged in water kept at 50°C before going to the shake flask in order to avoid temperature changes in the liquid phase; in this manner it is possible to ensure a constant temperature during the experimental run of gassing out in the shaken flask or in microfermenter. After this first phase, the gas above the liquid is again replaced
by air. Then the oxygen tension is registered on a recorder in a number of experimental conditions. The dynamic response of the process can be described by the following material balance equation [9]:

$$\frac{dC}{dt} = K_L a \cdot (C^* - C) \quad (1)$$

where:

C	= oxygen concentration in the liquid phase	(mg/L)
C*	= oxygen concentration in steady state conditions	(mg/L)
Kla	= oxygen mass transfer coefficient	(min^{-1})
t	= time	(min)

The integration of the equation (1) considering the equilibrium concentration of oxygen with the air (C^*) and the initial conditions ($C = C_o$ at $t = t_o$) gives:

$$\ln\left(\frac{C^* - C_o}{C^* - C}\right) = K_L a \cdot (t - t_o) \quad (2)$$

The value of the oxygen transfer coefficent $k_L a$ was evaluated by regression analysis using Mezaky's method (1968). The measures by the oxygen electrode can be made with no serious problems because its time constant is small compared with the time constant of the oxygen transfer into the shake flask. In this way the influence of the electrode response time on the probe output can be neglected. In the experimental tests carried out in presence of cotton plugs the same experimental procedure above reported was used. In this case, the purpose of the tests is the determination of the overall oxygen mass transfer coefficent (k_{Lo}). The dinamic response of the process can be described always by equation (6) but with a different meaning of the mass transfer coefficient:

$$\frac{1}{K_{Lo}} = \frac{1}{\frac{1}{Kl \cdot a} + \frac{1}{Kw}} \qquad (3)$$

$$K_w = \frac{K'_w}{V_G}$$

where :

K_{Lo}	= Overall mass transfer	(min^{-1})
Kl a	= mass transfer of the liquid phase	(min^{-1})
Kw	= overall mass transfer of the cotton closure	(min^{-1})
K'w	= mass transfer of the cotton closure	(m^3/min)
Vg	= volume of the gas phase into the flask	(m^3)

In fact, with a cotton closure, there are two obstacles to the oxygen mass transfer that are in series. The analysis of the main factors, that influence the oxygen mass transfer in the shaken flask tests at 20°C has been carried out using a three level full factorial designs. The factors investigated were: mixing conditions (rpm), temperature (°C), weight of closure (g), liquid holdup (mL) and the geometry of the shaken flasks. The factors and the levels investigated are reported in Table I. The same experimental design has been performed in the lab-scale microfermenter tests considering areation rate, stiring conditions and temperature as factors (Table II).

Table I - Factors and levels investigated in the shake flask tests: T=20°C.

Factors		Levels		
		-1	0	+1
Stirring	rpm	150	200	250
Liquid Volume	mL	100	125	150
Weight of the closure	g	3.0	5.0	7.0

Table II - Factors and levels investigated in the Lab-scale reactor tests: V=5 L.

Factors		Levels		
		-1	0	+1
Stirring	rpm	200	300	500
Air flow rate	L/h	10	30	50
Temperature	°C	25	31	37

Mathematical Models

The values of the oxygen mass transfer coefficients were correlated using an empirical model with dimensionless parameters; taking in mind some physical considerations, the factors, whether operative or geometric, that could influence the oxygen transfer were postulated as [8,11]:

$$k = \alpha' \cdot fz^{a'} \cdot D_E^{b'} \cdot D_T^{c'} \cdot D_i^{d'} \cdot \mu^{e'} \cdot \rho^{f'} \cdot Do_2^{g'} \cdot D_B^{h'} \tag{4}$$

where:

fz	= rotating shaker velocity	(s^{-1})
D_E	= apparent diffusivity of the cotton closure	(m^2/sec)
D_T	= diameter of the neck flask	(m)
D_i	= diameter of the gas-liquid interface in the flask	(m)
D_B	= diameter of the shake flask bottom	(m)
μ	= viscosity of the liquid phase	(Kg / m s)
ρ	= liquid density	(Kg / m^3)
DO_2	= oxygen diffusivity in the liquid phase	(m^2 / s)
a, b, c, d, e, f, g, h	= empirical constants	

Utilising Buckingham's theorem and considering shaken flasks having the same geometry and cultural media at constant temperature the (1) can be expressed using 3 dimensionless parameters as in the follow expression:

$$Sh = \alpha \cdot \mathrm{Re}^a \cdot \left(\frac{D_i}{D_B}\right)^b \cdot \left(\frac{\rho \cdot D_E}{\mu}\right)^c \tag{5}$$

where:

$$Sh = \frac{k \cdot D_i^2}{Do_2} \qquad \mathrm{Re} = \frac{\rho \cdot fz \cdot D_i^2}{\mu} \qquad Sc = \frac{\rho \cdot Do_2}{\mu}$$

It should be noticed that in the Reynolds number the diameter at the interface has been used instead of the bottom flask diameter. The use of the bottom diameter D_B in the Reynolds number is justified for bulk mixing but not for the mass transfer across the interphase. The diameter of the interphase gives, at a constant shaker table speed, a different interphase movement, from rocking to swirling. The consequence for the mass transfer rate is obvious.

The values of the oxygen mass transfer coefficents, obtained by experimental runs in lab-scale fermenter, were correlated using, usually, an empirical model [9]:

$$Kv = a \cdot \left(\frac{Pg}{V}\right)^b \cdot \left(\frac{Q}{S}\right)^c \tag{6}$$

where:

Pg	= power in the aerated system	(W)
V	= liquid hold-up in the vessel	(m^3)
Q	= air feed	(m^3 /h)
S	= cross-area of the fermenter	(m^2)
Kv	= oxygen mass transfer	(min^{-1})
a,b,c	= empirical constants	

Results

Oxygen Mass Transfer Tests

The oxygen mass transfer tests were carried out in shake flasks and in microfermenter. For these systems the factors and the levels investigated are shown in Table I and II respectively. The experimental planning was performed as a three-level full factorial design. In this manner it is possible to determine main and interaction effects among the investigated factors and, furthermore, it is possible to discriminate the linear and the quadratic component for each effect, because for each of them, there are two degrees of freedom available [13].

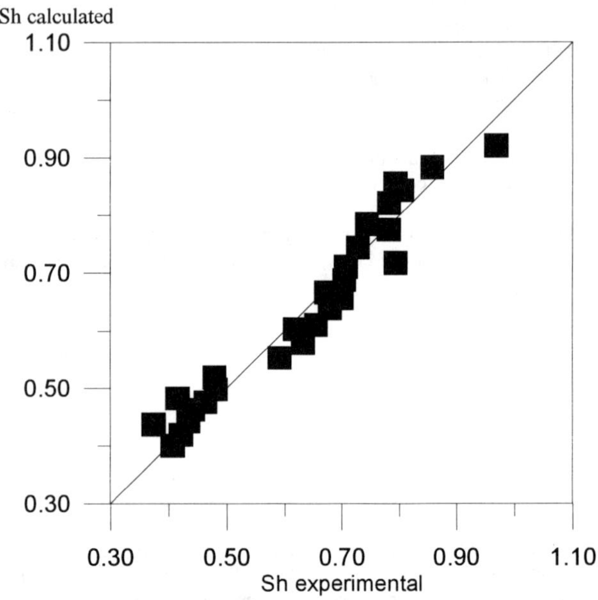

Figure 2 - Sh calculated vs. Sh experimental (model 5)

The experimental results were elaborated by ANOVA analysis (not shown here). From the analysis of these results obtained with shake flasks it is possible to observe that:

1) the only significant effects are due to liquid holdup, stirring and weight of the cotton closure. In other words the effects of the investigated factors are well evaluated with respect to the experimental error.
2) the liquid volume in the flask presents a positive effect on the oxygen mass transfer coefficients. Increasing the liquid volume there will be an increase of the mass transfer coefficient. These results should be explained considering that, as shown in the relationship 3, increasing the liquid volume decreases the gas-liquid interface then a reduction of the overal mass transfer take place. But at the same time there is also a reduction of the gas volume in the flask and then an increase of the oxygen mass transfer it will be possible. This possible explanation is supported by experimental evidence that without cotton closure a negative influence on the oxygen mass transfer due to a large liquid volume is found [11].
3) the increasing of the stirring conditions increases the oxygen mass transfer coefficients. This obvious result is linked to the reduction of the boundary layer in the gas-liquid interface (for both liquid and gas sides). An unavoidable increase of the mass transfer due to the increase of the interface gas-liquid area is possible because all the tests were carried out in shaken flasks without baffles and, then without wortex

formation prevention. However, the main goal of this study was to determine the effects of the main factors involved in bioleaching tests considering their influence on the overall oxygen mass transfer coefficients.

4) a significant effect due to the weight of the cotton closure was found. Some bioleaching tests confirmed that the rapid bioreductive processes occour in shake flake tests in which the oxygen limitation was more effective but not anaerobic. In fact narrow-neck shake flasks with high density cotton closures produce faster bioreductive processes compared with bioleaching tests carried out in wide-neck shake flasks with low density cotton closure.

5) the three investigated factors don't seem to interact among each other. This is important in the further study because it is possible to use fractional factorial experiments instead of full design in order to perform few experimental tests in the parameter estimation of the proposed empirical models.

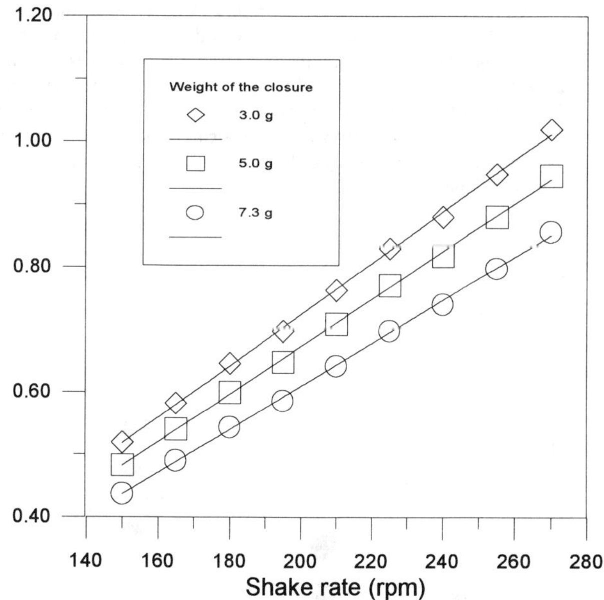

Figure 3 - Effect of the cotton closure weight on the Sh number for different Re number

The experimental results obtained in these tests were utilized to determine the parameters of the proposed empirical model. To determine its dimensionless parameters it is necessary to evaluate the apparent diffusivity (D_E) of the cotton closure of the shake flask. Naturally this transport parameter is connected to the weight of the cotton plug. For this reason an empirical relationship was found to correlate the apparent diffusivity (D_E) and the weight of the cotton closure (Cw) performing gassing-out tests without liquid phase into the shake flasks. The experimental results indicated that the empirical model can be expressed as follow :

$$\frac{Kw}{Vg} = -(8.005 \pm 1.283) \cdot 10^{-3} \cdot Cw + (0.1470 \pm 0.0056) \qquad (7)$$

where:

$$D_B = \frac{K'w}{Ac / L}$$

where Ac is the cross-section of the neck-flask and L is the length of the cotton closure. The empirical parameters of the relationship (5) were found using the model (7). The comparison between experimental and calculated values of the Sherwood number are also reported in Fig.2. As an example, in Fig.3 it is possible to observe the effect of the weight of the cotton closure in the empirical model where the Sh number is plotted vs. Re number. The empirical model obtained by regression analysis of the experimental results is:

$$Sh = (2.545 \pm 1.780) \cdot 10^{-6} \cdot \mathrm{Re}^{1.12 \pm 0.07} \cdot \left(\frac{\rho \cdot D_B}{\mu} \right)^{0.524 \pm 0.088} \cdot \left(\frac{Di}{D_B} \right)^{-2.425 \pm 0.476} \qquad (8)$$

The experimental results obtained in lab-scale fermentor runs were obtained as a function of the air flow rate and stirring. The results were utilized to correlate the oxygen mass transfer coefficient using the relationship (6). In Fig.4 the semi-empirical model values are plotted against the experimental results at 25°C. Considering the experimental results in shake flask tests it is possible to observe that similar oxygen mass transfer coefficient values are obtained in the lab-scale reactor tests at 300 rpm and 10-30 L/min (0.033-0.100 vvm). After this phase of gassing out tests it is possible to evaluate in advance how the parameters of both system (shake flasks and lab-scale reactor) influence the oxygen mass transfer coefficients. Then, it is possible to carry out bioleaching tests at similar values of oxygen mass transfer coefficients for both systems.

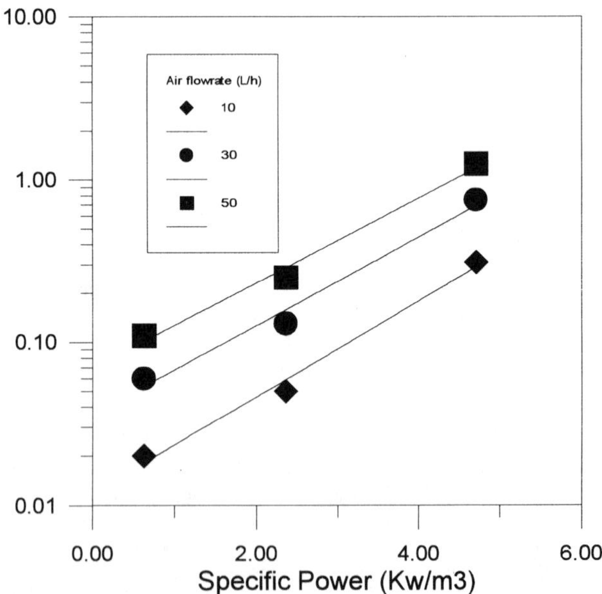

Figure 4 - Fitting of the model (11) in gassing-out microfermenter tests at 25°C (V=5 L).

Bioleaching Tests

The usual factors measured in a tipical bioleaching experiment are reported in Fig.5; in this case the run was carried out in a narrow-neck shake flask with a standard cotton closure at 30°C, 200 rpm and filled with 100 mL of cultural media. The progress of the bioleaching can be qualitatively observed considering the bleaching of the ore present in the flasks; this fact is due to the $MnCO_3$ formation during the process. In fact the reactions can be summarized as follow:

$$24\ MnO_2 + C_{12}H_{22}O_{11} + 48\ H^+ \longrightarrow 24\ Mn^{2+} + 12\ CO_2 + 35\ H_2O \qquad (9)$$

$$Mn^{2+} + CO_2 + H_2O \longrightarrow MnCO_3 + 2\ H^+ \qquad (10)$$

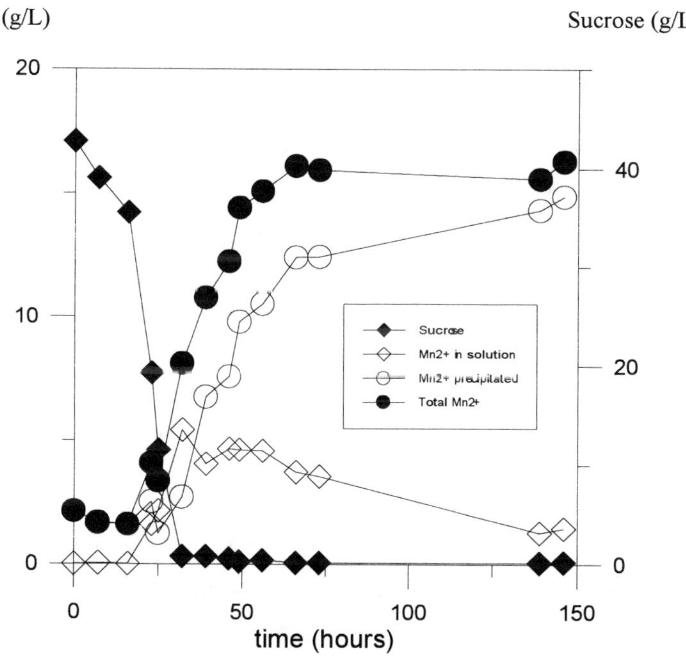

Figure 5 - Tipical experimental results obtained in shake flask bioleaching tests

Several experimental conditions were tested in order to determine the main factor that can influence the manganese extraction yield such as temperature, mixing conditions, media composition, pulp content and so on [6,7]. Some significant differences between bioleaching tests carried out in wide and narrow-neck shake flasks were found. In particular the bioleaching tests indicated qualitatively that a faster bioreductive process took place in narrow-neck compared to wide-neck shake flask tests.
Large differences in the extraction yield were also observed comparing the behaviour of the lab-scale reactor compared to the shake-flask tests.
So, an extraction yield depending of the experimental system utilized was found: significant differences can be shown between the shaken flask system and fermentation reactor lab-scale. In the last system it is possible to control the oxygen transfer into cultural medium manipulating the mixing conditions (rpm) and the air flow rate sparged into it. In the shaken flask the oxygen supply may become limited if the oxygen demand exceeds the oxygen tranfer capacity through the shake flask closure or/and the gas-liquid interface.

In Fig.6 it is possible to see how the aeration rates influence the bioleaching of MnO_2: the best experimental conditions are obtained at 0.1 vvm. In fact the bioreduction of the manganese didn't occour either in presence of high-air supply or in absence of aeration. It is possible to conclude how microaerobic conditions are needed in order to obtain the quantitative bioreduction of the MnO_2: this situation happens spontaneously in shake-flask experiments, being a limiting oxygen system. This fact may also explain the observed difference on the extraction yield between narrow-neck and wide-shake flasks; in fact it is possible to show how the oxygen mass transfer coefficients are larger in the wide-neck flask with respect to the narrow-neck ones.

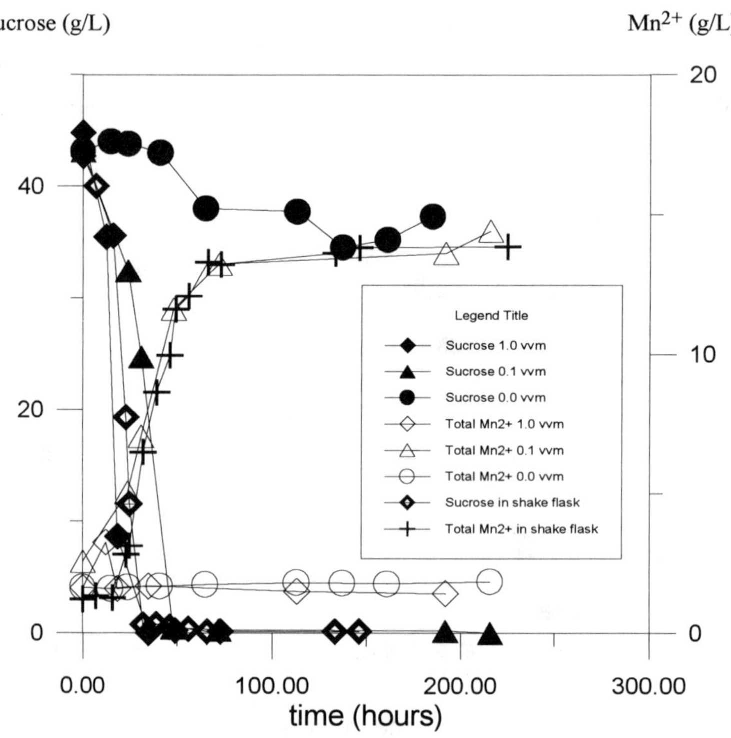

Figure 6 - Effect of the air flow rate on the MnO2 bioleaching: a comparison between lab-scale reactor (5L) and shake flask tests (100 mL).

Conclusions

In this work a dimensionless model to correlate the oxygen mass transfer coefficent in shake flask tests has been proposed: several experimental conditions were tested in order to evaluate the main factor that influences this coefficient. Therefore a parameter estimation of an usual model to correlate the oxygen mass transfer in lab-scale reactor with mechanical mixing was performed. These relationships were employed to explain the different experimental results obtained in the bioleaching of MnO_2 carried out using a heterotrophic mixed culture both in shake flask tests and in the lab-scale reactor. It was observed how microaerobic conditions are necessary in order to obtain the reductive process of the MnO_2. Futhermore , the relationships found can be utilized in the scale-up and/or scale-down problems in microbiological studies.

Acknowledgements

The authors are indebted with Mr. Marcello Centofanti for his helpful collaboration in the execution of the experimental work. This study has been financially supported by the EEC in the ambit of the contract MA2M CT91 0047 and by MURST 40%.

References

1. Boogerd F.C., Bos P., Kuenen J.G., Heijnen J.J., Van der Lans R.G.J.M., "Oxygen and carbon dioxide mass transfer and the aerobic, autotrophic cultivation of moderate and extreme thermophilies: a case study related to microbial desulfurization of coal", Biotechnoly and Bioengineering, Vol.35, (1990), 1111-1119.

2. Ehrlich H.L., "Manganese oxide reduction as a form af anaerobic respiration", Geomicrobiology Journal, 5, (3/4) (1987), 423-431.

3. Silverio M., Madgwick J.C.. "Biodegradation of manganese dioxide by purified leaching microorganisms", Australas. Inst. Min. Metall., 290 (7), (1985), 63-64.

4. Srimekanond A., Thangavelu V.J. and Madgwick J.C.. "Thermophilic bacterial leaching of manganese dioxide", Journal of Microbiology, 10, (1993), 217-220.

5. Toro L., Abbruzzese A., Paponetti B., Veglio' F., "Biometallurgy for manganese and copper ores", in Advances in Fine Particle Processing, (J. Hanna, A. Attia Eds., Elsevier Science Publisching Co., Inc., New York, 1989), 441-451.

6. Toro L., Veglio' F., Terreri M., Ercole C. and Lepidi A., "Manganese bioleaching from pyrolusite: bacterial properties reliable for the process", FEMS Microbiology Reviews , 11, (1993), 103-108.

7. Veglio' F., Terreri M. and Toro L., "Factorial experiments in the development of a pyrolusite bioleaching process using heterotrophic cultures", in Biohydrometallurgical Technologies, (A.E. Torma, J.E.Wey and V.I. Lakshmanan Eds., TMS, Warrendale PA, 1993) 1, 269-276.

8. Veglio' F. and Toro L., "A scale-up problem in the bioleaching of pyrolusite using heterotrophic microorganisms", Sixth European Congress on Biotechnology, Firenze (Italy), (13-17 June 1993), 4, th034.

9. Aiba S., Humphrey A.E., Millis N.F., Biochemical Engineering, (second edition, Accademic Press, Inc., New York & London, 1973), 224-238.

10. Van Suijdam J.C., Kossen N.W.F., Joha A.C., "Model for oxygen transfer in a shake flask", Biotechnology and Bioengineering, , Vol.XX, (1978), 1695-1709.

11. Veglio' F. and Toro L., "Coefficiente di trasferimento di ossigeno in beute agitate", Industria Chimica e di Processo (ICP), (Gennaio 1993), 50-56.

13. Davies O.L., The Design and Analysis of Industrial Experiments, (Longman Grop Limited London, second edition, New York, 1979).

IV.

BIOSORPTION

IMMOBILIZATION OF MICROBIAL CELLS FOR METAL ADSORPTION AND DESORPTION

C.Ercole, F.Vegliò*, L.Toro*, G.Ficara*, A.Lepidi

Department of Basic and Applied Biology, University of L'Aquila, 67010 Coppito, Italy;
*Department of Chemistry, Chemical Engineering and Material, 67040 Monteluco di Roio, L'Aquila, Italy.

Abstract

Microorganisms are known to concentrate metal species from aqueous solution and to accumulate them within the structure.

Manganese biosorption by microorganisms was examined using free and immobilized bacterial and algal cells. Data on manganese uptake by cells of different cultural age and cells grown in media enriched with phosphate, nitrate or sucrose are reported. These data show that bacterial cells of Arthrobacter sp. exibited higher metal-uptake capacity than the algal biomass of Chlorella and Scenedesmus for the uptake of manganese. The nutrient and cultural age affected the metal up-take capacity of the biomass.

Mineral Bioprocessing II
Edited by David S. Holmes and Ross W. Smith
The Minerals, Metals & Materials Society, 1995

Introduction

Several organisms, including cultivated plants, are able to concentrate heavy metals to a very large extent. Microbial cells, microbial microcolonies and large microbial populations have been described as the agents of several kind of metal dislocation and bioaccumulation [1,2,3,4]. The effects of such microbial activities are connected with events of local, regional and planetary concern [5].

The quality of deep and surface waters (from both an environmental and technological ground) is determined by the quantities and qualities of dissolved and solubilized organic and mineral compounds [6]. Among such compounds, heavy metals deserve a relevant role due to their toxic effects on living beings in the natural contexts or in the civil and agroindustrial ones (drinking, irrigation and bathing water). Disturbances are also found in the industrial use of waters polluted with heavy metals (washing, cooling etc). Fungal and other microbial biomass have been studied as a tool for removing heavy-metals from water and wastes.

Living and dead microbial cells of both procariotic and eukaryotic nature, are able to remove relevant amounts of heavy metals by mechanisms such as physico-chemical reaction, specific binding and transportation by proteins, metabolic transformation etc [7,8].

Physico-chemical adsorption, ion exchange complexation and/or microprecipitation of heavy metals are relatively rapid and eventually reversible by means of cheap and easy procedures.

The present paper deals with the study of some microbial strains bioaccumulating manganese. Free and immobilized cells are adopted for the study. When free cells are used, experiments are specifically intended to clarify the physiological and biochemical aspects of metal absorption as well as to establish the basis for further developments of the procedure of cell immobilization and their use for practical purposes.

Two groups of microorganisms (the prokariotic Arthrobacter sp. and the eukariotic Chlorella and Scenedesmus) have been taken into consideration in the preliminary study and rough measurements of the basic parameters (culture growth, biomass yield, metal absorption). The immobilization of the cells is only studied with cells of the more promising strains.

Studies on manganese biosorption for technological scopes have already been reported for cleaning surface and pipe water and protecting cooling industrial devices from mineral deposits and encrustments due to water evaporation [9,10,11].

Materials and Methods

Microorganisms: a bacterium (Arthrobacter sp.) and two green microalgae (Chlorella and Scenedesmus) were made use of in the present study.

Cultivation of bacteria : the strain of Arthrobacter, was isolated from a soil sample collected near Monteluco di Roio (L'Aquila, Italy). The following minimal medium (MM) was used for the cultivation of the cells and maintainance (gxl^{-1}): glucose 50; NH_4NO_3 2.5; KH_2PO_4 1.5; $CaCl_2$ 0.1; $MgSO_4$ 1.0; yeast extract 0.5. MM was enriched with extra quantities of nitrate, phosphate and sucrose (rising to 5, 3 or 100 gxl^{-1} respectively) to modulate the physiological status of the cells. Batch cultures were performed in 250 ml Erlenmeyer flasks filled with 100 ml of medium; the incubation of the flasks was done in an orbital rotatory shaker at 250 rpm, at 37°C. Cultures were harvested after 5 days of incubation during the exponential growth phase or alternatively 29 days, long after attainment of the stationary growth phase.

Cultivation of algae : green microalgae of the genera Chlorella and Scenedesmus were isolated from heavily polluted waters around Florence [12]. Cells were grown in a medium with the following composition in gxl^{-1}: $NaNO_3$ 1.5; $K_2HPO_4*3H_20$ 0.04; $MgSO_4*7H_20$ 0.075; $CaCl_2*2H_20$ 0.036; Na_2CO_3 0.02; $C_6H_8O_7$ 0.006; $FeNH_4(SO_4)_2*12H_2O$ 0.006; E.D.T.A. 0.001; 1ml of microelements was added to the medium [13]. The salts were dissolved and the final pH of the solution was adjusted at 7.5. Batch cultures were performed in 250 ml Erlenmeyer flasks filled with 50 ml of medium; incubation of flasks was carried out at room temperature and under day light.

Harvesting and use of cells

Cells of Arthrobacter, Chlorella and Scenedesmus were harvested by centrifugation and washed twice with sterile distilled water (SDW). For Mn^{++} uptake by free cells, the pellets were resuspended in SDW to the standard cell density of 0.2g/ml (d.w.). Aliquots (0,8ml) of this cell suspension were mixed with 0.2 ml of the Mn^{++} solution at different concentration.

Cells remained in the Mn solution for different times at room temperature (Chlorella and Scenedesmus) or 37°C (Arthrobacter).Then cells were removed by centrifugation at 3500xg for 15 min at 21°C.

Mn left unabsorbed from the solution was measured by Nuclear Magnetic Resonance (NMR) [14].

Harvesting preparation and use of immobilized cells

Cells of Arthrobacter grown for 5 days in the shaker, were collected by centrifugation and washed twice with SDW. The supernatant was discharged and the cells were again resuspended in SDW to the cell density of 0.02 g/ml (d.w.). 1ml of such a suspension was mixed with 2ml of solution containing 3% sodium alginate. The mixture was then thoroughly mixed and dropped into a solution of $CaCl_2$ (5%). After 1.5 hours the Ca-alginate beads were washed with SDW and suspended in the Mn^{2+} solution at the proper concentrations. The Mn uptake by immobilized cells was measured after 0.5, 1.0, 2.0 4.0, 22.0 hours of exposure.

Analytical determinations

Mn^{++} was determined as a function of the variation of T1 relaxation time of water solution using a spectrometer Brunker SXP 2-100 at a working frequency of 26 MHz and a temperature of 25°C. The modified null method was used to measure the T1 relaxation times [15]. Mn^{++} in acqueous solution lowers the T1 relaxation times of water. Therefore T1 values of the test solutions at Mn^{++} concentrations (mM): 5.9, 7.4, 8.8, 11.8, 14.8, 17.7, 19.2 have been measured. A calibration curve was used to relate experimental T1 values to Mn^{++} concentration of the samples. Alternatively, Mn^{++} in solution was determined using an atomic absorption spectroscope.

Results and Discussion

Biosorption of manganese by cells was examined as a function of a)different cultural age; b) physiological status of the cells grown in media enriched with phosphate, nitrate. sucrose; c) amount of Mn^{++} bounded by cells.

Different cultural age of the cells. Fig. 1 depicts the Mn uptake as a function of the Mn concentration in the test solution when Arthrobacter cells are grown in MM and harvested during the exponential growth phase (5 days). From Fig. 1 it turns out that Mn^{2+} uptake is constant in that it is not influenced by the Mn concentration of the test solution.

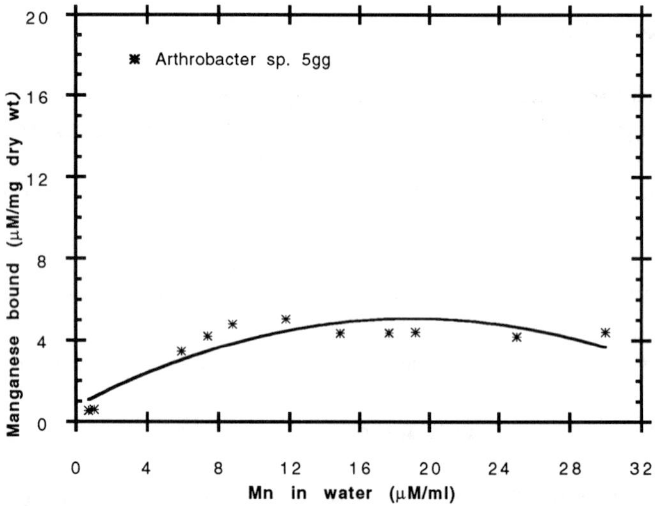

Figure 1 - Manganese uptake by Arthrobacter versus manganese concentration of the test solution. Cells were grown in MM and harvested at 5 days.

If cells of Arthrobacter are harvested during the stationary phase (29 days), the Mn uptake is a function of the concentration of the Mn in the solution (data are not shown).

Fig. 2 shows the Mn uptake by cells of Chlorella harvested at 5 days and 39 days. In this graph, a maximum of bioaccumulation is shown when cells are harvested at 5 days. On the contrary cells of Chlorella harvested at 39 days are able to remove Mn^{2+} no matter what its concentration in the range of the present investigation.

In Fig. 3 the Mn uptake by cells of Scenedesmus harvested at 5 and 39 days, is reported. Data show that Mn uptake is independent from the concentration of the solution at the same

behaviour.

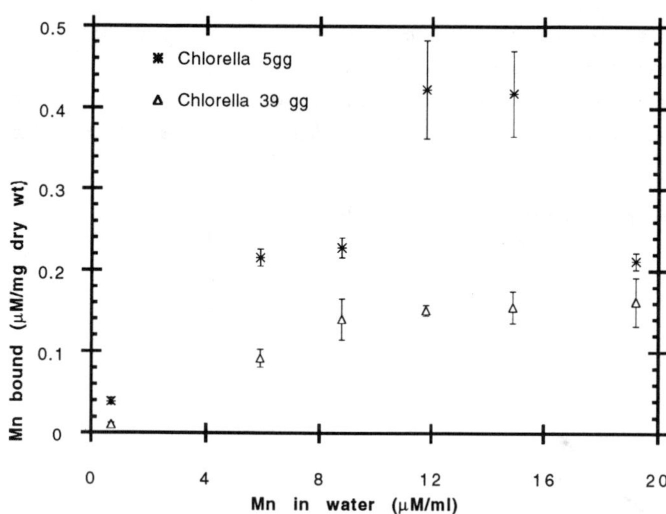

Figure 2 - Manganese uptake by Chlorella versus manganese concentrations of the test solutions at 5 (*), 39 (Δ) days.

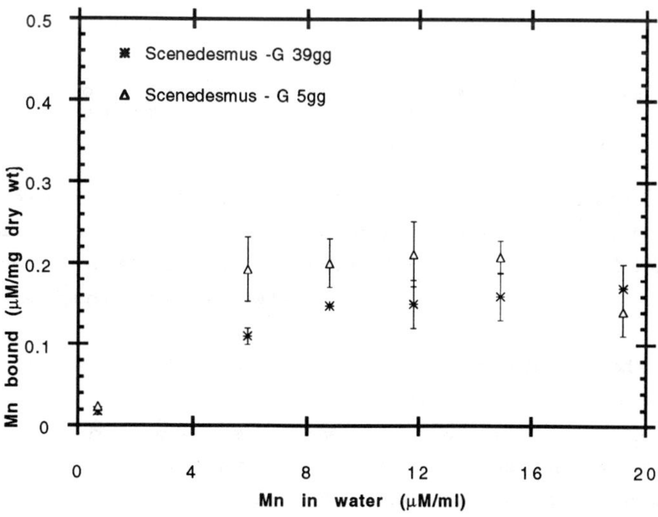

Figure 3 - Manganese uptake by Scenedesmus versus manganese concentrations of the test solutions at 5 (Δ), 39 (*) days.

Phisiological status of the cells grown in media enriched with phosphate nitrate or sucrose. Fig. 4 shows Mn^{2+} uptake by cells of Arthrobacter grown on MM and MM enriched with phosphate, sucrose and nitrate. Figure 4 highlights that the cells grown on MM, MM+sucrose and MM+phosphate bioaccumulate a constant amount of Mn, independently of the increasing Mn^{2+} concentration in the solution.

On the contrary, the amount of Mn^{2+} taken up by bacterial cells grown in MM supplemented with extra nitrate are highly sensitive to the Mn concentration present in the solution (Fig.4).

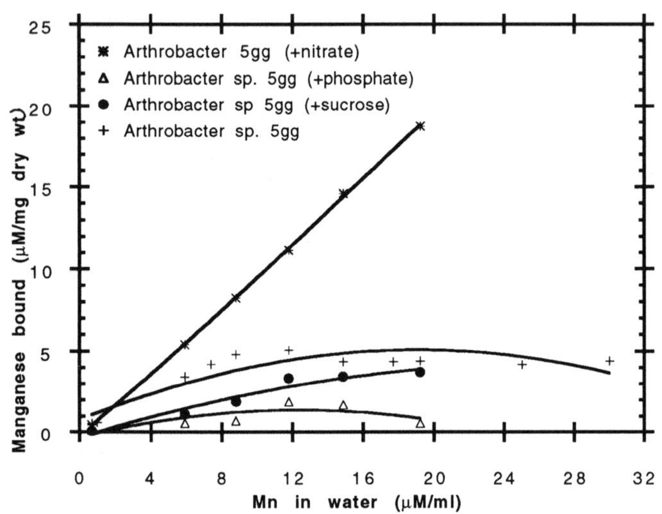

Figure 4 - Manganese uptake by Arthrobacter versus manganese concentrations of the test solutions at 5 days. Cells were grown in MM (+), MM+nitrate (*), MM+phosphate (Δ), MM+sucrose (o).

Amount of Mn⁺⁺ bounded by cells. As shown in in Fig.1 the amount of Mn^{2+} uptake is 5 μM/mg dry wt when cells of Arthrobacter are grown in MM and suspended in a solution containing 12μM of Mn^{2+}. On the other hand, according to Fig.4, a maximum loading of 0.3 or 0.15 μM/mg dry wt is measured when cells grown in MM enriched with sucrose or phosphate are employed.

This plot shows that the maximum measured is 18.7 μM/mg dry wt of Mn^{2+} bound to cells grown in MM enriched with nitrate and suspended in a solution containing 19.2 μM of Mn^{2+}.

The data on Mn^{2+} uptake by Chlorella harvested at 5 and 39 days, as reported in Fig. 2, show that quantities of the ion bounded to the cells is respectively 0.44 and 0.16 μM/mg dry wt with cells suspended in a solution containing Mn^{2+}12 μM.

As shown in Fig. 3 the Mn uptake by Scenedesmus was as high as 0.2 and 0.16μM/mg dry wt if 5 and 39 days harvested cells are employed. The results obtained with immobilized cells indicate that: (i) the uptake is almost completed with 30 minutes of contact between the solution and the cell containing beads. This fact is likely to be associated with the saturation of binding sites present around or inside the cells. In a second stage, with the gradual occupation of these sites, the uptake becomes less effective; (ii) the total Mn^{2+} uptake by immobilized cells is a little bit lower than the one obtained with cells suspended in the ion solution. It turns out that 137mg/ gr dry wt are bioaccumulated if a 535mg/l of Mn solution is employed; (iii) the yield is directly proportional to the quantity of the beads (indicated in the table as the volumeof the beads) but not to that of the biomass. Probably, this indicates that the biomass around the beds takes part in the process of bioaccumulation. On the other hands the biomass inside the beads uptake Mn in smaller measures, probably because the diffusion of the ions inside the beads is less. (Tab.1).

Test N	A (mL)	B (mg/L)	C (mg/mL)	Xtot (mg)	Mn2+ (mg/L)	q after 30' (mg/g)	Yield (%)
1	5	109	9.66	48.3	30	31.1	25.0
2	10	109	9.66	96.6	40	20.7	32.3
3	5	535	9.66	48.3	133	137.7	23.1
4	10	535	9.66	96.6	215	111.3	35.1
5	5	109	32.43	162.2	28	8.6	22.7
6	10	109	32.43	324.3	42	6.5	33.9
7	5	535	32.43	162.2	151	46.6	26.6
8	10	535	32.43	324.3	204	31.5	33.7

Legend:

A	Vol. beads	mL	5	10
B	Mn2+	mg/L	100	500
C	X	mg/mL	9.7	32.4

note: X= biomass content in the beads (mg/mL)
q = Mn2+ uptake/ g biomass
Yield = (Cin-Cfin)/Cin x 100
Mn2+ = Mn2+ uptake after 30 minutes
Xtot= total biomass content in the flasks

Table 1 - Bioaccumulation of manganese by immobilized Arthrobacter

As concluding remarks we can say that the data on the ability of different kinds of cells to remove the Mn^{++} ion from water are promising enough. The present research has been developed with the aim to establish a suitable microbial strain, proper culture conditions in terms of composition of the medium and culture age of the cells as well as the cells concentration, in both free and immobilized form in order to provide some basic information for a process development. Previous investigation have shown that uptake of manganese by bacterial and algal cells is a rapid and reversible reaction [14].

Further data are being collected on the yield and kinetics of Mn++ adsorption and desorption with the double aim of building up a process for water cleaning and recovery precious metals.

Acknowledgements

The autors would like to tank Dr. Angela Gasbarro, Dr. Antonella Nardini, Dr. Monia Felice and Mr. Marcello Centofanti for their helpful collaboration in the analytical measures, during work.

References

1."Biomineralization: processes of iron and manganese," Microbial accumulation of iron and manganese in different aquatic environments:An electron optical study, H.C.W.Skinner et al. eds.(Catena Verlag supplement 21, 1992),115-131.

2. G.W.Garnham, G.A.Codd, and G.M.Gadd, "Uptake of cobalt and cesium by microalgal-and cyanobacterial-clay mixtures," Microb. Ecol. 25 (1993),71-82.

3. N.Friis and P.Myers-Keith, "Biosorption of Uranium and lead by Streptomyces longwoodensis," Biotech.and Bioeng. 28 (1986),21-28.

4. A.C.A. Da Costa and S.G.F.Leite, Metals biosorption by sodium alginate immobilized Chlorella homosphaera cells," Biotech. Lett. 13 (8) (1991),559-562.

5. W.S.Reeburgh, "Rates of biogeochemical processes in anoxic sediments,". Ann.Rev.Eart Planet Sci.11 (1983),269-298.

6. C.R.Myers and K.H.Nealson, "Bacterial manganese reduction and growth with manganese as the sole electron acceptor," Science 240 (1988),1319-1321.

7. N.Kuyucak and B.Volesky,"Biosorbents for recovery of metals from industrial solutions," Biotech. Lett. 10 (27) (1988),137-142.

8. F.E.Brinckman and G.J.Olson,"Chemical principles Underlying bioleaching of metals from ores andf solid wastes, and bioaccumulation of metals from solutions," Biotech. and Bioeng.Symp. 16 (1986),35-43.

9. L.I.Sly, M.C.Hodgkinson and V.Arunpairojana,"Deposition of manganese in a drinking water distribution system," Appl. and Env.Microb. 56 (3) (1990),628-639.

10.M.J.Parking and I.S. Ross,"The specific uptake of manganese in the yeast Candida utilis," Jour.Gen.Microb. 132 (1986),2155-2160.

11. T.Peitchev and V.Semov,"Biotechnology for manganese removal from groundwaters," Wat.Sci.Tech. 20 (3) (1988),173-178.

12. E.Capolino, et al. "Cadmium toxicity and bioaccumulation in microalgae and cyanobacteria" (Paper presented at the 9th ISB, Troia, Portugal 9-13 September 1991).

13. M.Felice, "Bioassorbimento di Manganese da parte di diversi microrganismi" (Thesis, L'Aquila University, 1992),10-22.

14. G.Patrizio et al. "NMR study of Mn^{2+} bioaccumulation and recovery from a bacterial strain," Physica Medica 9 (1) (1993),25-28.

15. S.K.Ghosh, E.Tettamanti, and A.Panatta "A convenient way of measuring nuclear spin-lattice relaxation times based on the null method," J.Phys. E:Sci.Instrum. 13 (1980),84-84.

THE USE OF IMMOBILISED YEAST CELLS FOR HEAVY METAL REMOVAL FROM WASTEWATERS

John R. Duncan, Dean Brady and Peter D. Rose

Department of Biochemistry and Microbiology, Rhodes University,
Grahamstown, 6140, South Africa

Abstract

Biotechnology based processes for removal and/or recovery of these metals were investigated and yeast was chosen as the organism of choice in this study since it is a readily obtainable and inexpensive source of biomass in the local context. By immobilizing the yeast in polyacrylamide gel and packing this material into columns, Cu^{2+}, Co^{2+} or Cd^{2+} could be removed from influent aqueous solutions yielding effluents with no detectable heavy metal, until breakthrough point was reached. Cross-flow microfiltration (CFMF) units were successfully utilized to remove metal ions from solution. Yeast cells were blended into a stream of heavy metal-laden water in a reaction vessel with a HRT of 10 minutes. The yeast-metal suspension was then passed to the filter (flux rate: 9.2 x 10^2 $\ell/m^2/hr$). In the use of copper a removal efficiency of 65% was determined with cation binding of 20 g copper per kg dry mass of yeast. Using a packed system yeast cells were loaded onto the hollow fibre membrane filter at 0.42 kg/m^2 yeast dry mass/filter surface area and flux rates stabilized at 12.5 x 10^2 $\ell/m^2/hr$ before addition of heavy metal to the water. Removal efficiencies of 65-80% were recorded. These levels were sustained at flow volumes of over 3095 ℓ/kg dry wt cells without breakthrough at the above metal/yeast mass ratio. The passage of metal-laden influent through a series of sequential CFMF bio-accumulation systems allowed for further reductions in the levels of copper, cadmium, and cobalt in the final effluent than that afforded by a single bioaccumulation process. Serial bioaccumulation systems also allowed for partial separation of metals from dual metal influents. More than one elemental metal cation could be accumulated simultaneously and in greater quantities than when a single metal was present in the effluent (Cu^{2+} 0.43 mmol, Cu^{2+} + Cd^{2+} 0.67 mmol, and Cu^{2+} + Co^{2+} 0.83 mmol per g yeast dry mass when the initial concentration of each of the metal species was 0.2 mmol/ℓ). Co-accumulation of two different metal cations allowed higher total levels of bioaccumulation than found with a single metal.

Mineral Bioprocessing II
Edited by David S. Holmes and Ross W. Smith
The Minerals, Metals & Materials Society, 1995

Introduction

Water availability is a major factor limiting further industrial and urban development in South Africa (1,2). High levels of contamination has resulted in the substantial elevation of heavy metal levels in the public water system and the general environment globally (3) to a point where drinking water quality criteria may be exceeded (4). The technique of selective bioaccumulation of heavy metals by microbial systems offers a possible approach for the remediation of contaminated water bodies. The yeast *Saccharomyces cerevisiae* has previously been shown to accumulate heavy metals from water (5-8). As a waste product of numerous fermentation industries it may be obtained both in large quantities and inexpensively, providing a reliable source for large scale application in bioremediation processes.

While heavy metal bioaccumulation is well known, problems in immobilization have rendered large scale engineering of bioaccumulation systems impractical. Gel immobilization has been used with some success. Gel immobilization of the fungus *Rhizopus arrhizus* has been particularly successful in the accumulation of Cd^{2+}, Cu^{2+}, Fe^{3+}, Mn^{2+}, Pb^{2+} and Zn^{2+}, and in the recovery of uranium from ore bioleach (9-13). Uranium also has been recovered from both fresh and sea water by *Streptomyces viridochromogenes* and *Chlorella regularis* trapped in polyacrylamide (14). Gel immobilization on a large scale can, however, be prohibitively expensive and limits to rates of diffusion within a gel is a major problem.

A practical alternative may involve the use of cross-flow (tangential flow) membrane filtration, which is widely used in cell separation and processing (16). Cross-flow micro-filtration (CFMF) has particular advantages in that it operates at low pressures and with high flux rates. Cross-flow systems have been demonstrated to be useful in harvesting and maintaining *S. cerevisiae* cells and cultures (16-18).

In this study the use of polyacrylamide and CFMF for cell immobilization to retain *S. cerevisiae* biomass during heavy metal bioaccumulation has been explored.

Materials and Methods

Gel Immobilisation

The method of immobilization used essentially was that of Chibata *et al.* (19) and was as follows: 10 g wet mass of *S. cerevisiae* was suspended in 20 ml physiological saline (0.15 mol/ℓ NaCl) at 8°C: acrylamide monomer (7.5 g) plus *N,N'*-methylene-bisacrylamide (0.4 g) was dissolved in 24 ml deionized water and cooled to 8°C: the monomer solution and cell suspension were thoroughly mixed together. Added to this mixture was 1 ml of 2.5% *N,N,N',N'*-tetramethylethylenediamine (TEMED), and 5 ml of 1% ammonium persulfate. The temperature of the solution/suspension was maintained below 50°C during the exothermic polymerization process so as to not damage the biomass. The immobilized biomass was passed through a 30-mesh (500 μm) sieve. This yielded thin threads of yeast-cell-containing polyacrylamide gel, 5 g wet mass of which was subsequently placed in deionized water and then poured as a slurry into a chromatography column.

Experimental conditions were as follows: The flow rate was 1 ml/min: the fraction volume, 10 ml: column height, 10 cm: column volume, 20 ml: and temperature, 20°C ± 2°C. The metal stock solutions were 200 μmol/ℓ solution of each metal chloride in 5 mmol/ℓ HEPES buffer, pH 7.2 a buffer that does not chelate heavy metal cations. The metal salts used were $CuCl_2$ (Merck). $CdCl_2$ (PAL) and $CoCl_2.6H_2O$ (Merck). The ethylenediaminetetraacetic acid (EDTA) solution for cation elution after bioaccumulation was 30 ml of 1.0 mmol EDTA/ℓ in 5 mmol HEPES/ℓ buffer, pH 7.2. Buffer (30 ml) was used to wash the column before and after the EDTA procedure.

Microfiltration Systems

An alternative apparatus was assembled using a microfiltration hollow fibre membrane (CFMF) system. This apparatus did not remove copper from solution. This system was used for harvesting of the yeast biomass after bioaccumulation of copper. Alternatively, the biomass was initially harvested onto the membrane and bioaccumulation occurred when buffered copper solutions were filtered through the biomass-packed membranes. A polypropylene microfiltration hollow fibre membrane was used. Pore size was 0.1 μm, P_i was maintained below 20 ± 5 kPa. The yeast cell suspensions were circulated around the exterior of the hollow fibres and the permeate passed into the lumen. 14 g of yeast was suspended in buffered solution and punped into the filter cartridge. Immediately buffered $CuCl_2$ stock solution (200 μmol/ℓ) was added to the reservoir in such a manner as to maintain a constant volume in the reservoir during the period of the experiment. This was compared to a parallel experiment where a similar mass of cells in suspension was packed into the filter cartridge by harvesting prior to addition of metal solution to the reservoir. This was achieved by diverting the permeate outlet back into the reservoir until packing was completed.

Serial Microfiltration

S. cerevisiae cells (14 g wet mass) were suspended in a 2 ℓ solution of 200 μmol/ℓ $CuCl_2$ buffered with 5 mmol/ℓ Tris HCl buffer, pH 7.2. After 15 min reaction time the yeast was harvested onto a hollow fibre membrane cartridge (System I) by partially restricting the reject flow (from 250 to 150 ml/min). Constant addition of further copper stock solution (200 μmol/ℓ $CuCl_2$) maintained the influent reservoir at its initial volume (2 ℓ) while permeate (100 ml/min) was passed into a second reservoir (0.5 ℓ), to which 14 g wet mass of yeast was added. This second yeast suspension and subsequent permeate from System I was pumped onto a second membrane cartridge (System II) which was similar to the first. Therefore the effluent from the first accumulation system was used as the influent for the second system and the permeate rates of the two systems was constantly matched. In further experiments an influent solution containing 200 μmol/ℓ $CuCl_2$ and either 200 μmol/ℓ $CdCl_2$ was contacted with fresh yeast biomass and the levels of the heavy metal cations in the permeates monitored.

Results

Gel Immobilisation

The bioaccumulation preference was Cu^{2+} > Co^{2+} > Cd^{2+}. The level of metal accumulated is presented in Table I.

Table I Accumulation of Metals by Polyacrylamide Immobilized Yeast

Influent pH during accumulation	Metal bioaccumulation (nmol metal/mg dry mass yeast		
	Cu	Co	Cd
7.2	480	100	133
5.6	110	87	87

Microfiltration System

Results of metal accumulation from heavy metal-containing solutions indicate that this method can be extremely effective in metal removal (Figure 1). The metals Cr^{3+}, Cu^{2+} and Pb^{2+} were accumulated in large quantities, while Cd^{2+} was accumulated to a lesser extent.

Removal efficiencies of 65-80% were recorded and these levels were sustained at flow volumes of over 3 000 ℓ/kg dry weight of cells without breakthrough. In the case of copper, a metal binding of 20 g Cu^{2+}/kg dry mass of yeast was found.

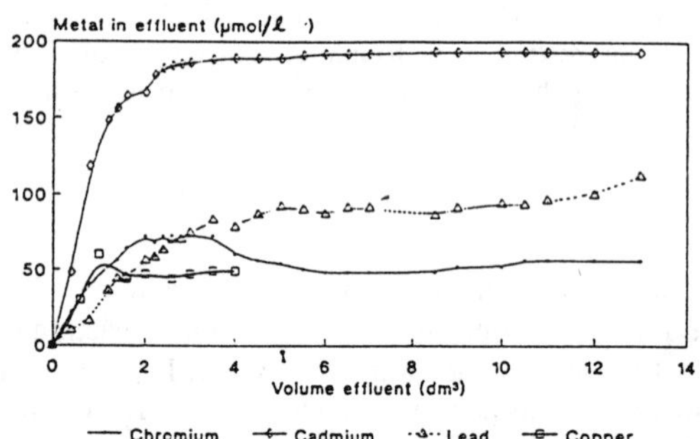

Figure 1: Bioaccumulation of metals from solution using cross-flow microfiltration cartridges. Cu^{2+} accumulation was only monitored for the first 4 hr. (Flux rates were maintained at 9.2 x 10^2 ℓ/m/hr).

Serial Microfiltration

The serial filter configuration is depicted in Figure 2.

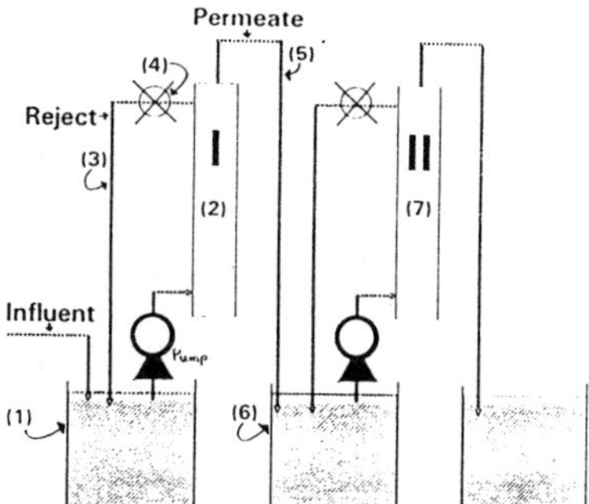

Figure 2: Schematic representation of the microfiltration equipment configuration. Points 1 and 2 are permeate sample points for Systems I and II respectively.

The use of serial heavy metal accumulation by biomass was demonstrated to be effective (Figure 3). The saturation of the cells with copper in Systems I and II is indicated by a steady increase in effluent copper. System I accumulated 0.26 mmol/g (16.6 mg) Cu^{2+} per g dry mass with an average flux rate of 2.9×10^2 ℓ/hr/m^2.

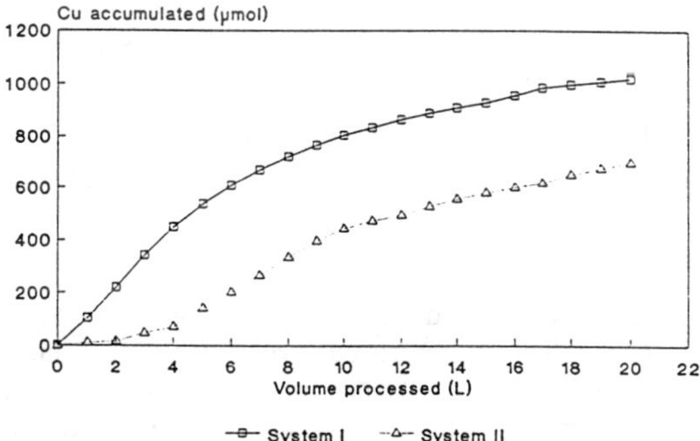

Figure 3: Copper bioaccumulation by yeast retained by application of serial cross-flow microfiltration. (Influent Cu^{2+} was 200 μmol/ℓ).

Inclusion of a second heavy metal in solution with the Cu^{2+} permitted accumulation of the two metals simultaneously (for example Cu^{2+} and Co^{2+}, Figure 4 and 5 respectively). The total quantity of metal accumulated in this case was greater than when a single metal was present in the effluent, i.e. Cu^{2+} alone was bound only to the limit of 0.43 mmol, while combinations of metals (Cu^{2+} and Cd^{2+} and Cu^{2+} + Co^{2+}) allowed for a total accumulation of 0.67 mmol and 0.83 mmol metal per g yeast dry mass respectively. In each case the total accumulation of Cu^{2+} decreased and saturation occurred more rapidly (compared to Figure 3). There is also an indication that Cu^{2+} displaces the other metals, particularly Co^{2+} from System I to System II, by cation competition. This is obvious in that accumulation of Co^{2+} is less in System I compared to System II (Figure 5) when in competition with Cu^{2+}. This process may allow for methods of partially purifying recovered metals from mixed metal solutions.

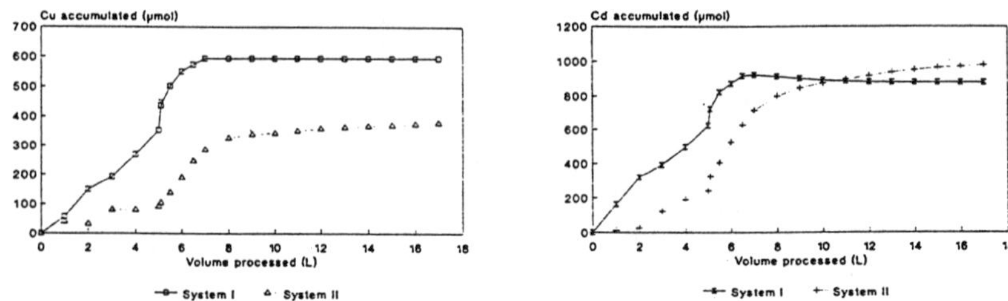

Figure 4: Simultaneous accumulation of Cu^{2+} (a) and Cd^{2+} (b) from a solution containing the $CuCl_2$ (200 μmol/ℓ) and $CdCl_2$ (200 μmol/ℓ).

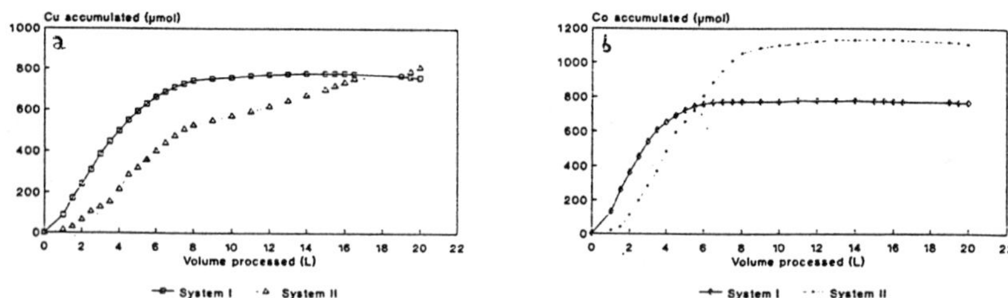

Figure 5: Simultaneous accumulation of Cu^{2+} (a) and Co^{2+} (b) from a solution containing Cu^{2+} (200 μmol/ℓ) and Co^{2+} (200 μmol/ℓ).

Discussion

The advantages of cell immobilized during metal accumulation over cell suspensions include improved biomass retention, improved levels of biomass reuse, higher biomass concentrations, and high flow rates. Unlike certain membrane systems, there may be only very limited clogging in continuous-flow systems involving immobilized biomass, (20) although limits to rates of diffusion may present a problem.

Immobilized *Rhizopus arrhizus* biomass has also been shown to accumulate Cd^{2+}, Cu^{2+}, Fe^{3+}, Mn^{2+}, Pb^{2+}, and Zn^{2+} (10). The quantity of Cd^{2+} accumulation was determined by a number of external factors: low pH or temperature diminished the level of Cd^{2+} accumulation but low levels of alkali metals did not affect the system (10). In the present study each of the three metal cations investigated were accumulated to a lesser degree at lower pHs.

In the immobilized column experiments > 99% removal of metal from solution (200 μmol/ℓ) was obtained before the breakthrough points were reached. For Cu^{2+}, removal of more than 70% was achieved over the total volume of 800 ml effluent passed through the column. Immobilized yeast column systems behave according to theoretical equilibrium plate theory and are therefore likely to be capable of removing metal cations more completely from solution than would a free cell suspension, which would permit only a single equilibration. The limitation of the efficiency of bioaccumulation is therefore only determined by the physical size of the column. The theoretical plate lengths depend on such factors as the biomass type, the diffusion coefficient in the immobilized biomass, the metal ion species and concentration and the flow rate. A limitation of the column immobilized system may also be the cost of such systems.

An alternative system of immobilization may be the use of cross-flow microfiltration (CFMF) membranes. These have been widely used in cell separation and processing (15-18). This system has particular advantages in being able to operate at low pressures and with high flux rates.

A single microfiltration cartridge packed with 14 g wet mass of yeast cells was found to be effective in metal removal in experiments where the filter unit was packed with yeast during simultaneous yeast harvesting or where the cells were packed onto the membranes before addition of heavy metal containing water. Flux rates were maintained at a high level and 65-80% removal efficiencies over large throughput volumes. Potential for the application of CFMF is yeast based bioaccumulation processes was thus demonstrated as a promising immobilized system.

The use of serial batteries of CFMF-based yeast bioaccumulators was found to reduce the concentration of toxic heavy metals in waters significantly. This system could also ensure that particulate metals and metals associated with microorganisms or colloids would be simultaneously removed. Serial CFMF bioaccumulation also permitted simultaneous accumulation of two metals and resulted in a greater total metal accumulation than with a single metal in the effluent. It further appears that there is some displacement of metals from one system to the other which, if further developed, may allow for the partial purification of metals during recovery from mixed metal solutions. Further experimentation and experience with CFMF may allow for improvement of the bio-accumulation process.

The use of serial membrane based bioaccumulation systems could thus improve the effectiveness of the process of metal bioaccumulation and CFMF-based bioaccumulation processes are potentially less expensive than gel-immobilized biomass and would allow for more rapid adsorption and desorption of cations. Moreover CFMF-based processes offer systems that could feasibly be engineered on a large scale.

Acknowledgements

The authors wish to thank the Water Research Commission for funding for this project. The hollow fibre microfiltration membranes were supplied by Dr. E.P. Jacobs of the Institute for Polymer Science, Stellenbosch University, RSA.

References

1. P.E. Odendaal, "Water research needs in South Africa," Proc. 2nd Anaerobic Digestion Symp, Bloemfontein, (1989), 10-21.

2. H.N.S. Wiechers, "Sewage purification in South Africa - Quo Vadis?" Water SA, **15** (1989), 141-146.

3. J.O. Nriagu and J.M. Pacyna, "Quantitative assessment of worldwide contamination of air, water and soils by trace metals," Nature (London), **333** (1988), 134-139.

4. M.J. Pieterse, "Drinking-water quality criteria with special reference to the South African experience," Water SA, **15** (1989), 169-178.

5. A. Nakajima and T. Sakaguchi, "Selective accumulation of heavy metals by microorganisms," Appl. Microbiol. Biotechnol., **24** (1986), 59-64.

6. P.R. Norris and D.P. Kelly, "Accumulation of cadmium and cobalt by *Saccharomyces cerevisiae*," J. Gen. Micro., **99** (1977), 317-324.

7. P.R. Norris and D.P. Kelly, "Accumulation of metals by bacteria and yeasts," Dev. Ind. Microbiol., **20** (1979), 299-308.

8. D. Brady and J.R. Duncan, "Bioaccumulation of metal cations by *Saccharomyces cerevisiae*," Appl. Microbiol. Biotechnol., **41** (1994), 149-154.

9. R. Ileri, F. Mavituna, M. Parkinson and M. Tucker, "The use of biosorption for uptake of low level contaminants by immobilized cells," Abst. 5th Euro. Conf. Biotechnol., Copenhagen, (1990), 83.

10. D. Lewis and R.J. Kiff, "The removal of heavy metals from aqueous effluents by immobilized fungal biomass," Environ. Technol. Lett., **9** (1988), 991-998.

11. M. Tzezos, R.G.L. McCready and J.P. Bell, "The continuous recovery of uranium from biologically leached solutions using immobilized biomass," Biotechnol. Bioeng., **34** (1989), 10-17.

12. M. Tsezos, "The development modelling and pilot plant testing of a new continuous metal recovery process using immobilized biomass," Abst. 5th Euro. Conf. Biotechnol., Copenhagen, (1990), 76.

13. M. Tsezos, "Recovery of uranium from biological adsorbents, - desorption equilibrium," Biotechnol. Bioeng., **26** (1984), 973-981.

14. A. Nakajima, T. Horikoshi and T. Sakaguchi, "Recovery of uranium by immobilized microorganisms," Eur. J. Appl. Microbiol. Biotechnol., **16** (1982), 88-91.

15. R.S. Tutunjan, "Cell separations with hollow fibre membranes," Comprehensive Biotechnology, (Vol. 2. M. Moo-Young, (ed.), Pergamon Press, Oxford), (1985), 367-381.

16. J.A. Scott, "Application of cross-flow filtration to cider fermentations," Process Biochem., **23** (1988), 146-148.

17. J-L Uribelarrea, J. Winter, G. Goma and A. Pareilleux, "Determination of maintenance coefficients of *Saccharomyces cerevisiae* cultures with cell recycle by cross-flow membrane filtration," Biotechnol. Bioeng., **35** (1990), 201-206.

18. R.K. Warren, D.G. Macdonald and G.A. Hill, "Cross-flow microfiltration of *Saccharomyces cerevisiae*," Process Biochem., **26** (1991), 337-342.

19. I. Chibata, T. Tosa and T. Sato, "Immobilized aspartase-containing microbiol cells: Preparation and enzymatic properties," Appl. Microbiol., **27** (1974), 878-885.

20. G.M. Gadd, "Heavy metal accumulation by bacteria and other microorganisms," Experientia, **46** (1990), 834-840.

EFFECT OF DIFFUSION PROPERTIES ON THE URANIUM SORPTION BY CHITOSAN POLYMERS

M. JANSSON-CHARRIER, E. GUIBAL*, B. DELANGHE and P. LE CLOIREC

Ecole des Mines d'Alès
Laboratoire Génie de l'Environnement Industriel
6, avenue de Clavières
30319 ALES CEDEX - FRANCE

Abstract

Chitosan is a well-known natural amino-polymer which has proved efficient in removing metal ions from dilute solutions. The uranium sorption is diffusion controlled and it occurs within a thin layer at the surface of the polymer. Sorption performances of chitosan could be increased by a substitution of functional groups on the chitosan backbone : the grafting of oxo-2-glutaric acid allowed glutamate glucan to be obtained. This modification of the chitosan promoted the formation of new chelating sites. Another impact of the modification was the change in the polymer structure which favoured metal ion diffusion in the whole mass of the sorbent. In order to estimate the influence of the modification of its structure, the chitosan was recrystallized by a dissolution/precipitation method. The performances of both polymers were compared regarding in terms of their diffusion properties. The recrystallization enhanced the mass transfer in the solid, although the process was not optimized. Scan Electron Microscopy and X-ray Energy Dispersed Analysis reveals the concentration gradient, thus confirming and the previous conclusion. The recrystallization of the chitosan produced behaviour somewhere between raw and substituted chitosan.

Mineral Bioprocessing II
Edited by David S. Holmes and Ross W. Smith
The Minerals, Metals & Materials Society, 1995

Introduction

For the last thirty years biological processes have been making a great contribution to the treatment of mineral wastes or wastewaters. Bioleaching is a promising field where biological techniques are involved[1-2]. Adsorbents of biological origin, such as bacteria or fungi, are also used in the removal of metal ion from diluted industrial wastewaters.[3-9] Bioaccumulation can involve various mechanisms including active or passive phenomena. For an industrial application passive phenomena appeared more interesting, taking advantage of physico-chemical properties of such sorbents, in terms of their cell walls and main constitutive polymers. Chitin and the deacetylated form, chitosan, are natural polymers present in large proportions in fungal cell walls or crustacean shells. These polymers exhibit high sorbent capacities for heavy metals.[10-11] The improvment of sorbent properties was obtained by grafting functional groups onto the chitosan backbone.[12-14]

The present study focuses on the removal of uranyl by chitosan and its modified forms : glutamate glucan (resulting from a substitution of oxo-2-glutaric acid), and recrystallized chitosan (obtained by dissolution and further reprecipitation). Kinetics are studied according to a major parameter, the particle size of the polymer. The discussion of diffusion mechanisms deals with intraparticular and external diffusion. Conclusions of the macroscopic study are compared with experimental data obtained with Scanning Electron Microscopy coupled with an X-Ray Energy Dispersed Analysis, in order to obtain the concentration gradient in the polymer.

Experimental section

Products

Chitosan was obtained (Fluka AG) as white, yellow tinged, polydisperse flakes of high molecular weight 2 10^6, with a deacetylation percentage of 80 %. Other reagents : oxo-2-glutaric acid and sodium cyanoborohydride (Fluka AG) were analysis grade. The modified polymers were produced according to a procedure similar to that proposed by Muzzarelli et al. [12-13], also used by Saucedo et al. [14] which consists of a two-step mechanism. First, keto-imination of the polymer by dissolution of chitosan in organic acid and a second step which leads to the reduction of this product. 10 g of chitosan (as 0.4 mol) were dropped into 400 ml of an aqueous solution of oxo-2-glutaric acid, with an acid to chitosan molar ratio of 1.5. The solution was stirred for 12 to 24 hours, a pale yellow viscous solution was obtained. The pH was adjusted to 4.5 with a molar solution of NaOH. Sodium cyanoborohydride (5 g in 100 ml of demineralized water) was dropped into the keto-imine solution drop by drop in order to reduce the keto-imine; simultaneously the pH was adjusted to 6.5. A white, slightly viscous product was obtained: glutamate glucan. The polymer was dried at ambient temperature and was later washed by soxhlet (acetone as solvent) for 3 hours. The dry product was sieved into four fractions: G1, G2, G3, G4 corresponding to the following respective particle diameter limits : 0-125 µm, 125-250 µm, 250-500 µm, 500-1000 µm. The particles were assumed to be spherical, and the external surface was considered as equivalent to the specific surface. The density of the polymer was evaluated at 1.4 g ± 0.1 g·cm^{-3}. Figure 1 shows monomer units of chitosan and glutamate glucan, and Table 1 presents some characteristics of both sorbents.

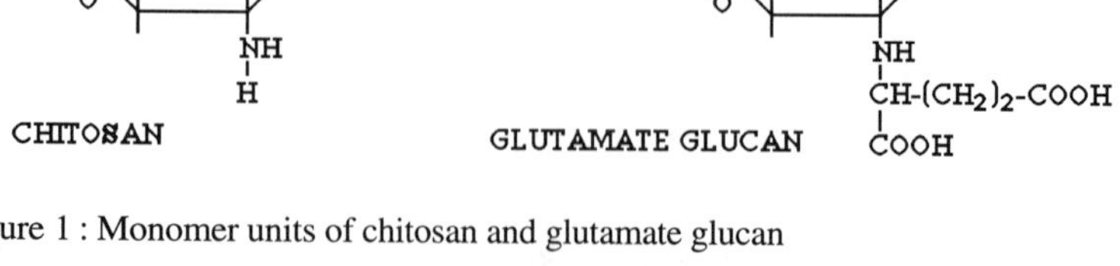

Figure 1 : Monomer units of chitosan and glutamate glucan

Polymer	% Free Amine	% Acetamide	% Subst. Amine	pK_a(s)
Chitosan	85	15	-	6.2
Glutamate glucan	60	15	25	5.1 - 8.6

Table 1 : Some characteristics of chitosan and glutamate glucan polymers.

The chitosan is also recrystallized, according to the method developed by Strusczyk.[31-32] Three simple steps are involved. The chitosan is dissolved in acetic acid solution (2 % w/w), for 4 hours at a low temperature. Undissolved suspended particles are eliminated. The mixture is set for 3 hours at 70 °C in an oil bath. At the end of this thermic step, the solution is rapidly frozen in an ice bath. The solution is set at room temperature under mild agitation, and the pH slowly adjusted to 8 by adding sodium hydroxide (3 % w/w). In order to purify the sorbent, the suspension is filtered, washed with ethanol and acetone in a soxhlet apparatus, and finally with demineralized water. After gentle drying, the powder is ground and separated into the four previously described particle sizes.

Sorption studies

Weighed amounts of polymer were dropped into solutions containing a known (25 $mgU \cdot l^{-1}$) concentration of uranyl nitrate, with a solid (mg) to liquid (ml) ratio of 0.1. The pH of the initial solution was adjusted to 5, with molar solutions of NaOH. The contact of the polymer with the solution inducing a change in pH, it was corrected by adding micro-volumes of HNO_3 or NaOH. The agitation of the solutions in beakers was maintained at a speed of 600 rpm with a magnetic stirrer (Variomag). Kinetics were performed by withdrawing samples at regular times. The collected sample was filtered through a Whatman cellulose filter (pore diameter : 1,2 µm). The arsenazo III colorimetric method was used to determine the uranium concentration in solution. The analysis was performed on a Shimadzu UV-160A at a wavelength of maximum absorbance fixed at 650 nm. The standard curve was linear between 0 and 30 $mgU \cdot l^{-1}$. The concentration q of metal in the polymer was obtained by a mass-balance between solid and liquid according to the formula :

$q = V(C_o - C_t)/m$ where V, m, C_o, C_t are the volume of the solution, the mass of polymer, the concentrations at time t=0, and at t.

Sorption kinetics : theory

Kinetics

The adsorption of metal onto a particle of polymer can be considered as proceeding through four steps :

- diffusion of the solute from the bulk solution to the boundary layer
- diffusion from the boundary layer to the surface of the sorbent
- diffusion of the metal within the particle : pore or solid diffusion mechanism
- adsorption at a site; other mechanisms such as precipitation can be involved in kinetic control.

The first step is regarded as non limiting : the agitation provided by the stirrer allows a homogenous distribution of the metal in the solution. The fourth step is quasi-instantaneous ; consequently the second and third steps are the rate limiting steps.

In order to predict the main controlling step, modelization of sorption kinetics is experimented on single models of external and intraparticular diffusion. A better and real simulation would take into account both models simultaneously.

A simple model of single diffusion, assuming hypotheses such as intraparticle diffusion and metal concentration at the solid surface negligible for time t -> 0, is used to obtain the initial rate of transfer for the external mass transfer resistance model. The classic mass transfer equation applied to the initial rate gives the evolution of the metal concentration in solution according to :

$$\frac{d}{dt}\left(\frac{C_t}{C_o}\right)_{t \to 0} = -\beta_L S$$

with β_L the external mass transfer coefficient and S the specific surface.

The initial rate is given by the initial slope of C_t/C_o.
The evaluation of the intraparticle diffusion coefficient takes into account the evolution of the solid concentration with the square root of time. Weber and Morris,[17] McKay and Poots,[18] applying classical diffusion Fickian's laws demonstrate that q varies with $(Dt/r^2)^{0.5}$. The intraparticle diffusion rate parameter , k, is obtained from the slope of the linear regression of q versus $t^{0.5}$.

Results and discussion

The influence of intraparticular diffusion leads to the study of the influence of the particle size of the sorbents. This work is presented in terms of both sorption kinetics by determination of diffusion properties and physico-chemical identification of the sorption location by S.E.M. and micro-analysis.

Sorption kinetics

Chitosan kinetic behaviour is characterized by an increase in the time necessary to reach the equilibrium. With substituted form the equilibrium is achieved within 24 hours, with original chitosan it is complete after 2 to 3 days. The modification of the chitosan involves dissolving the material in organic acid, and the subsequent reduction must increase the porosity of the sorbent. So metal ion species diffusion through the layers constituting the polymer is favoured. Conversely the chitosan is fibrous and has a stratified structure characterized by a poor ability to diffuse, and equilibrium is reached after a long contact time.
Particle size influences the sorption mechanism essentially in kinetic rate. The time necessary to reach equilibrium increases with increasing particle size of the polymer : the saturation of internal layers by solute sorption occurs over a longer period (Figure 2). Uranium sorption kinetics are strongly dependent on the particle size of the polymer : the importance of diffusion mechanisms in the control of sorption rate cannot be ignored. The linearization of q versus $t^{0.5}$ is represented, for low contact time, by a line which includes the origin. An intercept of this line equal to zero is interpreted by the reduction of the thickness of the external layer covering the polymer. Two ranges can be distinguished : for short time of contact the initial rate is estimated according to diffusion through the polymer, a second linear portion of the sorption kinetic curve is observed, indicating another possible diffusion or precipitation mechanism.[19] The results presented lead to the conclusion that the external layer is negligible and that external mass transfer is not rate limiting (Figure 3). The study of agitation would demonstrate that these parameters are ineffective in the kinetic control of uranium sorption by Glutamate glucan and that mass-transfer within the film is not rate limiting.[20] Equilibrium in uranium sorption by chitosan occurs at a longer contact time and final metal ion concentration in solution depends on particle size of the polymer : sorption is located on the external surface of the polymer or in a layer of restricted thickness. The simulation of sorption kinetics by the Morris and Weber model gives a positive ordinate intercept, instead of an intercept at the origine as obtained with glutamate glucan or a negative intercept which can be explained by an external mass-transfer control. The restricted diffusion of metal ions in chitosan can explain the poor fit obtained between experimental and calculated points.
The improvement of sorption properties by modification of chitosan can be explained by the grafting of new functional groups or by the increase in the porosity of the sorbent which allows the whole mass of polymer to be in contact with the metal ion. In order to predict the influence of the porosity, the recrystallization, which involves a dissolution and further reprecipitation, is studied in terms of kinetic and equilibrium performances. Figure 5 shows that for low particle sizes recrystallized chitosan exhibits similar behaviour to that of glutamate glucan : modification of the polymer structure seems to be predominant in the improvement of sorption properties.

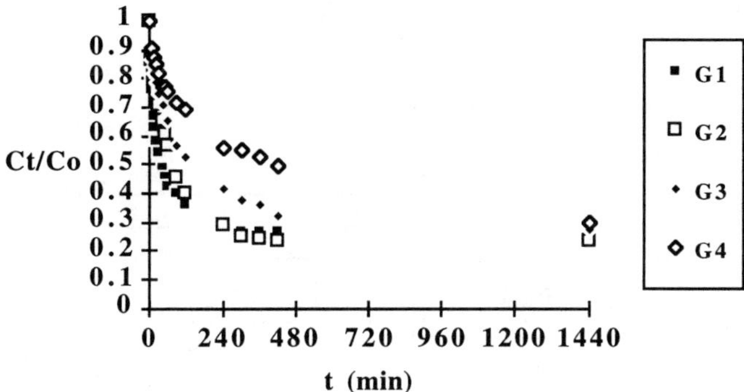

Figure 2 : Effect of particle size on uranium sorption by glutamate glucan : External diffusion (C_o = 25 mg·l^{-1}, Stirrer speed = 600 rpm, pH 5, T = 20 °C)

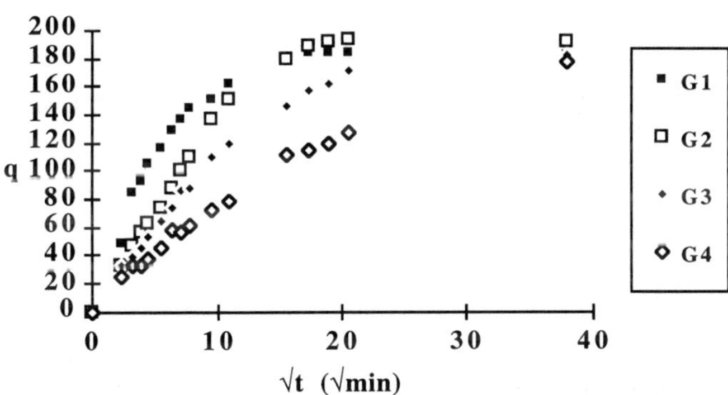

Figure 3 : Effect of particle size on uranium sorption by glutamate glucan : Intraparticular diffusion (Morris and Weber model) (C_o = 25 mg·l^{-1}, Stirrer speed = 600 rpm, pH 5, T = 20 °C)

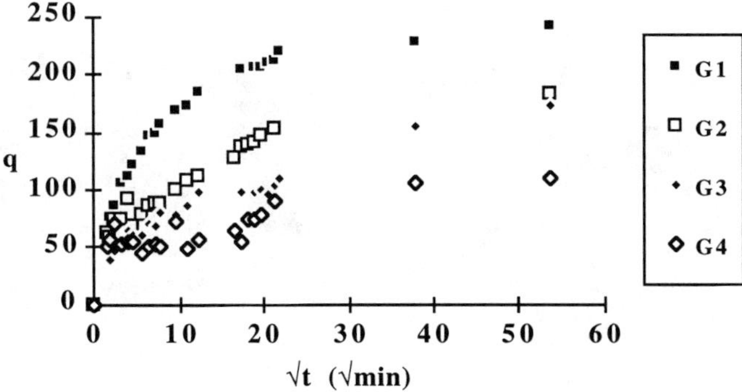

Figure 4 : Effect of particle size on uranium sorption by chitosan : Intraparticular diffusion (Morris and Weber model) (C_o = 25 mg·l^{-1}, Stirrer speed = 600 rpm, pH 5, T = 20 °C)

The results obtained from the parameters and sorbents studied show the predominance of intraparticle diffusion in the rate control of the sorption mechanism. External mass transfer could be disregarded within the conditions under which the experiments were carried out, the conclusion could be different with agitation speed lower than those used here.

Parameters					Diffusion models	
Polymer	pH	C_0 ($mgUl^{-1}$)	Agitation (rpm)	Particle size	External mass transfer rate	Intraparticle diffusion rate
Glut. glc.	5	25	600	G1	$4.3\ 10^{-2}$	18.6
"	5	25	600	G2	$2.7\ 10^{-2}$	13.8
"	5	25	600	G3	$2.5\ 10^{-2}$	11.2
"	5	25	600	G4	$2.0\ 10^{-2}$	7.7
Recr. chit.	5	25	600	G1	N.S.	21.3
"	5	25	600	G2	N.S.	15.3

Table II : Effect of various parameters on the sorption kinetics : external mass transfer and intraparticle diffusion rates (units previously cited - N.S. : non significant)

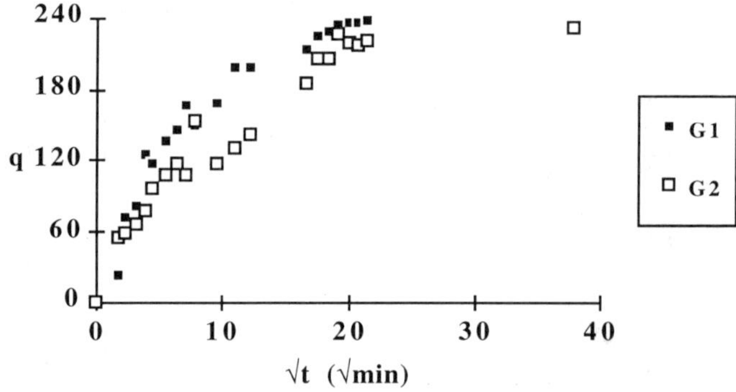

Figure 5 : Effect of particle size on uranium sorption by recrystallized chitosan : Intraparticular diffusion (Morris and Weber model) ($C_o = 25\ mg{\cdot}l^{-1}$, Stirrer speed = 600 rpm, pH 5, T = 20 °C)

These results in terms of both predominance of intraparticular diffusion in the rate control and influence of porosity in the overall sorption performances, are confirmed by a Scan Electron Microscopy and X-ray Energy Dispersed analysis. This complementary study allows the section and the distribution of uranium in these sections to be presented (Figures 6-8).

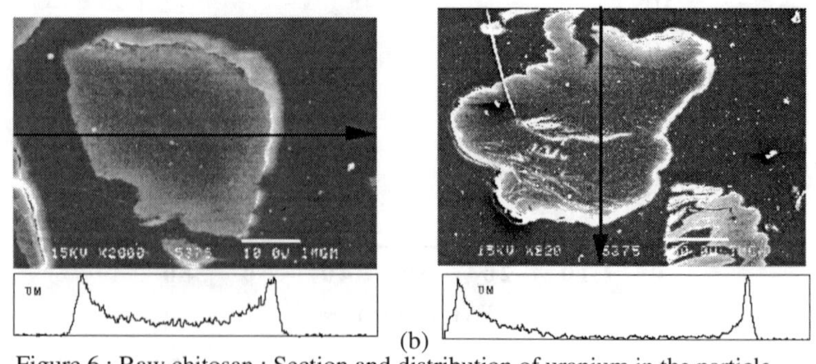

(a) (b)

Figure 6 : Raw chitosan : Section and distribution of uranium in the particle (a) G1 particle size, (b) G3 particle size

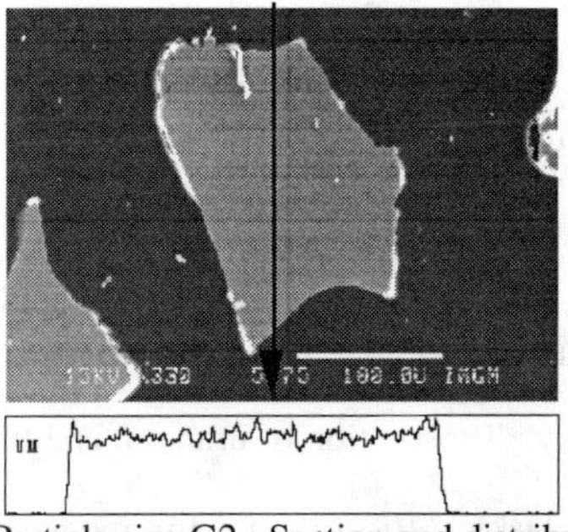

Figure 7 : Glutamate glucan - Particle size G2 : Section and distribution of uranium in the particle

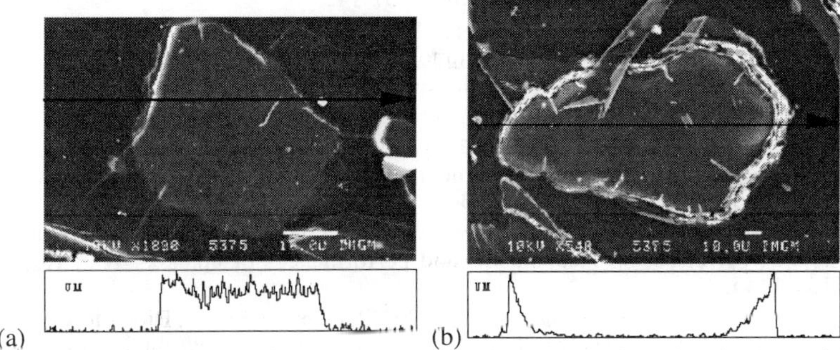

Figure 8 : Recrystallized chitosan : Section and distribution of uranium in the particle
(a) G1 particle size, (b) G3 particle size

The SEM and X-Ray Energy Dispersed Analysis of glutamate glucan, raw and recrystallized chitosan demonstrate that glutamate glucan and recrystallized chitosan exhibit similar behaviour in terms of diffusion of metal ion in the polymer. For low particle sizes the whole mass of the sorbent contributes to the metal ion sorption, for higher particle sizes the recrystallization of chitosan is not complete and the distribution of uranium in the section is not homogeneous. For raw chitosan, increasing the size of the polymer produces a reduction of the percentage of the polymer section saturated by the pollutant.

Conclusion

It appears that uranyl sorption, which involves mechanisms of both complexation and precipitation, is largely an intraparticle process. Its kinetics are favoured by a great porosity of sorbent. The optimum treatment conditions are a function of pH, metal concentration, agitation and particle size of the polymer. The optimization of the process depends on kinetic and equilibrium constraints, and should be a compromise between these two approaches. Among the various models of diffusion in solid/liquid exchange, two models are preferred : external mass transfer resistance and intraparticle diffusion. Particle size has a significant effect on sorption kinetics. Uranium sorption by chitosan occurs in a thin layer as opposed to modified chitosan such as recrystallized chitosan of glutamate glucan. S.E.M. and X-ray Dispersed Energy analysis confirm this conclusion : sorption is located in the whole mass of glutamate glucan and in a wider layer of recrystallized chitosan than one observed on raw material.
The sorption performances of modified chitosans are significantly higher than those of pure chitosan. The most interesting factors of these new chelating ampholytes are high porosity and hydrophilicity. These properties, allied to the high content of complexing groups (carboxylic or amino functions), explain the high kinetic rate of sorption. Chitosan needs a longer contact time to reach equilibrium at a lower capacity of sorption.

References

1. W. Sand, T. Gehrke, R. Hallmann, K. Rohde, B. Sobotke and S. Wentzien, "In-situ bioleaching of metal sulfides : the importance of Leptosirillum ferrooxidans", Biohydrometallurgical Technologies, Vol. I Bioleaching processes, A.E. Torma, J.E. Wey & V.I. Lakshmanan Eds., T.M.S., Proceedings of International Biohydrometallurgy Symposium, Jackson Hole, Wyoming, U.S.A., August 22-25 (1993) 15-28.

2. C.L. Brierley and J.A. Brierley, "Microbiological processes in recovery of metals from ores". Processes and Fundamental Considerations of selected hydrometallurgical systems, (New York, NY: M.C. Khun, 1981), 64-68.

3. M. Tsezos and B. Volesky, "Biosorption of uranium and thorium," Biotechnol. Bioeng., 23 (1981), 583-604.

4. N. Kuyucak and B. Volesky, "Biosorbents for recovery of metals from industrial solutions," Biotechnology Letters, 10 (2) (1988), 137-142.

5. R.G.L. McCready and V.I. Lakshmanan, "Review of bioadsorption research to recover uranium from leach solutions in Canada," Immobilisation of ions by biosorption, ed. H Eccles and S. Hunt (Chichester : Ellis Horwood Ltd. - Society of Chemical Industry, 1986), 219-226.

6. A. Nakajima and T. Sakaguchi, "Selective accumulation of heavy metals by microorganisms," Appl. Microbiol. Biotechnol., 24 (1986), 59-64.

7. C.L. Brierley, "Bioremediation of metal-contaminated surface and groundwaters," Geomicrobiology Journal, 8 (1990), 201-223

8. E. Guibal, Ch. Roulph and P. Le Cloirec, "Uranium biosorption by a filamentous fungus *Mucor miehei* : pH effect on mechanisms and performances of uptake," Wat. Res., 26 (8) (1992), 1139-1145.

9. K. Bosecker, "Biosorption of heavy metals by filamentous fungi", Biohydrometallurgical Technologies, Vol. II Fossil energy, Materials, Bioremediation, Microbial physiology, A.E. Torma, M.L. Apel & C.L. Brierley Eds., T.M.S., Proceedings of International Biohydrometallurgy Symposium, Jackson Hole, Wyoming, U.S.A., August 22-25 (1993) 55-64.

10. G. McKay, H.S. Blair and A. Findon, "Sorption of metal ions by chitosan," Immobilisation of ions by bio-sorption, ed. H Eccles and S. Hunt (Chichester : Ellis Horwood Ltd. for Society of Chemical Industry, 1986), 59-69.

11. G. McKay, H.S. Blair and A. Findon, "Equilibrium studies for the sorption of metal ions onto chitosan," Indian. J. Chem., 28A (1989), 356-360.

12. R.A.A. Muzzarelli, F. Tanfani and M. Emanuelli, "Chelating derivatives of chitosan obtained by reaction with ascorbic acid," Carbohydr. Polymers, 4 (1984), 137-151.

13. R.A.A. Muzzarelli, "Removal of uranium from solutions and brines by a derivative of chitosan and ascorbic acid," Carbohydr. Polymers, 5 (1985), 85-89.

14. I. Saucedo, E. Guibal, J. Roussy, Ch. Roulph and P. Le Cloirec, "Uranium sorption by glutamate glucan : a modified chitosan - Part I : Equilibrium studies," Water S.A., 19, 2 (1993) 113-118.

15. H. Struszczyk, "Microcrystalline chitosan. I. Preparation and properties of microcrystalline chitosan," J. Appl. Polym. Sci., 33 (1987) 177-189.

16. H. Struszczyk and O. Kivekäs "Microcrystalline chitosan. Some areas of application," Brit. Polym. J., 23 (1990) 261-265.

17. W.J. Weber and J.C. Morris, "Advances in water pollution research," (Proc. Intl. Conf. on Water Pollution Symp. Pub. Div., Pergamon Press, Oxford, Sept. 1962) (2) 231-266.

18. G. McKay and V.J.P. Poots, "Kinetics and diffusion processes in colour removal from effluent using wood as an adsorbent," J. Chem. Tech. Biotechnol., 30 (1984), 279-282.

19. K.S.L. Lo and J.O. Leckie, "Kinetic studies of adsorption-desorption of Cd and Zn onto Al_2O_3/solution interfaces," Wat. Sci. Technol., 28, 7 (1993) 39-45.

20. E. Guibal, I. Saucedo, J. Roussy, Ch. Roulph and P. Le Cloirec, "Uranium sorption by glutamate glucan : a modified chitosan - Part II : Kinetic studies," Water S.A., 19, 2 (1993) 119-126.

V.

FOSSIL FUELS AND BYPRODUCTS: BIOCONVERSION, PROCESSING AND BIOREMEDIATION

APPLICATION OF BIOCHEMICAL INTERACTIONS IN FOSSIL FUELS

Mow S. Lin and Eugene T. Premuzic

Biosystems and Process Sciences Division
Department of Applied Science
Brookhaven National Laboratory
Upton, NY 11973

Abstract

Certain extreme environments tolerant microorganisms interact with heavy crude oils by means of multiple biochemical reactions, asphaltenes, and bituminous materials. These reactions proceed via pathways which involve characteristic components of oils and coals such as asphaltenes, and in the chemically related constituents found in bituminous coals. These chemical components serve as markers of the interactions between microorganisms and fossil fuels. Studies in which temperature, pressure, and salinity tolerant microorganisms have been allowed to interact with different crude oils and bituminous coals, have shown that biochemically induced changes occur in the distribution of hydrocarbons and in the chemical nature of organometallic and heterocyclic compounds. Such structural chemical rearrangements have direct applications in monitoring the efficiency, the extent, and the chemical nature of the fossil fuels bioconversion. Recent developments of chemical marker applications in the monitoring of fossil fuels bioconversion will be discussed.

Mineral Bioprocessing II
Edited by David S. Holmes and Ross W. Smith
The Minerals, Metals & Materials Society, 1995

Introduction

Biochemical and geochemical reactions convert biogenic material continuously to fossil fuels. Similar reactions can be employed to process various fossil fuels provided the processes are fast and economical and compete with conventional processes. Over geological periods of time, the chemical and biochemical conversion rates occurring in reservoirs are too slow for practical use. However, under controlled and optimal conditions, biochemical processes can be used as cost efficient upgrading of fossil fuels.

Conventional coal cleaning methods remove most inorganic sulfur, e.g., pyrites, by mechanically crushing and grinding of coals, followed by removal of the impurities by different physical methods (1,2). Such processes are capital intensive and only remove the inorganic portions of the total sulfur content. Compared to chemical and physical methods, biodesulfurization of pyrite in coals has been studied predominantly on a laboratory scale (3,4). Although recently, a pilot plant has been built for testing in Italy (5).

By conventional cleaning methods the least treatable sulfur is the organic sulfur found in polar, tar, and asphaltic macromolecular parts of coal or heavy crude oils. It is believed that the rigid structures are due to sulfide, disulfide, and thiophenic type bonds, cross-linked to various hydrocarbons within these macromolecular matrices (6).

Small molecules, like dibenzothiophene (DBT), are often used as model compounds for studying organic sulfur degradation reactions. Bacterial strains and their enzymatic pathways for DBT degradation have been reported (7,8,19). However, such experimental strategy when applied to macromolecular matrix of coals and heavy crude oils, is more difficult to follow. However, in complex matrices, chemical markers can be identified by which the overall biochemical conversion process can be monitored.

Earlier reports on the fungal degradation of lignite suggested a biological coal liquefaction process (9,10). The resultant coal liquid, however, was a mixture of highly oxygenated compounds of low heating value (11). Therefore, extensive biodegradation may not be suitable for processing of fossil fuels. Limited, selective reactions aimed at heteroelement sites in the macromolecular matrix are beneficial and yield cleaner as well as lighter convenient fuels. For example, degradation of sulfur bonds leads to depolymerization of the macromolecular matrix. At Brookhaven National Laboratory (BNL), using the biomarker approach, other heteroatoms have been shown to be also involved in the depolymerization process (1,2). Such reactions may also be beneficial in downstream operations (hydrocracking, used in oil refineries, or in coal liquefaction). For example, during such processing, the catalysts gradually lose their activities through interactions with trace metals (Ni, V) present in the raw materials. Biochemical processes remove the adverse elements, and thus, extend the life of catalysts and reduce the disposal costs. Further, in oil refinery processes, strong acids such as hydrofluoric or sulfuric acids, used to remove nickel and vanadium from crude oils, are hazardous chemical reagents. Therefore, an alternative non-hazardous biochemical method may be useful. Preliminary results indicate that biochemical processing developed in this laboratory is promising in removal of trace metals.

Methods

Biotreatment of Coals and Oils

Details for the selection and culturing of bacteria used in this work have been reported elsewhere (1,2,12). Powdered coals (100-200 mesh) or oils were cultured at 25°C with mesophilic (BNL-4-23 and BNL-4-24) or at 65°C with thermophilic strains of microorganisms (BNL-TH-31 and NZ-5). Control samples were treated under identical experimental conditions, however, without inoculation. No attempt was made to sterilize the samples. The effect of indigenous bacteria was accounted for by changes in the control.

Pyrolysis Gas Chromatography Mass Spectroscopy (Py-GC-MS) and X-ray Absorption Near Edge Structure (XANES) Spectroscopy Analysis. Crude oil samples were dissolved in twenty volumes of chloroform. A 1 $\mu\ell$ sample was introduced into a GC split/splitless injector with a split ratio of 10 to 1. Coal samples were introduced by using a Chemical Data System (CDC) model 190 pyroprobe. The detailed procedures were published elsewhere (1,2,12) and will not be repeated here.

Ultraviolet-visible Spectroscopy and Total Sulfur Analysis. Oil emulsion samples were analyzed by using a Hewlett-Packard Model 8452 spectroscopy (1,2,12). Total sulfur content of coal and oil samples were analyzed by using a Perkin-Elmer Model 240C elemental analyzer with published procedure (13) and samples were also analyzed by combustion methods (Hoffman Analytical Laboratory, Golden, CO.)

Results and Discussion

Biotreatment of coals and oils involves heterogeneous reactions. For both types of substrates, reaction rates depend on the substrate surface area in contact with the microbial phase. In the case of oils, the oils are initially stirred in the culture medium, resulting in an oil emulsion. As the reaction proceeds, the formed emulsion exposes more surface area to the culture medium, thus accelerating the reaction. The extent of emulsification varied among the different strains used (13).

A number of bioproducts have been identified in the reaction mixture. These include emulsifying agents, such as alcohols and organic acids (12,14,15). Since a small amount of emulsifying agents may emulsify large amounts of oil, it is advantageous to encourage the production of emulsifying agents and thus enhance the reaction rates. In bioprocessing of coals, emulsification was not observed. Therefore, the reaction rates have to be enhanced by grinding the coal to a powder form, and thus, enlarge the surface area of the coal.

The biochemical mechanisms involved in bioprocessing of fossil fuels are selective and involve hereroatom sites. Thus, for desulfurization of oil studies, two Venezuelan high sulfur (4-6%) heavy crude oils were treated. Boscan crude is characterized as heavy because it is immature while Cerro Negro crude is heavy because it was biodegraded. The significance of "biodegraded" versus "bioconverted" material has been discussed elsewhere (1). For this discussion, it suffices to say that "biodegradation" means alteration of fossil fuel by indigenous organisms present in the reservoir over geological periods of time and bioconversion means biochemical conversion due to deliberately introduced microbial strains. From the measurement of the total sulfur content, biotreatment reduced up to 25% sulfur in Boscan crude and up to 29% sulfur even in the highly biodegraded Cerro Negro crude (16). Table 1 shows a similar result obtained on the biotreatment of crude oils and bituminous coals (17). In these experiments, thermophilic strains from the BNL collection were used and full advantage of fast reaction rates at higher temperatures has been taken.

Most sulfur species are contained in macromolecular structures of coal and heavy crude oils. Their analyses have been problematic for years. In this study, a destructive method of Py-GC-MS and a non-destructive method of XANES were used and compared in analyses of sulfur and sulfur structures. Pyrolysis of macromolecular substances, such as coals and heavy crude oils, is known to lead to formation of artifacts. However, recent studies of pyrolysis of fossil polymers at temperatures about 610°C, have indicated that products are formed mainly by cleavage of one bond at a time (18). An analogous single or multiple simultaneous cleavage at specific heteroatom sites may be the possible mechanism of biochemical degradation of organic sulfur compounds by thermophiles in which a stepwise enzymatic breakdown of organic sulfur structures occurs. For instance, the flash pyrolysis products of alkylthiophenes are derivatives of their corresponding alkylthiophene moieties in kerogen (18). In the present study, the treatment of lignite coal by thermophilic bacteria, BNL-TH-29, reduced the sulfides and C1 to C5 substituted thiophenes in the pyrolysate. The results indicated substantial degradation of sulfur linkages and thiophenic rings caused by the biotreatment (17). Similar, but to a lesser extent, results were also observed by treatment with strains of BNL-NZ-3 and BNL-NZ-5. The results are consistent with those for the total organic sulfur removal from other crudes already reported (1,2).

The biotreatment results for bituminous coal are somewhat different. Py-GC-MS analyses showed that there was substantial degradation of the thiophenic ring type compounds with minor changes in the sulfide species. The results are consistent with the XANES analyses shown in Table 1 (2,17). In this analysis, the bituminous Kentucky No. 8 coal is analyzed in terms of sulfur distribution of organic compounds against multiple sulfur standards. Each group of sulfur species in the coal was calculated as mol% of total sulfur species. A comparison of data for controls and untreated samples has shown that there was a reduction of more than 20% in sulfones and sulfates, with a 10% increase of thiophenic forms. This may be due to a 10% dissolution of total sulfur on heating of the Kentucky No. 8 coal control samples in culture medium at 65°C for seven days. An alternative possibility is that there were some intramolecular transformations of sulfur forms. The treatment of bituminous coal with both BNL-NZ-3 and BNL-NZ-5 resulted in a small reduction of thiophenic and sulfoxide sulfur forms, with small increases of sulfone and sulfate forms of sulfur. These results are similar to those obtained with aerobic oxidation of model compounds (8,19). Additional reduction of thiophenic forms was found in the biotreatment with BNL-TH-29 strain with variable levels of reduction in sulfoxide, sulfone, and sulfate forms. After biotreatment, there was a small increase of sulfide forms. Further exploration of these variations is warranted in order to understand the biochemistry of desulfurization of coals and similar materials. Results presented in Table I indicate that thiophenic forms of sulfur present in coal have been reduced and altered by biotreatment, and are consistent with those obtained by Py-GC-MS analysis (2,17).

Analogous to coal, heavy crude oils can also be treated for organic sulfur removal and analyzed by the same analytical techniques. In Figure 1, gas chromatography analysis (GC) has been carried out in the following manner: peaks detected at retention time (35 min) are the benzothiophenes, with increasing retention time (45 to 50 min) for dibenzothiophenes and multisubstituted alkyl dibenzothiophenes, respectively. The GC results indicate that the treatment by BNL-4-22 removed a substantial amount of the thiophenic compounds (35 to 50 min), as compared to the untreated Boscan oil. The results also showed that the BNL-4-23 treatment was not as effective as BNL-4-22 in removing high thiophenic compounds. This is consistent with the results shown in Table 1 for XANES analysis.

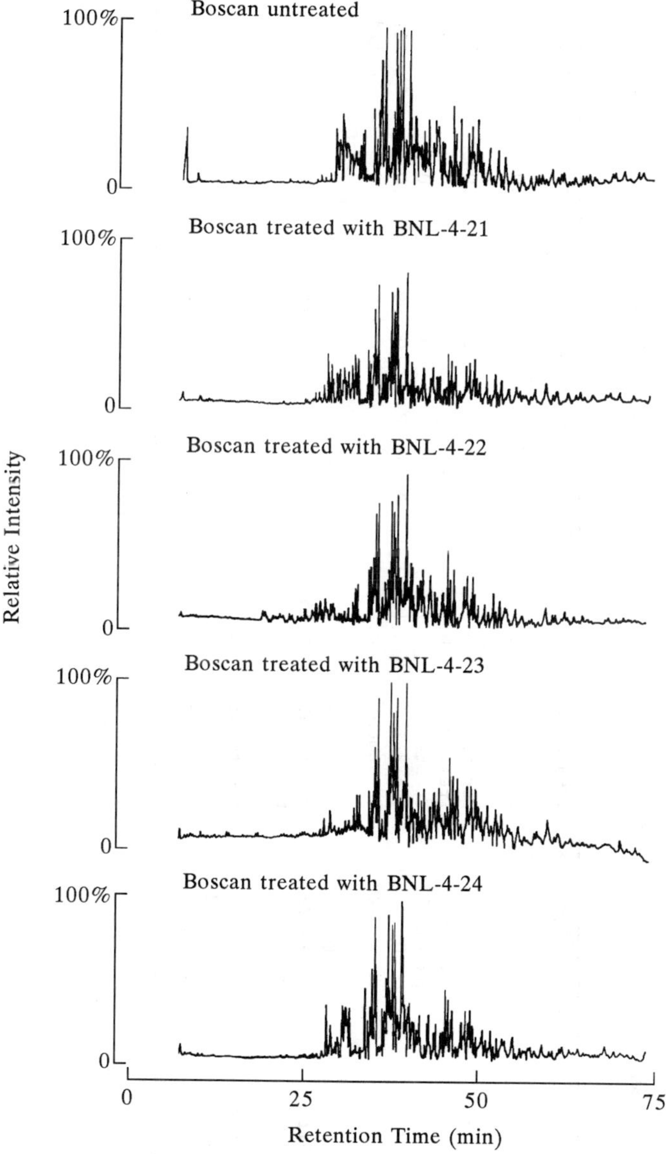

Figure 1. Gas chromatography analysis of organosulfur compound by Flame Photometric Detection (FPD).

Table I. XANES analysis of sulfides, thiophenes, and sulfoxide contents of untreated and treated fossil fuels

Fossil Fuels	Microorganisms	Treatment	Sulfides	Thiophenes	Sulfoxides	Sulfones	Sulfates
Kentucky No. 8	0	Untreated	0.262	0.397	0.133	0.074	0.234
Bituminous	0	Control	0.240	0.441	0.121	0.045	0.153
Coal	BNL-NZ-3	Treated	0.234	0.408	0.086	0.047	0.194
	BNL-NZ-5	Treated	0.208	0.296	0.089	0.053	0.179
	0	Control	0.184	0.303	0.109	0.055	0.239
	BNL-TH-29	Treatment	0.198	0.198	0.177	0.045	0.173
Boscan	0	untreated	0.198	0.738	0.064	-	-
Crude Oil	BNL-4-22	treatment	0.159	0.655	0.186	-	-
	BNL-4-23	treatment	0.121	0.743	0.135	-	-
Cerro Negro	0	treatment	0.147	0.781	0.072	-	-
Crude Oil	BNL-4-22	treatment	0.179	0.683	0.138	-	-
	BNL-4-23	treatment	0.103	0.713	0.184	-	-

*Measured in mole fractions.

"Biodegradation" of model organosulfur compounds (OSC) has been extensively reviewed (4,6,7,8,19). In general, the biochemical reactions of model OSC leads to a reduction of sulfur content due to the conversion of thiophenic compounds to hydroxy substituted rings, as well as open ring sulfones and/or sulfoxides. If similar biochemical reactions occur during the biotreatment of coals, the reduction in sulfur content may lead to a biochemical partial breakdown of coal structure. In some way, such "cracking" processes are similar to these observed in coal pyrolysis. Analysis of biotreated and untreated bituminous coal by Py-GC-MS shows that the biotreated coals contained more of small to medium molecular fractions when compared to high molecular weight residues. The effect is further pronounced in the biotreated lignite. It appears that the overall breakage of sulfide linkages leads to an extensive breakdown of crosslinking in lignite coal structures (2,17). Accepting that such mechanisms may be operative, then biochemical processing may be useful in the pretreatment of coal prior to the mild gasification of coal and could also be used in biodesulfurization of heavy crude oils.

Mechanisms favoring bond cleavages at heteroatom sites are furthermore supported by the changes occurring in the metal contents as shown in Table 2. Metal analyses of treated and untreated samples have shown that more than 90% of nickel and vanadium in heavy crudes can be removed by selected microbial strains (2,17).

Conclusions

Depolymerization of the macromolecular structure has obvious applications in future refinery and coal liquefaction processes. The reduction of trace metals in fuel processes will extend the useful life of catalysts and consequently reduce operation costs. However, to develop a large volume process, bench experiments need to be scaled up to pilot plant scale processes, to gather parameters for the design reactor types, mixing rates, mass transfer and reaction rates needed for engineering and cost-efficiency evaluations.

Acknowledgments

This work is performed under the auspices of the U.S. DOE Contract No. DE-AC02-76CH00016. We acknowledge B. Sylvester, L. Racaniello, and J. Yablon for laboratory assistance; W. Q. Zhou and J. Jeon for XANES analysis; and the contribution of Daniel Winn, who was supported by the Department of Energy, Office of Science and Education Programs, Office of Energy Research, as a participant in its Bioprocesses Program.

Table II. Impurity removing biotreatment

Fossil Fuels with Treatment	Impurities ppm				
	S	Ni	V	Sr	Mn
Untreated Cerro Negro Crude Oil	43700	246	493	9.6	15.8
CN/BNL 4-24	30000(-29%)	186(-25%)	276(-38%)		
CN/BNL 4-23	37400(-25%)	160(-35%)	214(-58%)	1.7(-82%)	1.8(-89%)
CN/BNL 4-22	32100(-27%)	12(-95%)	6(-99%)		
Kentucky No. 8 Bituminous Coal	11000	-	157	1400	168
Ky. 8/BNL 4-24	10340(-6%)	-	-	-	-
Ky. 8/BNL 4-23	8140(-26%)	-	-	-	-
Ky. 8/BNL TH-31	7590(-31%)	-	99(-37%)	1040(-26%)	41(-76%)

* The figure quoted in the parenthesis represent percentage removal relative to control.

- Not detectable.

References

1. Premuzic, E. T., Lin, M. S., and Manowitz, B. The significance of chemical markers in the bioprocessing of fossil fuels. Fuel Processing Technology 40, 227-239 (1994).

2. Lin, M. S., Premuzic, E. T., Manowitz, B., Jeon, J., and Racaniello, L. Biodegradation of coals. Fuel 72, 1667-1672 (1993).

3. Andrews, G. F., Stevens, C. J., Quintana, J., and Dugan, P. R. Oxygen transfer problems in coal processing bioreactors. Proc. 7th Pittsburgh Coal Conference, Pittsburgh, 1990, pp. 358-372, University of Pittsburgh, 1990.

4. Olsen, G. J., Finseth, D., Rhee, K., and Schaffstall, M. Laboratory studies of microbial desulfurization of Pittsburgh seam coal related to heap depyritization. Proc. Second International Symposium on the Biological Processing of Coal, 1991. Electric Power Research Institute, pp. 3-35 - 3-50, EPRI GS-7482, 1991.

5. Loi, G., Mura, A., Trois, P., and Rossi, G. The Porto Torres biodepyritization pilot plant: Light and shade of one year operation. Fuel Processing Techn. 40, 262-268 (1994).

6. Yen, T. F., Editor. The role of trace metals in petroleum. pp. 221, Ann Arbor Science, Ann Arbor, MI, 1975.

7. Kilbane, J. J. Sulfur-specific microbial metabolism of organic compounds. Resour. Conserv. Recyl. 3, 69-79 (1990).

8. Kodama, K., Umehara, K., Shimizu, K., Nakatani, S., Minoda, Y., and Yamada, K. Identification of microbial products from DBT and its proposed oxidation pathway. Agr. Biol. Chem. 37, 45-50 (1973).

9. Cohen, M. S. and Gabriele, P. D. Degradation of coal by the fungi polyporus versicolor and poria monticola. Appl. Environ. Microbiol. 44, 23-28 (1982).

10. Ward, B. Lignite-degrading fungi isolated from a weathered outcrop. Syst. Appl. Microbiol. 6, 236 (1985).

11. Scott, C. D., Strandberg, G. W., and Lewis, S. N. Microbial solubilization of coal. Biotechn. Prog. 2, 131-139 (1986).

12. Premuzic, E. T. and Lin, M. S. Prospects for thermophilic microorganisms in microbial enhanced oil recovery. Proc. Microbial Enhancement of Oil Recovery - Recent Advances, Norma, May 1990, E. C. Donaldson, Editor, Developments in Petroleum Science, Vol. 31, Chapter R-18, pp.277-96, Elsevier Sci. Publ., Amsterdam, 1991.

13. Perkin-Elmer publication. Instruction of sulfur analysis kit for the Model 240 Elemental Analyzer. Perkin-Elmer Co., Norwalk, CT, pp. 993-9417, 1985.

14. Premuzic, E. T. and Lin, M. S. Interaction between thermophilic microorganisms and crude oils - recent developments. Resour. Conserv. Recyl. 5, 177-284 (1991).

15. Premuzic, E. T., Lin, M. S., Racaniello, L. K., and Manowitz, B. Chemical markers of induced microbial transformations in crude oils. In Microbial Enhancement of Oil Recovery--Recent Advances, E. T. Premuzic and A. Woodhead, Eds., Vol. 39, pp. 37-54, Elsevier Science Publishers, Amsterdam, 1993.

16. Premuzic, E. T., Lin, M. S., Jin, J.-Z., Manowitz, B., and Racaniello, L. Biochemical alteration of crude oils in microbial enhanced oil recovery. Proc. Biohydrometallurgical Technologies, Jackson Hole, Aug. 1993, A. E. Torma, M. L. Apel, and C. L. Brierley, Editors, pp. 401-13, Minerals, Metals, and Materials Society, 1993. (BNL 48049).

17. Lin, M. S., Sylvester, B. J., and Premuzic, E. T. Biochemical desulfurization and demineralization of coals with mixed microbial cultures Proc. Second International Symposium on the Biological Processing of Coal, 1991. Electric Power Research Institute, pp. 79-85, EPRI GS-7482, 1991.

18. Sinningle D., J. S., Eglinton, T. I., Rijpstra, W. I. C., and de Leeuw, J. W. In Geochemistry of Sulfur in Fossil Fuels. W. L. Orr and C. M. White, Editors, ACS Symposium Series 429, pp.486-528, American Chemical Society, Washington, DC, 1990.

19. Van Afferden, M., Tappe, D., Beyer, M., Trüper, H. G., and Klein, J. Biochemical mechanisms for the desulfurization of coal-relevant organic sulfur compounds. Fuel 72, 16-35 (1993).

THE FLOOD/DRAIN BIOREACTOR FOR COAL AND MINERAL PROCESSING

G. F. Andrews and K. S. Noah

Center for Industrial Biotechnology and Process Engineering
Idaho National Engineering Laboratory
P.O. Box 1625, Idaho Falls, ID 83415-2203

Abstract

Large mineral particles can be bioprocessed in heaps, which are inexpensive to build and operate but produce low rates of leaching. Finely ground minerals can be processed in slurry bioreactors which give higher leaching rates but with higher costs. The flood/drain bioreactor is an alternative design that seeks to combine high rates with low costs while processing intermediate sized particles (approx. 1/4 in. x 30 mesh). A bed of solids is fluidized periodically by pumping in a bacteria/nutrient solution, providing excellent agitation, removing precipitates and other metabolic products off the solids, washing out fines, and bringing in fresh nutrients. Subsequent draining of the bed draws in fresh O_2 and CO_2 for microbial metabolism. In the coal depyritization experiments described here the fluidization also allows the dense mineral matter and large pyritic inclusions to settle to the bottom of the bed, giving a physical separation that complements the microbial pyrite degradation. Results from a laboratory-scale bed 10 ft high x 4 in. diameter for various pre-cleaned and run-of-mine coals show 50 to 80% pyrite removal in four weeks. Operational difficulties, including air-binding of the bed and the appearance of acidophilic protozoa, and ways of improving the sulfur removal rates are discussed.

Mineral Bioprocessing II
Edited by David S. Holmes and Ross W. Smith
The Minerals, Metals & Materials Society, 1995

Introduction

Virtually all minerals bioprocessing operations are carried out in one of two types of bioreactor. In the first, a bacterial culture/lixiviant solution is recirculated over a custom-built heap or a dump of waste ore (1,2). The advantage of this type is that it costs very little to build and operate. The disadvantage is that the particle size must be quite large to allow uniform flows of liquid and air, and large particles require long processing times due to the low rate of diffusion of reactants and products through the solid. Practical heaps usually have some arrangement to enhance the flow of air by natural or forced convection, but they contain a wide range of particle sizes and most have some "dead-zones." These may be regions that receive too little liquid flow, causing nutrient limitation or the accumulation of metabolic products (H^+, Fe^{3+}) to inhibitory levels or, they may be regions that receive too much liquid and become waterlogged, preventing the penetration of the O_2 and CO_2 essential for microbial metabolism. Waterlogging is particularly likely where there is a local accumulation of fine solids, and along the bottom of the heap where a phenomenon called the "capillary-fringe effect" prevents a packed-bed from draining down into a more permeable layer beneath it.

The second bioreactor type is the aerated slurry reactor, either air-lift or mechanically agitated (3,4). The solids in these reactors must be finely ground or they settle out, and reaction rates are correspondingly higher than in heaps. The obvious disadvantage is that slurry reactors cost far more than heaps to build and operate.

The question arises whether these are the only two possibilities. Can we not conceive of a different bioreactor configuration, treating intermediate-sized solids, and combining, in some sense, the lower costs of the heap with the higher rates of the slurry reactor. One possibility is the flood/drain bioreactor shown in Figure 1. A centrifugal pump, controlled by a timer, pumps the bacterial culture/luxiviant solution from a reservoir, up through a bed with sufficient force to fluidize the solids. As the liquid level rises in the column its velocity slows, due both to the natural characteristics of a centrifugal pump, and to a by-pass valve opened by the timer before the level reaches the overflow. Approximately one bed volume of liquid is allowed to overflow in order to remove the soluble metabolic products generated during the previous period. The overflow rate is so slow that only fine particles are washed out of the bed, to be removed in a settling tank. The pump is then turned off and the liquid falls back through the bed, drawing in fresh air containing the O_2 and CO_2 needed for the next period of microbial activity. The cycle is repeated when this activity is slowed by depletion of nutrients in the gas or liquid, or accumulation of inhibitory products in the bed.

The laboratory-scale column in Figure 1 is 13 ft high by 4 in. diameter. Scale-up would involve increasing the latter, so a commercial-scale reactor would be a lined pit, possibly segmented to allow sequenced-batch operation, with a gravel base containing the liquid inflow/outflow pipes, and suitable equipment for loading and discharging the solids.

The advantages of the flood/drain system are as follows. Forced aeration is achieved without the inherently inefficient and expensive process of air compression. Control of the cycle time gives an operating flexibility not available in heaps. Fluidization provides intense agitation of the whole bed, washing off any ferric hydroxysulfate precipitates produced by the microorganisms. "Dead zones" are unlikely, and even if they form they can only last one cycle time. The smallest

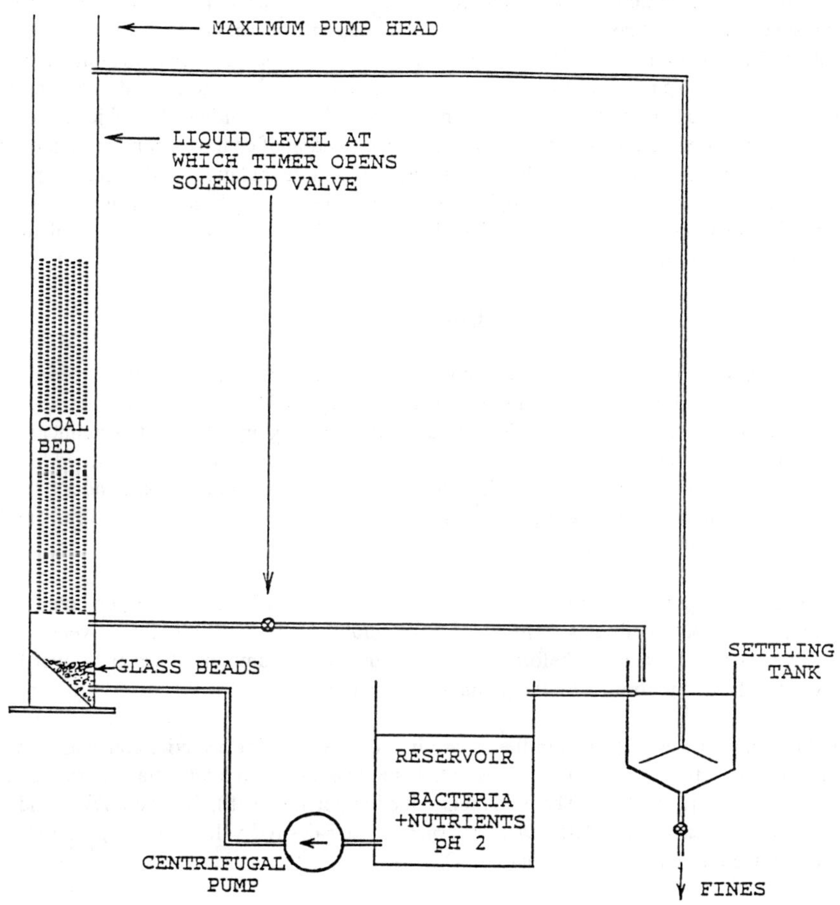

Figure 1 - The Laboratory-Scale Flood/Drain Bioreactor

particle left in the bed is controlled by the liquid overflow rate, and the bed actually stratifies during the flood/drain cycle, with the larger particles at the base of the bed and the smaller ones further up. The accumulation of large particles at the bottom reduces the "capillary fringe effect," reducing waterlogging in beds of small average particle size.

This stratification is particularly important in the beneficiation of coal. Pyrite and other minerals that must be removed are so much denser than the organic coal macerel (S.G. ~ 4 versus 1.3) that the settling velocity of a liberated mineral particle in water may be an order of magnitude larger than that of a "coal" particle of the same size. Thus, mineral-rich particles tend to accumulate near the base of the bed, from where they can be removed and discarded. The objective of the experiments reported here was to demonstrate the beneficiation of coal by a combination of this physical separation (analogous to that in a Batac Jig) and degradation of the smaller pyritic inclusions by a mixed culture of autotrophic, iron and sulfur oxidizing bacteria. It is possible that this density-separation effect, combined with microbial leaching, may be beneficial in the treatment of other mineral ores.

Experiments

Preliminary experiments on various coal types (5) resulted in the design of the experimental equipment shown in Figure 1. The series of experiments reported here (see Table 1) studied run-of-mine and cleaned Pittsburgh #8 coals supplied by the Consol Corp. The first experiment used a 3/8 in. x 28 mesh size fraction and a 1/25 HP pump but this combination was found to give inadequate fluidization. All other experiments were done with a 1/4 in. top size and a 1/15 HP pump. The maximum head for this pump was 21 ft, compared with the height of 15 ft from the pump to the column overflow.

No attempt was made to optimize the cycle time, which was kept at once per hour. The flood/drain cycle took 6 minutes; 2 minutes for the column to fill to the overflow level (the by-pass valve opened 15 seconds before); 3 mins for the desired amount of water to overflow; 1 minute for the bed to drain after the pump was turned off.

Each day samples were taken from the reservoir and measured for bacterial concentration by direct cell counts, total iron by atomic absorption spectroscopy, and total sulfate by the barium chloride turbidimetric method. The oxidation/reduction potential (Eh), the liquid level, and the pH were also checked. The latter were returned to their desired values (46 L, pH 2) by the addition of dechlorinated tap water and H_2SO_4.

Core samples were taken periodically from three sampling ports in the bed. The "Bottom" port was 4.5 in. above the support screen, but the "Middle" and "Top" ports were at different heights in different experiments (see Table 1). Previous work (5) had shown that the standard ASTM "forms of sulfur" procedure for removing the non-pyritic iron (boiling for 30 min in 4.8N HCL) was not always adequate for microbially-treated coals, probably due to the formation of complex ferric hydroxysulfate precipitates in the coal particles. It was therefore preceeded by a cold wash in 50 mL of 4.8 N HCL for 30 min. The iron and sulfate in the composited wash solutions were measured by atomic absorption and the barium chloride methods. They will be referred to as the "precipitate" iron and sulfur, although they actually include the sulfate sulfur and non-pyritic iron in the feed coal and that dissolved in the water on the coal samples, as well as the actual ferric precipitates. The pyritic sulfur in the washed coal samples was measured by the standard ASTM procedure.

TABLE 1. SUMMARY OF COLUMN EXPERIMENTS

Coal Type	Pittsburgh #8 ROM	Pittsburgh #8 ROM	Pittsburgh #8 ROM	Pittsburgh #8 Clean	Pittsburgh #8 Clean
Grind Size	3/8 in. x 28 mesh	1/4 in. x 28 mesh	1/4 in. x 28 mesh	1/4 in. x 28 mesh	1/4 in. x 28 mesh
Pump used HP: head ft	1/25:17	1/15:21	1/15:21	1/15:21	1/15:21
Wash procedures	0.1% $Na_4P_2O_7$	O.5 ml 85% H_3PO_4	0.5 ml 85% H_3PO_4	declorinated tap	H_2O
Days	41	27	30	56	59
Pyritic S(%)	1.27	1.18	1.18	1.04	1.20
Settled bed height (in.)	86	85	70.5	89.5	89.5
Dry weight in bed start/end (kg)	15/13.31	15/14.4	12.9/12	15/13.5	15/13.1
Initial Fe/Final Fe (gm/L)	0.27/0.83	0.16/0.59	0.41/0.80	0.6/1.7	1.1/2.6
Ash (%)	14.8	12.8	12.8	6.87	6.87
Product ash (%) Bottom 8 in.	23.7	33.9	32.3	13.0	13.8
Product ash (%) Remainder	7.6	5.1	5.2	5.9	5.6
Pyritic reduction (%) whole/whole-bottom 8 in.	58/58	41/49	49/62	35/46	56/67
Cell counts initial/highest (cells/ml x 10^7)	2.1/6.2	1.3/3.9	2/3.4.9	1.9/6.2	2.1/8.4
Sampling port heights middle/top in	55/86*	55/91*	41/74	33.5/86.5	41/86.5

*Cores taken by boring vertically into the top of the bed.

The first and second experiments in Table 1 were started with an inoculum consisting of 8 L of iron-grown *Thiobacillus ferrooxidans*, 1 L of iron-grown *Leptospirillum ferrooxidans* and 1 L of tetrathionate-grown *T. thiooxidans*. The inoculum for all other experiments was prepared as follows. At the end of each experiment, the contents of the settling tank (15 L or 37% of the total liquid volume) were drained and replaced with an equal volume of wash water which (by by-passing the reservoir) was used to wash the product coal in the column. The wash water was then mixed with the contents of the reservoir and used to start up the next experiment. The main objective of this procedure is to remove the fines from the system and produce a clean product coal with low levels of "precipitate" sulfur. It also washes off some fraction of the bacteria, Fe^{+++}, and H^+ generated by the microbial activity in each experiment for re-use in subsequent experiments.

Several operational problems were overcome in these experiments. The coal was initially loaded dry, but first attempts to fluidize the bed resulted in air-binding and the bed rising as a solid plug. This is a "scale-down" problem that would be less severe in a large-diameter column, and was associated with the presence of fines that were difficult to "wet." It was solved by pre-wetting the coal and loading it into the column as a slurry. Another problem caused by fines was their tendency to filter out during the drain cycle, forming an impermeable layer on top of the bed that prevented proper draining. This was resolved by correct positioning of the overflow and timing of the flood/drain cycle. Some early experiments had been plagued by the appearance of acidophilic protozoa in the reservoir, that caused a catastrophic decline in bacterial numbers (5). These protozoa never appeared in the experiments reported here, presumably because some unknown compound leaching from the coal inhibited their growth.

Results and Discussion

Product Coal Washing and the Concentrations of Bacteria and Iron

After repeated experiments with the coal washing/inoculation procedure described above the microbial culture should be well adapted to the coal, and the concentrations of bacteria and dissolved iron should reach a quaesi-steady state. The data in Table 1 show that this quaesi-steady state was not reached. Concentrations of iron and cells in the reservoir increased as expected during an experiment. However, the initial and final concentrations were slowly increasing from experiment to experiment. Note that the cell numbers remain well below the level ($\sim 10^{10}$ cells per gram of pyrite) usually considered optimum for coal pyrite degradation. However, measurements of oxidation/reduction potential (Eh) showed that virtually all of the dissolved iron in the reservoir was in the ferric state, so the pyrite oxidation rate was not limited by the activity of the iron-oxidizing bacteria. The implication that sulfur oxidation is rate-limiting should mean that the process can be accelerated by increasing the concentration of either the sulfur oxidizing bacteria (for the "direct leaching" mechanism) or dissolved iron (for the "indirect leaching" mechanism). The data in Table 1 support this idea, the pyrite reduction attributable to the bacteria generally increasing with the initial concentrations of bacteria and iron. (The drop in reduction in the first "cleaned coal" experiment can be explained by the time needed by the culture to acclimate to the new feed).

The coal washing/inoculation procedure described here simulates how a commercial process would be run, but is far from being optimized and the achievable rate of microbial pyrite degradation may be considerably higher than that in these experiments. There is, for example, no guarantee that 37% replacement of the liquid with each batch is best. A lower percentage (in sanitary engineering terms a higher mean cell residence time) would increase the quaesi-steady concentrations of Fe^{3+} and bacteria, which would be beneficial, but also the concentrations of organics and other ions leached from the coal, which may be inhibitory. Another open question concerns additives in the wash water. It seems correct to acidify this water in order to prevent the formation of precipitates when it contacts the high (Fe^{+++}) and (SO_4^{--}) water on the coal. But what acid should be added? Organic acids are expensive and inhibit *Thiobacillus*, while the common mineral acids add undesirable anions (Cl^-, NO_3^-, SO_4^{--}) to the coal. Phosphoric acid was thought best because it adds an essential nutrient and stimulates the desorption of *Thiobacillus* from pyrite surfaces. However, the amount of phosphoric acid needed to reduce the pH of tap water to 2 was far greater than the amount needed for microbial growth, and would itself have produced iron phosphate precipitates and inhibit pyrite degradation. Small amounts were tested but gave no advantage over dechlorinated tap water. Pyrophosphate was tested once because it gives improved desorption of bacteria, but once again the required concentrations were excessive. Acid cation exchange is recommended for pre-treatment of wash water in practice.

Bed Stratification and Ash Removal

The height of the settled bed provides an easy indicator of the stratification of the solid particles. In a well-mixed bed the smaller particles can occupy the interstices between the larger ones, giving a high solids holdup and thus a short bed. In a stratified bed the large particles are separated from the small particles, this void-filling can no longer happen, and the bed occupies more space. Thus, in the last experiment shown in Table 1 the settled bed height increased from its initial value of 89.5 in. to 102.5 in. in the first day, followed by a slow decline to 98 in. over 60 days due to abrasion and wash-out of fines. This observation suggests that bed stratification is compete after 10 to 20 flood/drain cycles.

The total mass loss from the bed during these experiments averaged 9% (Table 1). However, some of this loss was due to the solubilization of the pyrite by bacteria and dissolution of the basic minerals (carbonate etc.) under the acidic conditions of the process. The actual loss of heating value due to abrasion and wash-out of fines was more like the 4 to 5% suggested by the bed height data. Only the insoluble mineral matter (quartz etc.) shows up in the measured "Ash in the Bottom 8 in.," so the numbers in Table 1 actually understate the efficiency of the physical separation. The 33% figure for the second and third experiments (stratification was poor in the first experiment due to an inadequate pump) was evident as the rocks etc. associated with run-of-mine coal visibily accumulated at the bottom of the bed. Making a product (everything above the bottom 8 in.) with only 5% ash from run-of-mine coal in a single step is excellent performance. Even the pre-cleaned coal used in the last two experiments was significantly upgraded, generating a waste (the bottom 8 in.) with at least twice the insoluble ash of the feed.

A major factor in determining the physical separation performance is the size distribution of the solids. A reasonable criterion is that the smallest liberated mineral particle should have a higher settling velocity than the largest pure organic coal particle. For the particles of interest settling velocity is proportional to $(p - p_w)d^{1.6}$ where p = solids density, p_w = density of water and d = particle size. If $p = 3.5$ for mineral matter and $p = 1.3$ for organic macerels this suggests that the ratio (largest particle)/(smallest particle) should be 3.8, too narrow a size fraction to be practical. The 1/4 in. x 28 mesh fraction used in the experiments has a ratio of 10, so the observed physical separation is better than expected and suggests that economical performance can be achieved.

Forms of Sulfur

The variation of pyritic sulfur in the core samples during the last experiment in Table 1 is shown in Figure 2. By the time of the first sample at day 4 the bed was completely stratified, the mineral-rich particles at the bottom of the bed containing more than three times the pyritic sulfur in the particles near the top. Thereafter there was a slow decrease in pyritic sulfur throughout the bed due to microbial oxidation.

After stratification particles near the top of the bed were small, so a significant fraction of the pyritic inclusions they contain were exposed to direct microbial attack at the particle surface. This explains why the pyritic sulfur in the "top" core samples decreased rapidly between days 4 and 18 (Figure 2). It then leveled off due to the many inclusions that are completely submerged in the coal particle where they are inaccessible to the bacteria. This submerged pyrite can be removed by the "indirect mechanism" of pyrite biodegradation (2), but a significant rate requires a higher concentration of ferric ion in the liquid than was achieved here. There is a similar sequence of events at the middle and bottom of the bed, but the particle sizes are larger here so a lower fraction of the pyrite is exposed to direct microbial attack.

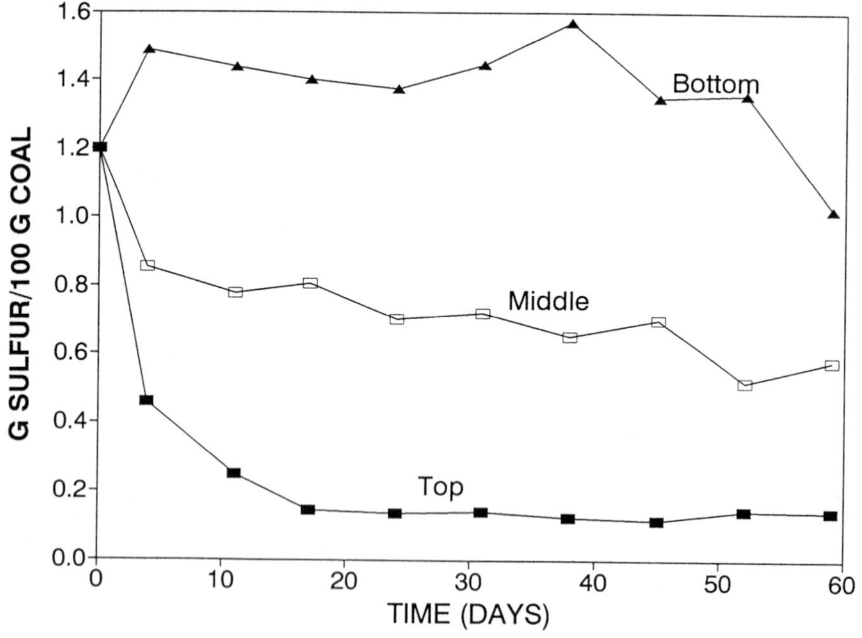

Figure 2 - Variation of Pyritic Sulfur in Core Samples

The "precipitated" sulfate (Figure 3) started out at the sulfate sulfur content of the feed coal and increased rapidly for the first 11 days due to the microbial production of sulfate. However, it never reached more than 10% of the pyritic sulfur, and later declined as microbial pyrite degradation slowed and ferric hydroxysulfate precipitates were washed off the coal. The precipitated sulfate was highest at the bottom of the bed, possibly due to the formation of sulfate salts by reaction between the basic minerals and the sulfuric acid in the liquid. The high levels at the top of the bed are believed to be related to the high surface/volume ratio of the small particles found there. As the liquid drains back down through the bed, colloid ferric hydroxysulfate precipitates would be filtered out by these small particles.

The Mass Balance

Previous work suggested a number of uncertainties associated with the application of the standard ASTM "forms of sulfur" procedures to microbial depyritization experiments.(5) These uncertainties make it desirable to close the mass balance; is the disappearance of pyritic iron from the coal balanced by the appearance of dissolved and precipitated forms? A similar balance could be done for sulfur, but is complicated by the fact that "pyritic sulfur" is not measured directly. The ASTM procedure actually measures pyritic iron and assumes a stoichiometry FeS_2, an assumption best avoided in the microbial oxidation of coal pyrite.

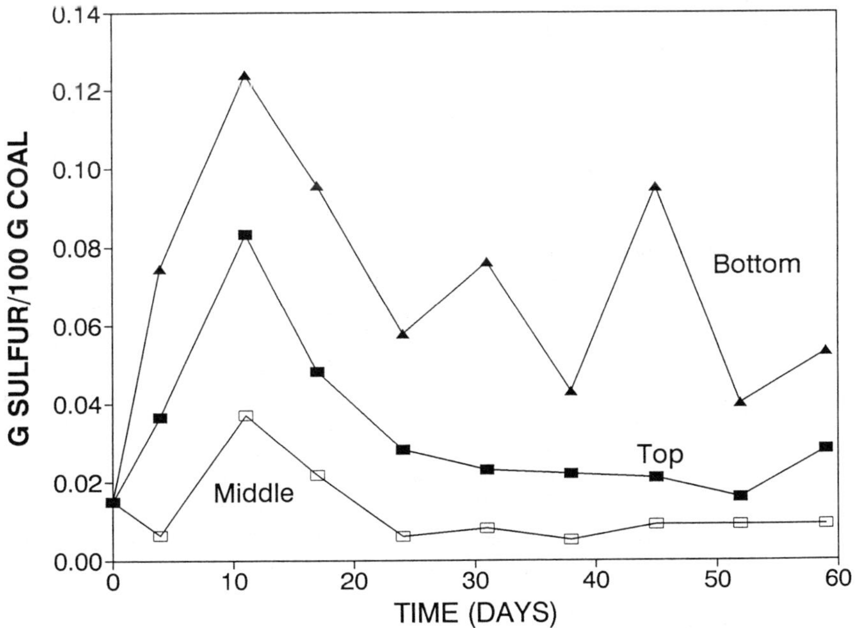

Figure 3 - Variation of Precipitated Sulfate Sulfur in Core Samples

Stratification of the bed greatly complicates these mass balance calculations. Core sample measurements of pyritic or precipitated iron (y) are given per gram of solids, but when the bed is stratified the bed density (ρ = gram of solids per cm^3 of bed) varies through the bed. How then can the average pyritic and precipitated iron be calculated from the measurements taken at the top, middle and bottom of the bed? The exact mathematical definition is

$$\text{Average } y = \frac{\int_0^H \rho y dz}{\int_0^H \rho dz} \tag{1}$$

The bed density was assumed to decrease linearly with height in the region $0 \leq z \leq A$ where the liberated mineral matter accumulates, and to be constant in the remainder of the bed, $A \leq z \leq H$, where the ash content of the coal is lower and more constant. Measurements of bed densities in the bottom 8 in. and the remainder of the bed at the end of the experiment, allowed the relevant densities to be determined in terms of A. The variation of the pyritic iron concentration with z, given by the core sample data was fit to a hyperbolic function. The integral in equation (1) could then be evaluated and the average pyritic iron calculated for any specified value of A. As expected, the results were found to be virtually identical for any reasonable value of this parameter ($0.2H \leq A \leq 0.5H$: depending on the ash content of the coal). The average precipitated iron was calculated in a similar manner, except that a hyperbolic function was inappropriate to describe the minimum found in the middle of the bed (Figure 3). The core sample data was therefore fit by two straight lines, one from Bottom to Middle and one from Middle to Top.

The average pyritic iron calculated by this procedure and plotted in Figure 4, decreased uniformly with time. Note that, although the pyrite reductions are given in Table 1 for the end of the experiment at 60 days, very similar values could be obtained in 24 days. At this time all of the exposed pyritic inclusions have been solubilized, and the oxidation of the remaining submerged inclusions by indirect leaching is a slow process. The average "precipitated" (i.e., HCl soluble) iron, also shown in Figure 4, stayed virtually constant at the level in the feed coal. This is in contrast to the results of most "heap" experiments (6), and illustrates the ability of the flood/drain system to wash precipitates off the coal as fast as they are formed. The "soluble" iron line in Figure 4 shows the mass of iron (dissolved and colloidal) that appeared in the water, converted to the same "per 100 gm coal" basis as the other measurements. It increased rapidly, due to microbial pyrite oxidation, after a lag period of approximately 2 days. This lag was much shorter than in previous experiments, showing that the product coal washing procedure described above was generating a microbial culture well acclimated to the coal.

To close the mass balance, all of the above measured forms of iron are added to give the total line in Figure 4. It is found to remain constant within the bounds of experimental error. The fact that it remains consistently below the total iron in the feed coal, rather than fluctuating randomly around it, does suggest a small systematic error. This is not due to the iron in the fines in the settling tank, which was checked and found to be negligible, and is probably a result of the approximations in the averaging procedure of equation (1).

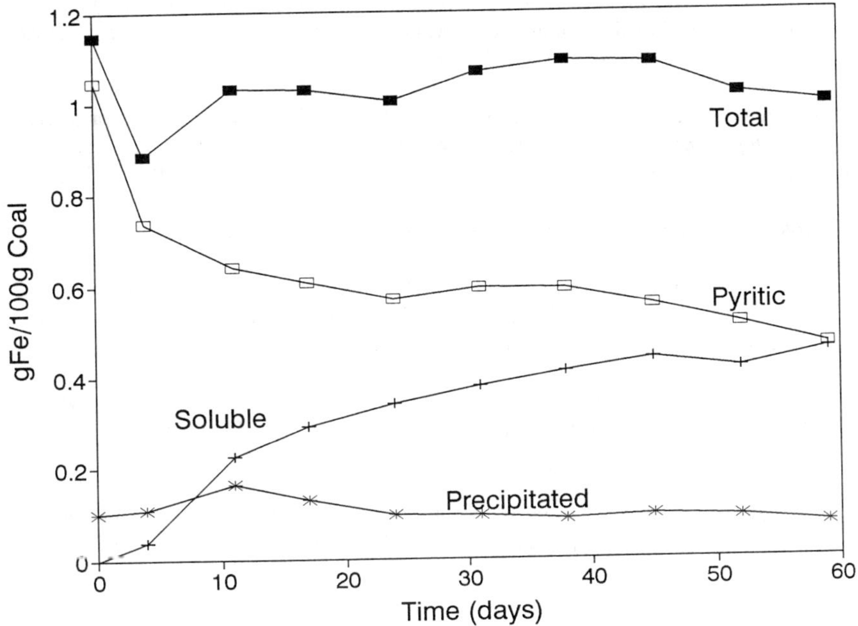

Figure 4 - Measured Forms of Iron and the Total Iron Recovery

Conclusions

The flood/drain bioreactor is a feasible alternative for the large-scale processing of solids. It falls between the common heap and slurry-type processes in terms of particle size, rates and costs.

For coal processing the repeated fluidization gives excellent physical separation, the denser, high-ash particles falling to the base of the bed from where they can be discarded. The combination of physical separation and microbial depyritization gives a high-quality product from run-of-mine coal in a single step. Further development requires acceleration of the microbial activity so that it works over a time scale similar to the physical separation. This involves optimization of the cycle time and the coal washing step so as to increase the concentrations of acclimated bacteria and ferric ions available to initiate the process on fresh coal.

Acknowledgments

This work was supported under contract no. DE-AC07-76ID1570 from the U.S. Department of Energy, Office of Advanced Research and Technology Development, Office of Fossil Energy to the Idaho National Engineering Laboratory/EG&G Idaho, Inc.

Literature Cited

1. Brierley J.A., L.Luinstra. *Biooxidation Heap Concept for Pre-Treatment of Refractory Gold Ore*. In Biohydrometallurgical Technologies. Torma A., Wey J., Lakshamanan V. (eds.). Vol. 1 p. 437-448. TMS (1993).

2. Olsen G. J., D. Finseth, K. Rhee, M.Schoffstall. *Laboratory Studies of Microbial Desulfurization of Pittsburgh Seam Coal Related to Heap Depyritization*. Proc. 2nd Int. Symposium on Biological Processing of Coal. P. 3-35. EPRI GS 7482 (1991).

3. Rossi G. *Microbial Depyritization of Coal*. Fuel 72:1581. (1993).

4. Hansford G., A. D. Bailey. *Oxygen Transfer Limitation of Biooxidation at High Solids Concentration*. In Biohydrometallurgical Technologies. Torma A., Wey J., Lakshamanan V. (eds.). Vol. 1 p. 469. TMS (1993).

5. Andrews G. F., K. S. Noah, A. W. Glenn, C. J. Stevens. *Combined Physical/Microbial Beneficiation of Coal in the Flood/Drain Bioreactor*. Fuel Processing Technology (in press).

6. Andrews G. F., C. J. Stevens, S. A. Leeper. *Heaps as Bioreactors for Coal Bioprocessing*. Applied Biochemistry and Biotechnol., 39/40:427. (1993).

MICROBIAL REMOVAL OF OIL FROM POLLUTED SOILS IN A PILOT

SCALE OPERATION IN THE TULENOVO DEPOSIT, BULGARIA

V.I. Groudeva[1], S.N. Groudev[2], G.C. Uzunov[2]
and I.A. Ivanova[1]

[1] Department of Microbiology, Faculty of Biology, University of Sofia, Sofia 1421, Bulgaria

[2] Research and Training Centre of Mineral Biotechnology, University of Mining and Geology, Sofia 1156, Bulgaria

Abstract

In the Tulenova deposit soils polluted with oil were treated by the heap technique using a mixed population of laboratory-bred oil-degrading microorganisms. The heap was irrigated with water to maintain the soil humidity in the range of 25-30 % and ammonium phosphate was added to stimulate the growth and oxidative activity of microorganisms. The soil was regularly ploughed up to enhance the natural aeration of the heap. In an experimental heap section initially containing 88 g oil/kg dry soil about 68 % of the oil was degraded within 5 months, while in an control heap section without ploughing up, irrigation and addition of ammonium phosphate only 6.4 % of the oil was degraded by the action of the indigenous oil-oxidizing microorganisms.

Mineral Bioprocessing II
Edited by David S. Holmes and Ross W. Smith
The Minerals, Metals & Materials Society, 1995

Introduction

Different heterotrophic microorganisms, both bacteria and fungi, related to a large number of taxonomic genera, are able to utilize oil hydrocarbons as sources of energy and carbon for their growth (1-5). Different biotechnologies for removing oil from polluted waters and soils have been developed on that basis (2,6-10) and some have been efficiently applied at an industrial scale. These biotechnologies can be divided into three main groups depending on the nature of the microorganisms being used:

- activation of the indigenous microflora in the polluted area by addition of nutrients (mainly in the form of mineral fertilizers);
- addition to the polluted area of oil-oxidizing microorganisms isolated from different natural biotopes and selected under laboratory conditions on the basis of the level of their oil-oxidizing ability;
- addition to the polluted area of genetically engineered microorganisms characterized by a pronounced oil-oxidizing ability.

In the last two cases, nutrients are also added to the polluted area to stimulate the activity of the microorganisms being used.

The process of the microbial cleaning depends not only on the level of the genetically determined oxidative ability of the relevant microorganisms but also on the character of the oil pollution and on the values of the physico-chemical parameters of the environment. These parameters, especially the pH, temperature, humidity of the soils being treated, the presence of oxygen and of some useful ions, exert a direct influence on the growth rate, number and physiological activity of the microbial populations. The maintenance of the environmental parameters at values, which are the optimum for the relevant microorganisms, is the main mechanism for enhancing the biodegration of oil.

The biotechnologies for cleaning oil pollution can be divided into three groups depending on the way of application: in situ, heap and reactor techniques, respectively.

Some data about the work of a pilot scale operation for microbial removal of oil from polluted soils in heaps in the Tulenovo deposit are shown in this paper. This deposit is located in Northeastern Bulgaria, very near to the Black Sea coast.

The oil is recovered through numerous wells which produce fountains of fluid containing brine and oil. The oil content in the fluid recovered from the different wells varies in the range of about 0.1 - 1 %. The oil is heavy, with a specific gravity of 0.939, rich in asphaltene-resinous substances (34.7 %), with a low content of paraffins (0.19 %) and with a high viscosity. The total ion concentration in the brine is about 2 - 3 g/l and the pH is in the range of 7.0 - 7.7.

The fluid from each well is collected in a separate vessel where the oil is separated from the brine as a result of their different specific gravities. The aqueous phase is removed from the relevant vessel and runs into the sea. In some cases oil escapes together with the water from the vessels. Several ponds having a surface of about 10 - 20 m^2 each and a depth of about 0.5 m have been constructed to collect the polluted waters and to prevent their running into the sea. In these ponds the oil is separated from the water. A considerable amount of oil in the deposit and the ground around the ponds is heavily polluted with oil.

Materials and Methods

It has been found that the oil from the deposit is degraded efficiently by some oil-oxidizing microorganisms under laboratory and field conditions (11-13). In 1993 polluted soil from locations around some ponds was collected and used to form a heap having the shape of a truncated pyramid. The heap was formed on a ground surface covered by a clay layer and was about 80 m long, 5 m wide and 0.5 m high. The heap was divided into several experimental sections which were treated in different ways. Control sections were also set up.

The soil in most heap sections was initially ploughed up to a particle size less than 30 mm to enhance the natural aeration inside the heap. Then such treatment was carried out periodically, at least once per month.

An artificially mixed microbial culture consisting of different oil-oxidizing microorganisms was added to the polluted soil (Table I). This culture was selected under laboratory conditions among a large number of different oil-oxidizing cultures on the basis of its higher oxidative activity toward the oil from the Tulenovo deposit. Each component of the culture attacks different components of the oil at different rates and the mixed culture as a whole degrades the oil by a strong synergetic effect. To prepare inoculum for treatment of the polluted soil in the heap, each component of the culture was grown in a fermenter containing the nutrient medium of Raymond (in the modification of Rozanova and Nazina (14)) with peptone as a source to carbon and energy. Late-log-phase microbial cultures containing about 10^9 cells/ml each were removed from the fermenters and were mixed together at different proportions to give the composition shown in Table I. The inoculum was added to the experimental heap sections in different amounts varying from 5 - 20 l/m^2.

The microbial inoculum was added to the heap together with brine from the deposit. The irrigation of the heap with brine was carried out to adjust the initial humidity

Table I. Composition of the Microbial Culture Inoculated into the Oil-Polluted Soil

Component	Relative portion in the inoculum, %	Maximum rate of degradation of oil from the deposit under laboratory conditions, mg/l.h
Bacillus sp. - 1	15	36.9
Bacillus sp. -2	15	33.0
Mixed culture of bacteria and yeasts	20	37.4
Sporosarcina sp.	15	20.3
Pseudomonas sp.	15	24.2
Mixed bacterial culture	20	42.4

of the separate experimental sections at different levels varying from 20 to 50 %. Then the humidity of the different sections was maintained at the desired levels by periodical irrigation with brine.

Ammonium phosphate dissolved in brine was added to the different heap sections in the amounts varying from 0.05 to 0.5 g per kg of dry soil. The different concentrations of this salt into the relevant heap sections were maintained by periodical additions. In some cases zeolite saturated with ammonium phosphate was added to the soil (in the amounts in the range of 2 - 5 kg/ton dry soil) to provide the microorganisms with ammonium and phosphate ions.

No laboratory-bred oil-oxidizing microorganisms were added to the control heap sections. However, the polluted soil contained its own indigenous microflora consisting of a large variety of microorganisms, including different oil-oxidizing bacteria and fungi. The control section No 1 contained soil initially broken to a particle size less than 30 mm but then no other activity was carried out in this section and it was maintained in its natural state. The soil in the control section No 2 was ploughed up and irrigated periodically to enhance the natural aeration and to maintain the humidity in the range of 25 - 30 %. Apart from these activities, in the control section No 3 ammonium phosphate was added periodically to stimulate the indigenous microflora.

Results and Discussion

The analyses of the microflora in the heap showed that the number of the oil-oxidizing microorganisms in the experimental sections was considerably higher than that in the control sections No 1 and No 2 and usually higher than that in the control section No 3 (Tables II and III). The microbial cenoses in the different experimental sections had similar species composition but differed from each other with respect to the number of cells related to the different species as well as to the levels of their oxidizing activities. It was found by using the method of DNA molecular probe analysis that the inoculated active strains were prevalent oil-oxidizing microorganisms in the microbial cenoses established in the experimental heap sections. In some cases the total number of oil-oxidizing microorganisms exceeded 10^6 cells/g dry soil. In the control section No 3 the total number of indigenous oil-oxidizing microorganisms was close to 10^6 cells/g dry soil during the summer months but their activity was lower than that of the inoculated laboratory-bred microorganisms in the experimental heap sections treated in the same way.

The regular ploughing up of soil, the addition of ammonium phosphate and the maintenance of the humidity in the Range of 25 - 30 % were essential factors for enhancing the microbial growth and activity. The temperature of the soil was also an essential factor in this respect. The highest rates of oil degradation were observed during the summer months (from July to September 1993) when the temperature inside the heap was in the range of 18 - 23°C. The process was efficient even at temperatures in the range of 10 - 15°C but practically stopped during the cold winter months (from December 1993 to February 1994) when the temperatures were often about 0°C or even lower. In some cases, however, the reduction of the biodegradation rate was not connected only with the low temperatures but also with the exhaustion of those oil fractions which were more amenable to microbial attacks.

Both the species composition and the number of microorganisms related to the separate species changed during the treatment (Table II). The number of microorganisms was higher during the summer months and then decreased steadily reaching its lowest values during the winter months. In March 1994 the number of microorganisms started to increase again and the biodegradation of oil was restored, especially in those heap sections in which oil hydrocarbons amenable to microbial attacks were still present. However, the addition of inoculum of the same composition to the heap was useful and enhanced the biodegradation to some extent.

On the other side, the number of some microorganisms as well as that of some higher organisms (protozoa, worms, insects, etc) increased simultaneously with the

Table II. Microflora and Some Essential Environmental Parameters in an Experimental Heap Section (No 7) for Microbial Removal of Oil from Polluted Soils

Indices	Date of sampling				
	2.7. 93°	1.9. 93	1.12. 93	1.4. 94	1.5 94°°
Microorganisms, cells/g dry soil:					
- Oil-oxidizers	5.10^3	2.10^6	3.10^1	4.10^3	1.10^4
- Aerobic hetero-trophic bacteria	5.10^6	8.10^7	5.10^2	2.10^6	9.10^6
- Oligocarbophiles	8.10^5	1.10^7	3.10^2	1.10^5	6.10^6
- Spore-forming bacteria	6.10^3	3.10^6	3.10^3	6.10^3	1.10^6
- Nitrogen-fixing bacteria	1.10^2	1.10^5	5.10^1	7.10^2	9.10^4
- Anaerobic heterotrophic bacteria	1.10^4	5.10^4	7.10^2	2.10^4	9.10^3
- Moulds	2.10^2	1.10^4	3.10^1	3.10^2	1.10^4
Temperature of the soil, °C	20-21	18-20	3-5	9-11	14-16
Humidity, %	33	31	25	25	28
pH	7.35	7.12	6.93	6.91	6.90
Oil content, g/kg dry soil	88	47	28	27	26
Oil content, % from the initial content	100	53.4	31.8	30.7	29.5
Oil removal, %	---	46.6	68.2	69.3	70.5

° Start of the treatment; °° End of the treatment.

Table III. Some Essential Characteristics of Different Heap Sections Subjected to Microbial Removal of Oil from Polluted Soils

Characteristics	Experimental sections		Control sections		
	No 9 (without zeolite)	No 10 (with zeolite)	No 1	No 2	No 3
Duration of treatment, months	5 (from 2.7.93 to 30.11.93)		10 (from 2.7.93 to 30.4.94)		
Initial oil content, g/kg dry soil	99	95	88	87	91
Final oil content, g/kg dry soil	33	30	82	64	48
Oil removal, %	66.7	68.4	6.4	26.4	47.3
Maximum oil degradation rate, g oil/kg dry soil.month	27	28	1.7	8	15
Ammonium phosphate consumption, kg/ton dry soil.month	0.6	0.45	--	--	0.6
Microbial counts during the most active phase of treatment, cells/g dry soil:					
- Oil-oxidizers	3.10^6	3.10^6	3.10^3	9.10^4	9.10^5
- Aerobic heterotrophic bacteria	1.10^8	3.10^8	1.10^5	3.10^5	1.10^8
- Oligocarbophiles	9.10^6	8.10^6	1.10^5	5.10^5	5.10^6
- Spore-forming bacteria	3.10^6	5.10^6	5.10^4	8.10^4	3.10^6
- Nitrogen-fixing bacteria	3.10^5	5.10^5	8.10^3	3.10^4	9.10^4
- Anaerobic heterotrophic bacteria	3.10^4	4.10^4	3.10^3	1.10^4	4.10^4
- Moulds	1.10^4	1.10^4	5.10^2	1.10^3	9.10^3

reduction of the oil content of the soil.

The presence of zeolite saturated with ammonium phosphate enhanced to some extent the rate of oil biodegradation and decreased considerably the consumption of this nutrient (Table III). The positive effect of the zeolite was connected with the maintenance of relatively constant concentrations of ammonium and phosphate ions in the soil solution due to the existence of a stable equilibrium between the dissolved and adsorbed forms of these ions. Furthermore, the zeolite improved the physico-mechanical properties of the soil.

In all heap sections the pH of the soil decreased during the treatment. This was due to the secretion of some acidic microbial metabolites (mainly CO^2 and organic acids).

The highest rate of microbial degradation of oil achieved in this study was about 28 g oil per month (Table III). In some experimental heap sections the oil content of the soil was reduced from the initial 90 - 100 g/kg dry soil to about 30 g/kg dry soil at the end of the treatment (April 1994), i.e. about 70 % of the oil was removed from the soil within 10 months. The residues, after treatment, consisted mainly of asphaltene-resinous substances.

In the control heap sections No 1 and No 2 the oil removal for the same period of time was 6.4 % and 26.4 %, respectively. In the control section No 3 the oil removal was 47.3 %. This was an indication that under optimum environmental conditions the oil degradation by the indigenous microflora can be efficient.

The data from this study showed that the oil-polluted soils from the Tulenovo deposit can be cleaned efficiently in heaps by using active oil-oxidizing microorganisms under conditions of a good aeration, optimum and constant humidity and presence of some essential nutrients.

Acknowledgements

The authors acknowledge the financial support of the Neft i Gas Company - Dolen Dubnik and the excellent assistance of Mr. V. Gamzakov and his colleagues from this company.

References

1. Atlas, R.M., Microbial degradation of pertroleum hydrocarbons: an environmental perspective, Microbiol. Rev., 45, 180-209, 1981.
2. Atlas, R.M., ed., Petroleum Microbiology, Macmillan Publishing Company, New York, 1984.
3. Leahy, J.G. and Colwell, R.R., Microbial degradation of hydrocarbons in the environment, Microbiol. Rev., 54, 305-315, 1990.
4. Kimura, B., Murakami, M. and Fujisawa, H., Utilization of hydrocarbon substrates by heavy oil-degrading bacteria isolated from the sea water of oil-degrading bacteria isolated from the sea water of oil-polluted Bisan Seto, Nippon Suisan Gakkaishi, 56, 771-776, 1990.
5. Venkateswaran, K., Iwabuchi, T., Matsui, Y., Toki, H., Hamada, E. and Tanaka, H., Distribution and biodegradation potential of oil-degrading bacteria in North Eastern Japanese coastal waters, FEMS Microbiology Ecology, 86, 113-122, 1991.
6. Chakrabarty, A.M., Genetically-manipulated microorganisms and their products in the oil service industries, Trends in Biotechnol., 3, 32-38, 1985.
7. Fox, J.L., Native microbes' role in Alaskan clean-up, Bio/Technology, 7, 852, 1989.
8. Prince, R.C., Clark, J.R. and Lindstrom, J.E., Bioremediation monitoring program of Alaskan oil spill, A publication of Exxon Research and Engineering, Annadale., 1990.
9. Pritchard, P.H. and Costa, C.F., EPA's Alaska oil spill bioremediation project, Environ. Sci. Technol., 25, 372-379, 1991.
10. Pritchard, P.H., Use of inoculation in bioremediation, Current Opinion in Biotechnology, 3, 232-243, 1992.
11. Ivanova, I.A., Analysis of the microflora of the Tulenovo oil deposit and possibilities for practical applicaton of its metabolic activity, Master of Science Thesis, University of Sofia, Sofia, 1993.
12. Groudeva, V.I., Ivanaova, I.A., Groudev, S.N. and Uzunov, G.C., Enhanced oil recovery by stimulating the activity of the indigenous microflora of oil reservoirs, In: Biohydrometallurgical Technologies, vol II, pp. 349-356, A.E. Torma, M.L. Apel and C.L. Brierley, eds., A publication of the Minerals, Metals & Materials Society, Warrendale, Pennsylvania, 1993.

13. Groudev, S.N., Goudeva, V.I., Uzunov, G.C., Ivanova, I.A., Stankov, I. and Gamzakov, V., Microbiological cleaning of soils from oil in the Tulenovo deposit, Minno Delo i Geologiya, No 2, 37-41, 1994.

14. Rozanova, E.P. and Nazina, T.I., Hydrocarbon-oxidizing bacteria and their activity in the oil strata, Mikrobiologiya, 51, 324-348, 1982.

BIOTREATMENT OF H_2S IN THE PETROLEUM INDUSTRY*

Hector M. Lizama** and Eric N. Kaufman

Chemical Technology Division, Oak Ridge National Laboratory, P.O. Box 2008, MS 6226, Oak Ridge, Tennessee 37831-6226

Abstract

Increasingly stringent limits on sulfur emissions have resulted in new challenges for the processing of H_2S in the petroleum industry. In general, conventional H_2S treatment processes are economical only on a large scale. For situations where H_2S are low but still unacceptable, biotechnological approaches may become an option. This paper describes two H_2S bioreactor design types based on available scientific literature. These are compared with conventional reactor systems. Also addressed are the inherent differences in the requirements of upstream (production) and downstream (refining) operations of the petroleum industry. It is concluded that the current state of the art in H_2S bioprocessing research needs further development for it to be considered a feasible alternative.

*Research supported by the AR&TD Program of the Office of Fossil Energy, U.S. Department of Energy under contract DE-AC05-84OR21400 with Martin Marietta Energy Systems Inc.

**Corresponding author.

Mineral Bioprocessing II
Edited by David S. Holmes and Ross W. Smith
The Minerals, Metals & Materials Society, 1995

Introduction

Hydrogen sulfide is a toxic, corrosive gas often associated with the production of petroleum. In addition, it is a byproduct of all current petroleum refining steps that involve the removal of organic sulfur by reductive processes. When H_2S is removed from a gas stream or generated from a petroleum refining step, it is generally converted to sulfur by way of the Claus process or variations of it (8). Figure 1 illustrates a typical Claus plant process scheme. Relatively pure H_2S passes through a combustion chamber where one third of the feed is oxidized to sulfur dioxide ($3H_2S + 1.5O_2 \rightarrow 2H_2S + SO_2 + H_2O$). The H_2S/SO_2 mixture then passes through a series of catalytic units and condensers where sulfur is formed ($2H_2S + SO_2 \rightarrow 3S + H_2O$). The number of reactor/condenser units determine the sulfur recovery which can reach as high as 99.5%. A typical Claus plant has a capacity of 100 tons per day (TPD) sulfur.

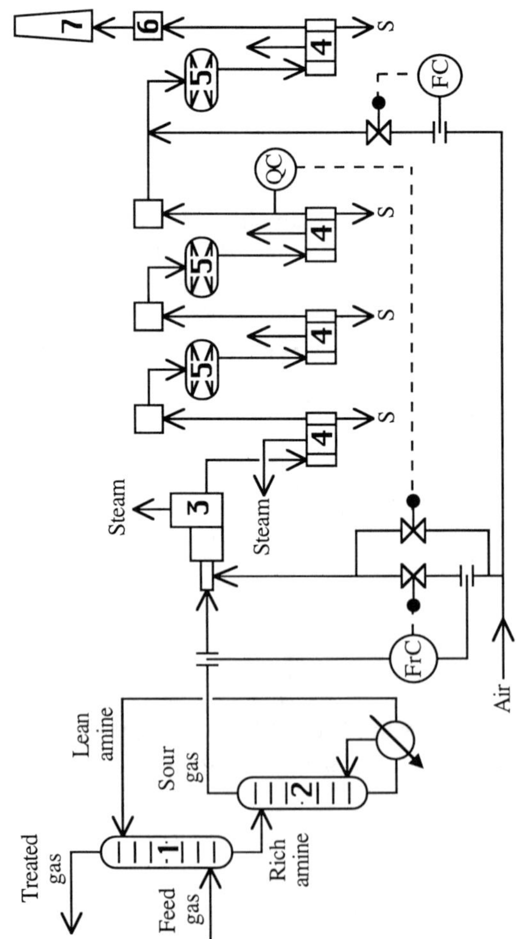

Figure 1. Claus plant with upstream amine scrubber. 1. absorber tower; 2. thermal regenerator; 3. combustion chamber; 4. condensers; 5. reactors; 6. incinerator; 7. stack.

During oil production, H_2S partitions with natural gas and must be reduced in content from several hundred parts per million (ppm) to less than 14 ppm. In situations where produced volumes are low and a Claus plant not economical, H_2S is commonly flared with a small amount of natural gas. This practice, however, is coming under increasing scrutiny due to the resulting sulfidic oxide (SOx) emissions. With increasingly restrictive regulations on flaring, there is a growing need for alternative low-cost technology for processing small or dilute volumes of H_2S.

The possibility of using microbial approaches to H_2S processing has been considered but to a limited degree. Sublette and coworkers (15) have carried out extensive studies with the bacterium *Thiobacillus denitrificans*, oxidizing H_2S completely to sulfate in a completely-stirred tank reactor (CSTR). Lizama and Sankey have used the bacterium *Thiobacillus thiooxidans* to partly oxidize H_2S to elemental sulfur and sulfate in a gas-stripping tower reactor (11). This paper makes a comparison of these two systems as they would be applied towards an industrial-scale process for the removal of H_2S.

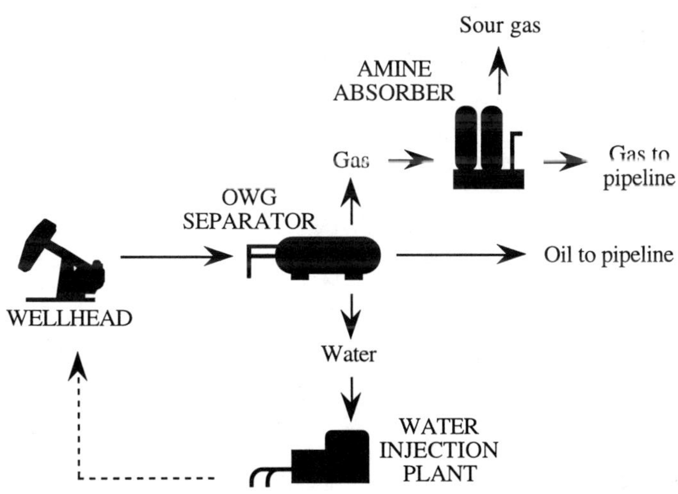

Figure 2. Petroleum production and the upstream generation of H_2S.

Criteria

To establish a criteria for design parameters, the nature and needs of the petroleum industry must be taken into account. This includes the upstream, or production, side as well as the downstream, or refinery, side. Oil production can be summarized by the scheme shown in Figure 2. Produced fluids originate at the wellhead and consist of crude oil, natural gas, and produced water. Production lines from several wells feed into an oil-water-gas (OWG) separator where the three phases separate by gravity. Produced water is sent to a water injection plant for subsequent injection back into the formation as part of waterflood process. Crude oil is usually fed directly to a pipeline. Sour natural gas, meaning it contains H_2S, must go through a sweetening step where H_2S is removed. This is carried out by passing the gas through an amine stripper (6), the sweetened gas being then sent to a compressor station and to pipeline. The concentrated H_2S stream generated by the amine stripper is usually small in volume. Fields on primary oil recovery, where oil flows out under its own pressure, are generally sweet. Once a field goes on secondary recovery, where water is injected to force out residual oil, the formation can become contaminated with sulfate reducing bacteria (SRB) which will cause progressive souring (12).

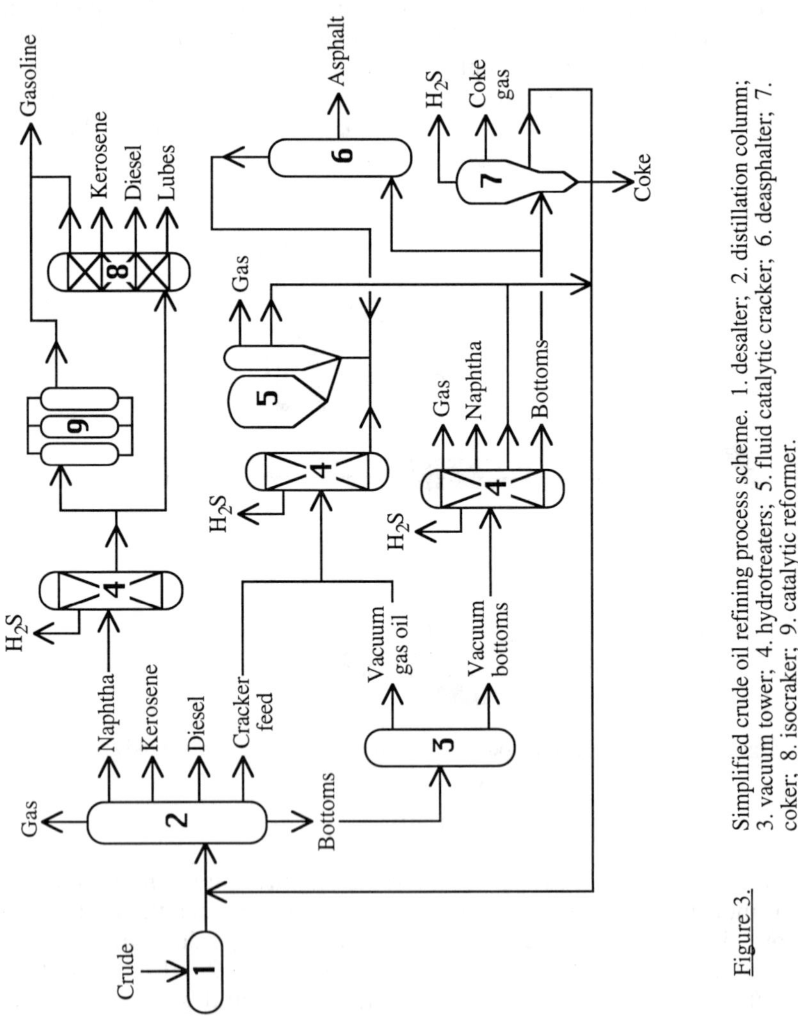

Figure 3. Simplified crude oil refining process scheme. 1. desalter; 2. distillation column; 3. vacuum tower; 4. hydrotreaters; 5. fluid catalytic cracker; 6. deasphalter; 7. coker; 8. isocraker; 9. catalytic reformer.

The refining of crude oil (7) is summarized by the process scheme shown in Figure 2. Crude passes through a desalter prior to undergoing distillation and fractionation into four streams, namely light and heavy naphthas, kerosene, diesel, and feedstock for the cracker unit. The distillation residuals, or bottoms, are condensed under vacuum to yield more feedstock for the cracker unit. The vacuum residuals consist of heavy cross-linked asphaltenic material high in sulfur and other heteroatoms that are used for asphalt or coke production. In catalytic cracking, the heart of the refining process, large hydrocarbon chains are split into smaller ones of greater value. The naphtha fractions undergo catalytic reforming or isocracking to make high-volume, low-value gasoline and low-volume, high-value lubes. As seen in Figure 2, H_2S is produced at the hydrotreating steps, whenever reduction of hydrocarbon molecules is required. The organic sulfur is liberated by combination with hydrogen which is introduced in various forms.

It is evident that the production and refining of crude oil have very different requirements. Whereas the upstream has to deal with low volumes, it is also constricted by available space and remoteness of location. In contrast, the downstream has to deal with large volumes but has ample capabilities to do so. In this work, a biological reactor system is envisioned which is capable of processing up to 1 TPD sulfur equivalent and releasing a gas stream with an H_2S content of 10 ppb (0.01 ppm). It is assumed that the H_2S feed has been concentrated by way of an amine scrubber.

Bioreactor 1.: Tower reactor

The first bioreactor candidate consists of a packed be tower reactor with countercurrent flow and liquid recycle (Fig. 4). The liquid recycle stream passes through a filtering unit to collect formed elemental sulfur. The biocatalyst consists of *T. thiooxidans* bacteria suspended in the recirculating liquid. The relevant design equations used for this system are:

(Ref. 10)
$$\frac{p}{RT} G_\mathrm{v} \partial Y_{\mathrm{H_2S}} = L_\mathrm{v} \partial C_{\mathrm{H_2S}} \qquad [1]$$

(Ref. 10)
$$\frac{1}{K_g} = \frac{1}{k_g} + \frac{H_{\mathrm{H_2S}}}{k_l} \qquad [2]$$

(Ref. 14)
$$\frac{RTk_g}{a_t D_g} = 2\left(\frac{\rho_g G_\mathrm{v}}{a_t \mu_g}\right)^{0.7} \left(\frac{\mu_g}{\rho_g D_g}\right)^{0.33} \left(\frac{1}{a_t d_\mathrm{p}}\right)^{2} \qquad [3]$$

(Ref. 14)
$$\frac{d_\mathrm{p} k_l}{D_l} = 25.1\left(\frac{d_\mathrm{p} \rho_l L_\mathrm{v}}{\mu_l}\right)^{0.45} \left(\frac{\mu_l}{\rho_l D_l}\right)^{0.5} \qquad [4]$$

(Ref. 2)
$$D_l = \frac{7.44 \times 10^{-8} T (X_a M_{\mathrm{H_2O}})^{0.5}}{\mu_l (V_{\mathrm{H_2S}})^{0.6}} \qquad [5]$$

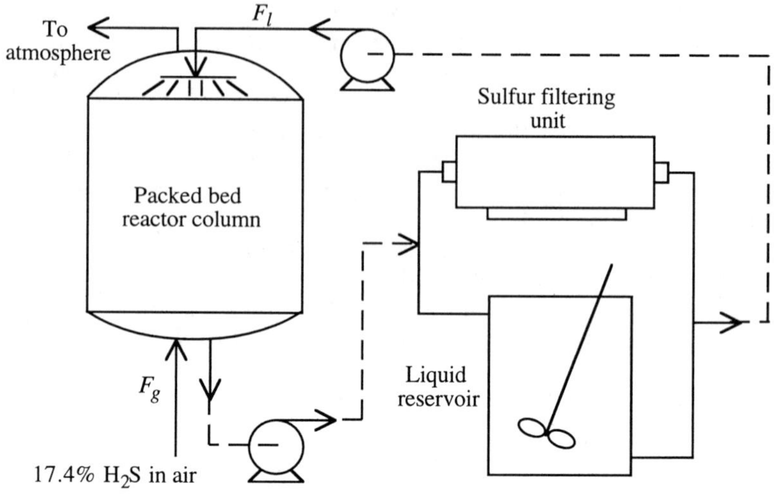

Figure 4. Oxidation of H_2S in a tower reactor with liquid recycle.

(Ref. 3)
$$D_g = \frac{1.86 \times 10^{-3} T^{1.5} \left({}^{1}\!/\!_{M_{H_2S}} + {}^{1}\!/\!_{M_{Air}} \right)}{p\left\{0.5\left[1.18(V_{H_2S})^{0.33} + \varpi\right]\right\}^2 \Omega} \tag{6}$$

(Ref. 14)
$$\frac{a}{a_t} = 0.31 \frac{(\sigma_{H_2O})^{0.5}}{h^{0.4}} \left(\frac{6.0 \mu_l L_v \rho_g}{a_t \sigma \mu_g} \right)^{0.392} \tag{7}$$

(Ref. 10)
$$h = \frac{pG_v}{aRT} \int_{Y_{H_2Sout}}^{Y_{H_2Sin}} \frac{1}{(-r''_{H_2S})} \partial Y_{H_2S} \tag{8}$$

(Ref. 10)
$$-r''_{H_2S} = \frac{1}{\dfrac{1}{K_g} + \dfrac{aH_{H_2S}C_{H_2S}}{v_{H_2S}}} pY_{H_2S} \tag{9}$$

(Ref. 13)
$$v_{H_2S} = \frac{\partial C_{H_2S}}{\partial t} = \frac{V_{\max(H_2S)} C_{H_2S}}{K_{s(H_2S)} + C_{H_2S} + {}^{(C_{H_2S})^2}\!/\!_{K_{is(H_2S)}}} \tag{10}$$

(Ref. 16)
$$v_{S^\circ} = \frac{\partial C_{S^\circ}}{\partial t} = \frac{\mu_{\max(S^\circ)} C_{S^\circ}}{K_{s(S^\circ)} + C_{S^\circ}} \tag{11}$$

(Ref. 10) $$-\frac{\partial N_{H_2S}}{\partial t} = \frac{1}{4}\frac{\partial N_{SO_4}}{\partial t} \qquad [12]$$

Y_{H_2Sin} is set at 17.36% H_2S in air so that there is an equimolar H_2S/O_2 ratio and the formation of elemental sulfur is favored. Gas and liquid flow rates are set so that $k_g > 95\%$ of K_g, making the H_2S mass transfer resistance come mainly from the gas film and therefore facilitating partitioning into the liquid phase. The packing of choice are ceramic Rasching rings of 13 mm in diameter (0.5 inch). Based on the available bench-scale data, a design configuration was arrived at which envisions a column 3m in diameter and 3m in height. A schematic of such a reactor is illustrated in Figure 5 and its parameters are listed in Table I.

Table I. Reactor parameters for tower reactor

Parameter	Value	Units
d_R	3	m
h	3	m
Voids	0.64	dimensionless
a_t	370	m^2/m^3
G_v	1.37×10^{-2}	mL/cm^2-sec
L_v	0.40	mL/cm^2-sec
K_g (H_2S)	4.99×10^{-7}	mol/cm^2-sec-atm
a	6.26	m^2/m^3
V_{max} (H_2S)	9.27×10^{-6}	M/sec
μ_{max} (S°)	3.31×10^{-5}	sec^{-1}
K_s (H_2S)	1.09×10^{-5}	M
K_s (S°)	3.20×10^{-3}	g/mL
K_{is} (H_2S)	0.15	M

H_2S bacterial kinetic values are from unpublished data (Imperial Oil, Calgary, Canada). Sulfur kinetic values are from Ref. 16.

Gas residence time in the proposed tower reactor is 3.9 hours; liquid residence time is four minutes. Even though the column has a bed volume of 21 m^3, it can only oxidize about 0.02 TPD equivalent sulfur. For a 1 TPD application, a total of 52 such columns would be required; 45 of these would be operational at any one time while the remaining seven reactors would be on maintenance downtime. The maintenance requirements are due mainly to the deposition of sulfur in the column bed which accumulates over time, resulting in plugging problems (11). An 80% efficiency is assumed for the sulfur filtering unit and the remaining sulfur is allowed to accumulate in the column to 20% of the bed voids. This results in a 12 day run time without affecting hydrodynamic behavior of the bed. The column is then run with only air as the gaseous feed, allowing the bacteria to oxidize the deposited sulfur fully to sulfide. Complete removal of sulfur is accomplished in two days resulting in a total cycle time of 14 days: 12 operating and 2 on regeneration. All of the reactors would feed into a common sulfur filtering unit and a small reservoir tank for the addition of make-up water and nutrients.

Bioreactor 2.: CSTR

The second bioreactor candidate consists of a completely stirred tank reactor with biomass recycle (Fig. 5). *T. denitrificans* bacteria make up the biocatalyst which is itself immobilized in a heterotrophic floc. Hydrogen sulfide in air is introduced into the reactor by sparged bubbles. The floc is separated from the liquid by a settling tank. Relevant design equations for this system are:

(Ref. 15) $$F_g = \frac{(-\bar{r}_{O_2})V_l}{Y_{O_2\text{in}}}\frac{RT}{p} \qquad [13]$$

(Ref. 1) $$P_0 = 0.035\rho_l(N_i)^3(d_i)^{3.7}w_i(n_i)^{0.8}(n_b)^{0.4}(w_b)^{0.3} \qquad [14]$$

(Ref. 1) $$P = 0.072\left[\frac{(P_0)^2 N_i(d_i)^3}{(F_g)^{0.56}}\right]^{0.45} \qquad [15]$$

(Ref. 1) $$\varepsilon_g = 0.52\left[\frac{F_g}{N_i(d_i)^3}\right]^{0.5}\left[\frac{\rho_l(N_i)^2(d_i)^3}{\sigma}\right]^{0.65}\left(\frac{d_i}{d_R}\right)^{1.4} \qquad [16]$$

(Ref. 2) $$u = \frac{F_g h_R}{V_l \varepsilon_g} \qquad [17]$$

(Ref. 1) $$k_l a = 0.0018\left(\frac{P}{V}\right)^{0.7} u^{0.3} \qquad [18]$$

(Ref. 2) $$\text{Re}_i = \frac{\rho_l N_i (d_i)^2}{\mu_l} \qquad [19]$$

(Ref. 2) $$\text{Fr}_i = \frac{(N_i)^2 d_i}{g} \qquad [20]$$

(Ref. 2) $$\text{Sc} = \frac{\mu_l}{\rho_l D_{g(O_2)}} \qquad [21]$$

(Ref. 2) $$\mathrm{Sh} = 0.13(\mathrm{Sc})^{0.33}(\mathrm{Re})^{0.75} \qquad [22]$$

Y_{H_2Sin} is set at 0.16% H_2S in air, a concentration low enough to limit substrate inhibition effects. Gas and liquid flow rates are respectively set so that there are no O_2 mass transfer limitations ($k_l a \gg \bar{r}_{O_2}/\{C^*_{O_2} - C_{O_2}\}$) and no biomass washout occurs. Scale-up regime was based on u_i since the floc is sensitive to shear. Based on the available bench-scale data (Sublette) and the scale-up criteria used, a design configuration was arrived at which envisions a reactor that is 2m in diameter and 6m in height with seven paddle impellers. A schematic of such a reactor is illustrated in Figure 5 and its parameters are listed in Table II.

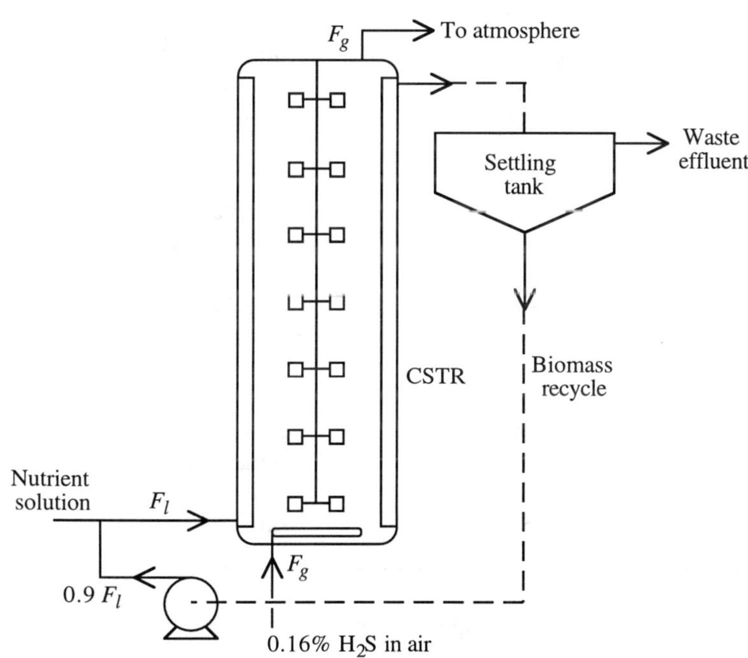

Figure 5. Oxidation of H_2S in a CSTR with biomass recycle.

The proposed reactor is some 19 m^3, holding almost 12 m^3 of liquid, but can only oxidize about 0.03 TPD equivalent sulfur. For a 1 TPD application, a total of 40 such reactors would be required; 33 of these would be operational at any one time while the remaining seven reactors would be on maintenance downtime. This follows the rule-of-thumb ratio of 75% operational, 25% maintenance for a continuously operated process. All of the reactors would feed into a common 350 m^3 settling tank with a residence time of three hours. The reactors themselves have a liquid residence time of 3.33 hours with a recycle ratio (ω) of 0.9. From this, the total water requirement per reactor is 8.4 m^3 per day; operational water requirement is more than 276 m^3 per day.

Table II. Reactor parameters for CSTR

Parameter	Value	Units
$-\bar{r}_{O_2}$	3.58×10^{-9}	mol/mL-sec
V_l	11.6	m^3
d_R	2	m
h_R	6	m
F_g	6.10	L/sec
F_l	0.97	L/sec
P_0	87403	dimensionless
u_i	68	cm/sec
N_i	0.11	rev/sec
d_i	67	cm
w_i	17	cm
n_i	7	blades
n_b	6	baffles
w_b	20	cm
P/V_l	668	KW/m^3
ε_g	0.61	dimensionless
u	0.32	cm/sec
k_la (O_2)	0.12	sec^{-1}
Re_i	5.40×10^4	dimensionless
Fr_i	7.96×10^{-4}	dimensionless
Sc	2.91	dimensionless
Sh	658	dimensionless

Discussion

Two bioreactor designs, a tower reactor and CSTR-type reactor, have been formulated for the aerobic oxidation of an H_2S gas stream. Both designs have very low capacities. In order to process 1 TPD sulfur equivalent, 45 tower reactors or 33 CSTRs would be needed. In terms of operating reactor volume this translates into 954 m^3 and 616 m^3, respectively. The requirements for process water are also very large; the tower reactor system recirculates more than 353 m^3 of water while the CSTR system uses 276 m^3 of water per day, holding almost 700 m^3 in its reactors and settling tank. The tower reactor system has a slight advantage in that it operates in a closed loop system, requiring no additional water except to make up that which is lost by evaporation. It also has the advantage of producing sulfur which can be sold to offset operating costs. The advantage of the CSTR system lies in using available established technology. Also, the requirement for operational reactor volume is significantly smaller.

Both designs involve ambient temperature and pressure operating conditions. Although this feature, characteristic of most bioprocesses, is advantageous, it does not compensate for the very large reactor volumes required. H_2S bioprocessing, then, does not appear to be a viable alternative at this time. Specifically, water requirements and reactor volume must decrease by an order of magnitude. It may also be necessary to improve the reaction rates by an order of magnitude. These issues may be overcome through the application of new engineering concepts to biocatalysts since research in this area has been limited. At the same time, some of the bioengineering applications already developed could be used to augment conventional processes already in place. This has been the approach taken by Sublette, who has successfully adapted his H_2S reactor to treat wastewater laden with caustic sulfides (9).

Prospects for H_2S bioprocessing is bound to improve due to two factors: increasing demand for crude with low sulfur content and increasing requirements for low sulfur emissions. In reality, worldwide crude sulfur content has remained relatively constant over the past decade and is not expected to increase substantially in the near future (5). As seen in Fig. 3, however, sulfur tends to accumulate in the heavy fractions and their products such as coke and asphalt. There is increasing pressure for these products to be relatively sulfur-free, particularly with coke which is used in other industrial processes. Furthermore, since desulfurization schemes are reductive in nature the need for H_2S processing will grow in parallel with sulfur removal.

Nomenclature

a	= interfacial area per unit bed volume
a_t	= bed surface area per unit bed volume
C_{H_2S}, C_{S°	= H_2S, sulfur concentrations in liquid phase
d_i, d_p, d_R	= impeller, packing, and reactor diameters
D_g, D_l	= diffusivities in air, water
F_g, F_l	= gas, liquid flow rate
Fr_i	= Froude number with reference to impeller (dimensionless)
G_v	= gas volumetric flow rate (mL/cm^2-sec)
h, h_R	= bed height, reactor height
H_{H_2S}	= Henry's law constant for H_2S (8856.61 mL-atm/mol, Ref. 5)
k_g	= gas phase mass transfer coefficient (mol/cm^2-sec-atm)
k_l	= liquid phase mass transfer coefficient (cm/sec)
K_g	= interfacial mass transfer coefficient (mol/cm^2-sec-atm)
K_{is}	= substrate inhibition constant (mol/mL)
K_s	= dissociation constant (mol/mL)
L_v	= liquid volumetric flow rate (mL/cm^2-sec)
M_{H_2O}, M_{H_2S}, M_{Air}	= molecular weights (H_2O: 18, H_2S: 34, Air: 29 g/mol)
n_i, n_b	= number of impellers, baffles

N_{H_2S}, N_{SO_4}	= moles of H_2S, SO_4^{2-}
N_i,	= impeller rotation rate
p	= total system pressure (1 atm)
P	= power (Watts)
P_0	= power number (dimensionless)
p_{H_2S}	= moles H_2S in gas phase
R	= gas constant (82.06 mL-atm/mol-°K)
Re_i	= Reynolds number with reference to impeller (dimensionless)
$-r''_{H_2S}$	= mass transfer rate (mol/cm^2-sec)
$\bar{r}_{O_2}$	= mean oxygen uptake rate (mol/mL-sec)
Sc	= Schmidt number (dimensionless)
Sh	= Sherwood number (dimensionless)
T	= temperature (298 °K)
v	= bacteria reaction rate (mol/mL-sec)
V_{H_2O}, V_{H_2S}	= molar volumes (H_2O: 18, H_2S: 33 mL/mol)
V_g, V_l	= gas, liquid volumes
V_{max}	= maximum bacteria reaction rate (mol/mL-sec)
w_i, w_b	= width of impellers, baffles
X_a	= H_2O association factor (2.6)
Y_{H_2S}, Y_{O_2}	= H_2S, O_2 concentration in gas phase (volume fraction)
ε_g	= gas holdup (volume fraction)
μ_g, μ_l	= viscosities (air: 1.83×10^{-4}, water: 8.90×10^{-3} g/cm-sec)
μ_{max}	= maximum specific growth rate (sec^{-1})
ρ_g, ρ_l	= densities (air: 1.30×10^{-4}, water: 1.00 g/mL)
Ω	= molecular collision integral from energy tables (Ref. 14)
σ	= surface tension (72 dynes/cm)
ϖ	= molecular collision diameter of air (3.617 Å, Ref. 3)

References

1. B. Atkinson, ed., Biochemical Engineering and Biotechnology Handbook (New York, NY: Stockton Press, 1991), 700-733.

2. J.E.Bailey and D.F. Ollis, Biochemical Engineering Fundamentals (New York, NY: McGraw-Hill Publishing Co., 1986), 470-494.

3. R.B. Bird, W.E. Stewart and E.N. Lightfoot, Transport Phenomena (New York, NY: John Wiley & Sons, 1960), 503-513.

4. J.R. Dosher and J.T. Carney, "Sulfur Increase seen mostly in Heavy Fractions of Lower-quality Crudes", Oil & Gas J., 92 (1994), 43-48.

5. B. Elvers, S. Hawkins and G. Schulz, Ullman's Encyclopedia of Industrial Chemistry (Weinheim, FRG: VCH Verlagsgesellschaft, 1990), 453.

6. Gulf Publishing Co., "Gas Processing Handbook '90", Hydrocarbon Processing, 69 (1990), 69-99.

7. Gulf Publishing Co., "Refining Handbook '92", Hydrocarbon Processing, 71 (1992), 135-211.

8. J.A. Lagas, J. Borsboom, G. Heljkoop, "Claus Process gets extra boost", Hydrocarbon Processing, 68 (1989), 40-42.

9. C.M. Lee and K.L. Sublette, "Microbial Treatment of Sulfide-Laden Water", Wat. Res., 27 (1993), 839-846.

10. O. Levenspiel, Chemical Reaction Engineering (New York, NY: John Wiley & Sons Inc., 1972), 409-441.

11. H.M. Lizama and B.M. Sankey, "Conversion of Hydrogen Sulphide by Acidophilic Bacteria", App. Microbiol. Biotechnol., 40 (1993), 438-441.

12. H.M. Lizama and B.M. Sankey, "On the Use of Bleach Soaks to control Bacteria-mediated Formation Souring", J. Pet. Sci. Eng., 9 (1993), 145-153.

13. H.M. Lizama and B.M. Sankey, "Oxidation of H_2S by *Thiobacillus thiooxidans* is inhibited by Substrate and Methane", Biohydrometallurgical Technologies, Vol. II, eds. A.E. Torma, M.L. Apel, C.L. Brierley (Warrendale, PA: The Minerals, Metals & Materials Society 1993), 339-348.

14. R.H. Perry and D.W. Green, Perry's Chemical Engineers' Handbook (New York NY: McGraw-Hill Publishing Co., 1984), 18.19-18.41

15. C. Ongcharit, Y.T. Shah, K.L. Sublette, "Novel Immobilized Cell Reactor for Microbial Oxidation of H_2S", Chem. Eng. Sci., 45 (1990), 2383-2389.

16. I. Suzuki, J.K. Oh, P.D. Tackaberry, and H.M. Lizama, "Determination of Activity Parameters in Thiobacillus ferrooxidans Strains as Criteria for Mineral Leaching Efficiency. I. Growth Characteristics and Effect of Metals on Growth and on Sulfur or Ferrous Iron Oxidation", Biominet Proceedings, ed. R.G.L. McCready (Ottawa, ON: CANMET SP87-10 1987), 179-209.

Enzymatic and Hydrodynamic Studies of a Liquid/Solid Fluidized-Bed Reactor for Coal Bioconversion*

Eric N. Kaufman**, Mark H. Little,
Charlene A. Woodward, and Charles D. Scott

Bioprocessing Research and Development Center
Oak Ridge National Laboratory
Oak Ridge, TN 37831-6226

Abstract

Biocatalysts allow the solubilization/liquefaction of coal at near ambient temperatures. This research has focused upon the chemical modification of enzymes to enhance their solubility and activity in organic media and on optimal reactor design for a biocatalyst coal liquefaction process. Modification of hydrogenase and cytochrome c using dinitrofluorobenzene (DNFB) or methoxypolyethylene glycol *p*-nitrophenyl carbonate (PEG-n) has effected increased solubilities up to 20 g/L in organic solvents ranging from dioxane to toluene. Use of these modified enzymes in a small fluidized-bed reactor (with H_2 sparge) resulted in greater than 40% conversion of bituminous coal in 24 h. Research using model compounds suggests that the conversion process may be in part due to splitting at methyl or ethyl bridges and perhaps saturation of ring structures. A new class of continuous columnar reactors will be necessary to achieve the high throughput and low inventory necessary for biocatalyst processes. The controlling mechanisms of particle transport in fluidized bed systems using very small coal particulates are being studied. This has included the hydrodynamic modeling of coal segregation in fluidized-bed reactors, with direct microscopic visualization using fluorescence microscopy. A summary of our previously published work on enzyme modification and reactor modeling and verification is presented.

---

* Research supported by the Advanced Research and Technology Development Program of the Office of Fossil Energy, part of which is managed by the Pittsburgh Energy Technology Center, U. S. Department of Energy under contract DE-AC05-84OR21400 with Martin Marietta Energy Systems, Inc.

** To whom all correspondence should be addressed.

Mineral Bioprocessing II
Edited by David S. Holmes and Ross W. Smith
The Minerals, Metals & Materials Society, 1995

Introduction

Advances in biotechnology are leading to novel bioprocesses for the conversion of coal to clean liquid and gaseous fuels. Coal represents our nation's most plentiful fossil energy resource and is the most economical fossil fuel on a cost-per-BTU basis *(1)*. Lower-rank coals (subbituminous and lignite) represent a substantial portion of U.S. fossil energy reserves but are underutilized due to increasing environmental concerns which require costly treatment to decrease acid rain and greenhouse gases. Coal may be converted into clean-burning liquid fuels by conventional catalytic processes. However, this type of coal liquefaction requires extremes in temperature and pressure, specific substrate ratios in order to meet the desired product spectra, and expensive catalysts which may be easily fouled *(2)*. There is increasing interest in utilizing biocatalytic systems (microorganisms or enzymes) for coal processing *(3)*. This is primarily because bioprocesses are conducted at modest temperatures and pressures in a relatively benign chemical environment and may provide a more specific product stream. Various fungi and bacteria have been shown to solubilize coal *(3, 4)*, remove pyritic sulfur *(5)*, and convert solubilized coal into liquid and gaseous fuels *(6, 7)*. A possible bioprocess schematic for coal liquefaction proposed by researchers at the Bioprocessing Research and Development Center at Oak Ridge National Laboratory is shown in Figure 1. The design and operation of such a biocatalytic process present obstacles in terms of both the chemistry of the catalytic interaction with coal and reactor design.

Due to the hydrophobic nature of coal, the chemical composition of the liquefaction product stream, and the increased penetration of coal allowed by organic solvents, it is desired to conduct the biocatalysis in an organic media. It has been demonstrated that various enzymes can be effectively used as particulate catalysts in conjunction with organic solvents for liquid/liquid interactions *(8-10)*. However, when contact with a solid substrate in an organic solvent is desired, the most efficient form of the enzyme is a homogeneous catalyst that is soluble in the organic solvent. Most proteinaceous materials have only trace solubility in anhydrous organics, especially nonpolar organic solvents *(11, 12)*. Since enzymes are naturally hydrophilic molecules, they must be chemically modified to increase their hydrophobicity, and this must be achieved without significantly decreasing their catalytic activity. It has been shown that polyethylene glycol activated with cyanuric chloride (PEG-c) can be attached to enzymes, resulting in increased organic solubilization *(13)*. Our recent work has demonstrated the use of the reagent 2,4-dinitrofluorobenzene (DNFB) to attach dinitrophenyl (DNP) groups to enzymes at their free amino groups and increase their solubility in organic solvents *(9, 10)*. DNFB and another modified polyethylene glycol, methoxypolyethylene glycol *p*-nitrophenyl carbonate (PEG-n), have been further investigated for enzyme modification with the subsequent study of conversion of coal to liquids enhanced by reducing enzymes with molecular hydrogen. The mechanisms for such a conversion have been studied by using model compounds.

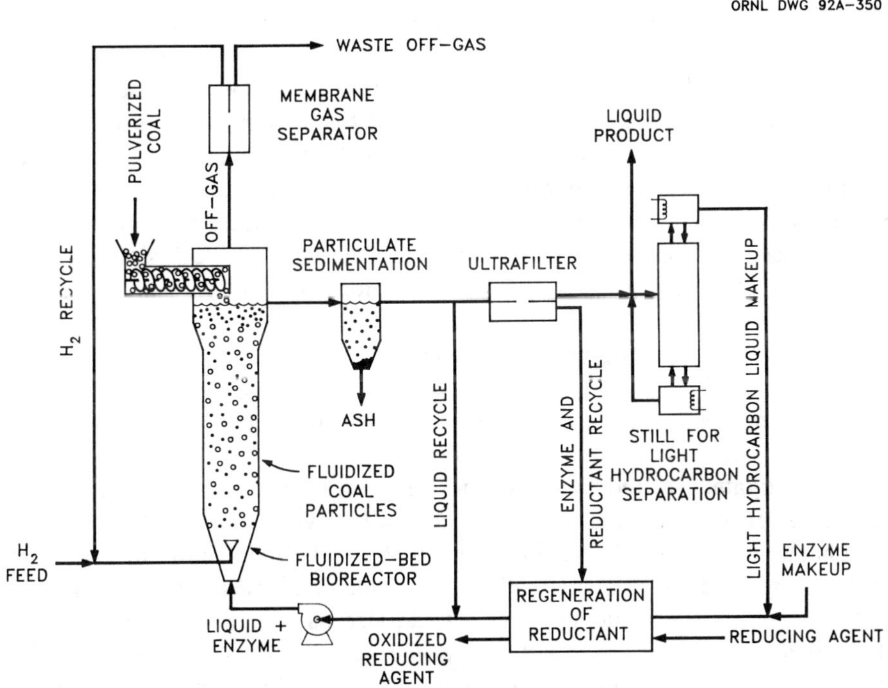

Figure 1 Possible bioprocessing system for the anaerobic solubilization of coal using modified enzymes in a process-derived liquid. Pulverized coal particles are continuously added to a liquid fluidized-bed reactor containing modified enzymes and a process-derived solvent. The reactor may be sparged with H_2. Undigested coal in the form of ash would be continuously removed from the reactor. Process liquid would be treated to recover modified enzymes from the product hydrocarbon stream. This hydrocarbon stream may be directly utilized as a liquid fuel product or may serve as the carbon source for further bioprocesses to produce other fuels or commodity chemicals. Various components of this process for coal liquefaction have already been demonstrated at the laboratory scale.

Liquid fluidized-bed reactors will be utilized in the envisioned biocatalytic process due to their efficient mass transport, ease of gas disengagement, and amenability to continuous operation. The efficient design, operation, and scaleup of such a reactor requires detailed knowledge of the reactor axial pressure drop, dispersion of the solid and liquid phases, particle size distribution as a function of axial position and flow rate, and solubilization chemistry and kinetics. While hydrodynamic issues have been well studied for conventional gas fluidized beds, little information has existed regarding liquid fluidized beds of particles with small size and density (low Reynolds number regimes). In conjunction with researchers at Washington State University, we have recently proposed a convection/diffusion model for the transient and steady state behavior of these systems *(14-16)*. This model is fully predictive and has been validated through its ability to forecast the overall bed height and transient pressure profiles caused by a step change in liquid flow rate *(17, 18)*. A fluorescence visualization method *(19, 20)* has been introduced to further validate the model's ability to predict particle segregation within the bed. Now validated on both a macroscopic and microscopic scale *(18, 20)*, this model will allow accurate design, operation, and scaleup of a fluidized-bed reactor for the bioconversion of coal.

Enzyme modification and coal conversion

Enzyme processes are catalyst-enhanced conversions that operate with reasonable specificity and at moderate operating conditions. The biocatalysts we have investigated are reducing enzymes that can utilize molecular hydrogen or other inexpensive reducing agents. Specifically, hydrogenase enzymes isolated from bacteria that are relatively oxygen stable (nickel-containing hydrogenases) are being studied for this application *(21)*. Such enzymes not only can utilize molecular H_2 as the reducing agent but can also use other artificial and natural electron acceptors and donors *(21)*. Some heme-bearing molecules, such as cytochrome c, are known to mediate the catalytic process, probably by enhancing electron transport *(12, 21)*. Reduced cytochrome c also appears to have some catalytic properties *(11, 12)*.

At least three biocatalytic approaches to the conversion of coal to liquids by reduction of the coal structure can be envisioned *(11, 12)*:

Coal + H_2 ---Hydrogenase---> Reduced Coal + H_2O (1)

Coal + H_2 ---Cytochrome c (?)---> Reduced Coal + H_2O (2)

Coal + Cytochrome c(H) --Hydrogenase-->

Reduced Coal + Cytochrome c + H_2 (3)

Equation (1) represents the oxidation of H_2 while the coal structure is being reduced, Equation (2) represents a similar reduction that may result from utilization of cytochrome c as the biocatalyst, and Equation (3) indicates that other reducing agents (such as reduced cytochrome c) may be utilized for this reaction.

Enzyme modification was carried out with either DNFB or PEG-n. Modification of the enzymes with DNFB was by a significant change in the technique originally developed by Sanger *(22)* and reported in detail elsewhere *(9)*. This technique results in the introduction of dinitrobenzene on free amino groups in the enzyme structure. Chemical modification with PEG-n was carried out by introducing the enzyme at a concentration of 5-10 mg/mL into an agitated, aqueous mixture buffered at pH 8.8 with a 0.1 M phosphate buffer that contained 30-100 mg/mL of the activated PEG. An anaerobic environment was maintained by H_2 overpressure. The reaction proceeded for 2 h, after which the pH was adjusted to 7.8. After addition of 5 mg/mL sodium dithionite to maintain reducing conditions, the solution was lyophilized, and the resulting solid material was stored under N_2 at 0^oC. In solubilization studies, the PEG-modified enzyme was contacted with the organic solvent of interest for 1 h at ambient conditions, after which the solid residue was removed by centrifugation.

Chemical modification of the enzymes has been found to increase the solubility in organic media, although severe modification may also result in loss of activity *(9)*. The modified enzymes in organic solvents enhance the conversion of coal to liquids, apparently by breaking aliphatic cross-linking and other reduction reactions. Enzymes modified with both DNFB and PEG-n have enhanced solubility in organic solvents ranging from levels as high as 20 mg/mL in relatively polar solvents, such as dioxane and pyridine, to less than 3 mg/mL in relatively nonpolar solvents, such as benzene and toluene (see *(10)*). Unfortunately, a deleterious effect on enzyme activity occurred as the level of DNFB reaction was increased, resulting in only 40% of the original at the highest level tested *(9)*. The interaction of PEG-n was slightly less effective in enhancing enzyme solubility in pyridine (~3 mg/mL) but was equally as effective in enhancing the solubility in the synthetic solvent and toluene. In addition, PEG-n was much more efficient at maintaining enzyme activity, with an order-of-magnitude increase in activity over enzymes modified by PEG-c.

The organic solvent also affects the enzyme activity, with irreversible loss especially noted in the more polar solvents. For example, even though the total quantity of modified enzymes in solution is greater in pyridine than in benzene, the remaining enzyme activity is significantly greater in the less polar organic *(23)*.

An important aspect of the conversion of coal to liquids is the cleavage of covalent bonds that make up the 3-D structure. Thermal disruption of these bonds requires relatively high temperatures, but it is anticipated that biological catalysts will allow such interactions to occur at modest operating conditions. Model compounds were chosen with methylene and/or ethylene bridges between aromatic moieties similar to those found in coal to help evaluate the mechanisms of enzyme-enhanced coal conversion. The degradation of 1,2-bis(4-pyridyl)ethane (BPE) and 1,2-bis(4-quinolyl)ethane (BQE) was studied in benzene at 30°C with 2-4 mg/mL of a mixture of DNP-hydrogenase and DNP-cytochrome c under a hydrogen atmosphere in shake flasks. Degradation of the coal model compounds and resulting products were monitored using gas chromatography (GC) on a fused silica capillary column. We found that up to 52% of these model compounds could be degraded over a 12-h period. This could indicate that the ethylene bridging was being disrupted. More complete tests were made with 1-[4-(2-phenylethyl)benzyl]naphthalene (PEBN), a more complicated model compound with both methylene and ethylene bridging (See Figure 2). Extensive GC analyses for the model compound, hypothesized reduction products of known composition, and actual products of model compound solubilization were made during the course of the enzyme-enhanced interactions (Figure 3). Previous studies *(24)* have shown that thermal degradation of this compound requires a temperature greater than 400°C, while carbon black can reduce the required temperature to 320°C. However, chemically modified biocatalysts have a measurable effect at temperatures as low as 30°C. The GC analysis of the reaction solutions indicated that the experimental reduction products included toluene, *p*-xylene, naphthalene, methylnaphthalene and bibenzyl, all of which would be expected if there was cleavage at the methylene and ethylene bridges. The presence of unidentified reduction products suggested that some interaction with the ring structures may have occurred. The two different hydrogenases produced different products, and we also observed a solvent effect with toluene producing the most products. The thermophilic hydrogenase from Pyrococcus furiosus was the most effective for interactions with PEBN. As expected, that enzyme could effectively operate at a higher temperature, with the maximum conversion occurring at 50°C (22%) but a noticeable decrease resulting at 75°C (5%). The optimum temperature for P. vulgaris hydrogenase was previously found to be 30°C *(11)*.

ORNL DWG 93A-648

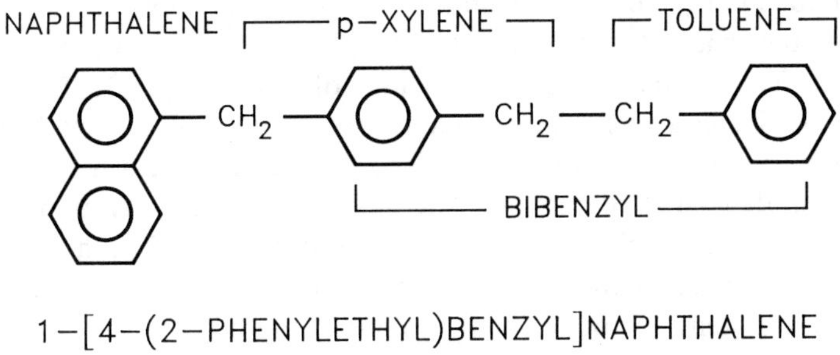

NAPHTHALENE — p-XYLENE — TOLUENE

CH_2 — CH_2 — CH_2

BIBENZYL

1-[4-(2-PHENYLETHYL)BENZYL]NAPHTHALENE

Figure 2 Structure of coal model compound 1-[4-(2-phenylethyl)benzyl]naphthalene, with both methylene and ethylene bridging between aromatic moieties similar to those found in coal. Figure adapted from *(23)*.

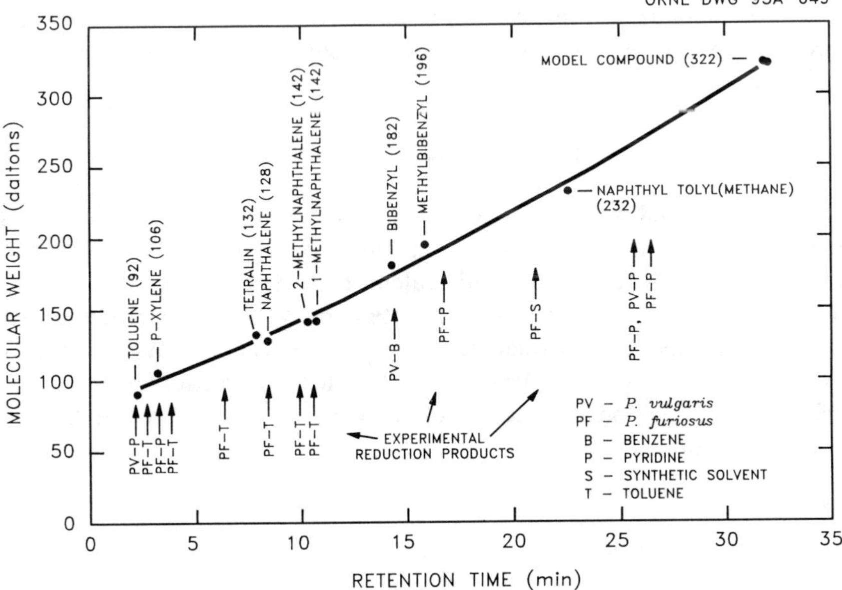

Figure 3 GC retention times of the model compound, 1-[4-(2-phenylethyl)benzyl]naphthalene, and some possible reduction products with indicated molecular weights as compared to the products of enzyme-enhanced reduction. The tests were made in shake flasks at 30°C under a hydrogen atmosphere for a 24-h period with 2-4 mg/mL of PEG-n-modified mixed enzymes (hydrogenase from P. vulgaris or P. furiosus with approximately equal amounts of cytochrome c). Figure adapted from *(23)*.

Mixed modified enzymes have been shown to enhance the conversion of coal to liquids in organic solvents under a hydrogen atmosphere *(9-12)*. We now know that the characteristics of the enzymes and the solvents affect the conversion. The increased light absorbance of the reacting fluid indicated that the enzyme-enhanced interaction occurs throughout a 24-h period but that the most rapid conversion reaction takes place during the first 8 h for experiments utilizing a single charge of enzyme. The most significant conversion of coal has been achieved in a fluidized-bed bioreactor utilizing pyridine at 30°C under a hydrogen atmosphere in which there were three additions of mixed enzymes (4 mg/mL of hydrogenase from P. vulgaris and 3 mg/mL cytochrome c modified with DNFB) at 0, 4, and 8 h *(9)*. One-half of a gram of bituminous Illinois No. 6 coal of 45 to 63 μm diameter was used in this small-scale study. The fluidized-bed reactor consisted of a 30-cm-long, truncated, jacketed cone with a 1.25-cm-diameter inlet and a 2.5-cm-diameter outlet. Hydrogen was sparged into the reactor at its outlet. In this case, the conversion of the coal was greater than 42% for the enzyme-enhanced test compared to less than 20% for the reference test. Approximately 55% of the resulting liquid product had a boiling point of less than 280°C.

Reactor design and modeling

Liquid fluidized-bed reactors will be utilized using either modified enzymes or whole microorganisms to achieve continuous coal conversion at a larger scale. The efficient design, operation, and scaleup of such a reactor require detailed knowledge of the transient and steady state expansion and segregation of coal particles within the bed. Collaborating with researchers at Washington State University *(14-18)*, we have proposed a transient convection/diffusion model which is a modification of that proposed by Kennedy and Bretton *(25)*. The governing equation is:

$$\frac{\partial C_i}{\partial t} = \frac{\partial}{\partial z}\left[D_i\left(\frac{\partial C_i}{\partial z} - \frac{C_i}{\rho}\frac{\partial \rho}{\partial z}\right) - U_i C_i\right] \quad (4)$$

where the coal charge is divided into classes (denoted by subscript "*i*") based upon size or density, and D_i is the dispersion coefficient of the class, U_i its segregation velocity, and C_i its fractional volumetric concentration. Time is represented by "*t*", ρ is the liquid density, and "*z*" is the axial position within the reactor. Equation (4) is written for each subclass "*i*", resulting in a system of coupled, partial differential equations. The boundary conditions are that there is no flux at the base of the column, where there exists a glass frit liquid distributor, and that there is no flux at the top of the column. The boundary condition at the top of the column may be modified to account for particle elutriation if desired. Key to solving this system of equations is the prediction of the dispersion coefficients and segregation velocities as functions of particle size, density, and void fraction and liquid superficial velocity, density, and viscosity. We have achieved this using the Richardson-Zaki relationship for the segregation velocity and our own correlation for the dispersion coefficient *(15, 17)*. With these correlations, we have a fully predictive model (with no adjustable parameters) which results in a system of equations which must be solved numerically for a given set of operating conditions. We have recently expanded our model to account for coal solubilization *(16)* and can now optimize for the size and rate of coal to be fed to the reactor.

We have conducted a series of experiments measuring bed heights, steady state and transient axial pressure profiles, and particle size distributions in order to verify the predictions of our model for a nonreacting system *(17, 20)*. We have introduced and validated a fluorescence visualization technique, which allows bed segregation data to be obtained directly and noninvasively in an operational reactor *(19, 20)*. The proposed method, by causing the continuous phase to fluoresce and observing the system using epi-illumination fluorescence microscopy, increases the contrast between the dispersed and continuous phases so that individual coal particles may be recorded and sized using video image digitization and analysis. This method enables direct visualization of discrete particulate entities at any volume fraction including close-packed. The microscope may be positioned at any axial location, allowing monitoring of the entire reactor. Further, the technique is noninvasive and does not require specialized reactor systems. By not altering the dispersed phase through direct labeling, and without appreciably altering the chemistry of the continuous phase, this technique may be used in situ in actual operating reactors.

Experiments conducted to verify our mathematical model were performed in a 4-ft-tall, 1-in-ID glass column with taps attached to pressure transducers in order to measure the mass of coal lying between given axial positions. Illinois No. 6 coal was wet-sieved into the desired size fractions ranging from 45 to 150 μm in diameter. Figure 4 represents the measured and model predictions for the concentration of 45 to 63 μm coal solids lying between various pressure ports as the liquid superficial velocity fluidizing the coal was stepped from 0.0057 to 0.0037 and then to 0.0109 cm/s. As can be seen, our model accurately predicts solids holdup within the reactor under a variety of operational conditions.

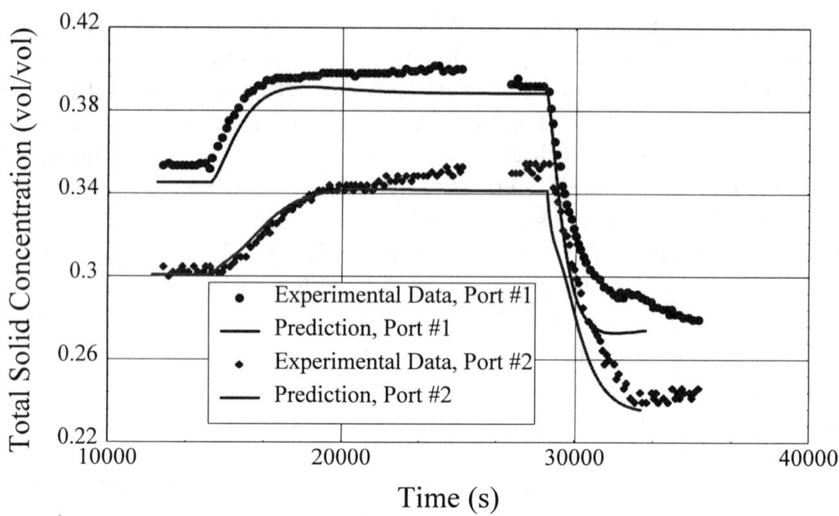

Figure 4 Model verification through its ability to predict transients in solids concentration in response to step changes in liquid velocity. One hundred grams of 45 to 63-μm coal were fluidized, and the weight of coal lying between various axial locations within the reactor was measured using pressure transducers. The liquid velocity was first lowered from 0.0057 cm/s to 0.0037 cm/s at 14,500 s and then raised to 0.0109 cm/s at 28,900 s. Our mathematical model of coal expansion and segregation accurately predicts the behavior of the fluidized-bed with no adjustable parameters. Figure adapted from *(17)*.

To study the ability to predict reactor performance on a microscopic scale, we have performed fluorescence visualization with a wide variety of coal types, sizes, and flow rates. Figure 5 demonstrates our ability to noninvasively measure particle size distribution as a function of axial position in a liquid fluidized bed of coal and our ability to predict this microscopic distribution. This size distribution was taken using a nonreacting bed of 40 g of 45 to 63 μm and 80 g of 106 to 150 μm Illinois No. 6 coal at a liquid superficial velocities between 0.011 and 0.019 cm/s. The top row of size distribution contours represents the experimental data, whereas the bottom row represents our model predictions. Research has demonstrated that our models are capable of predicting this microscopic behavior with no adjustable parameters *(18, 20)*. With this ability to accurately forecast reactor performance on both a macroscopic and microscopic scale, we expect to design, operate, and scale up a fluidized-bed bioreactor for coal solubilization.

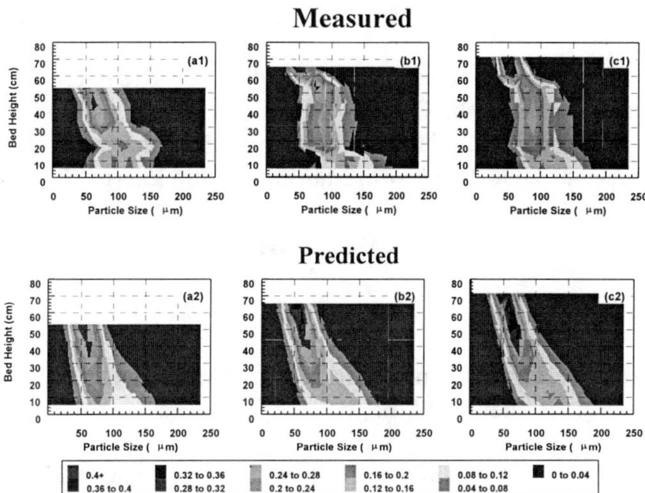

Figure 5 In situ measurement of coal size distribution in a fluidized-bed using fluorescence visualization and comparison to model predictions. Forty grams of 45 to 63 and 80 g of 106 to 150 μm coal were fluidized in our reactor at a superficial liquid velocity of 0.011 (a), 0.016 (b), and 0.019 (c) cm/s. In situ measurements of the coal size distribution as a function of axial location were performed as described previously *(19, 20)*. The patterns of coal size segregation as a function of position and flow rate were well predicted by our mathematical model *(18)*.

Conclusions

Techniques for chemically modifying enzymes to increase their solubility and activity in organic media have been introduced for the purposes of biologically mediated coal conversion. Conversion of both model compounds and bituminous coal has been demonstrated at the bench-top scale. A mathematical model of fluidized-bed expansion and segregation has been introduced to allow the design and operation of a larger-scale process. This model has been validated at the macroscopic level by its ability to accurately predict overall bed heights and transient pressure profiles measurements with no adjustable parameters and has also been validated at the microscopic level using fluorescence visualization of particle segregation. Continuing research is investigating the mutation of hydrogenase-producing bacteria to enable their use in organic environments and on staged reactor systems for the solubilization of coal followed by the fermentation of this liquid product to clean liquid and gaseous fuels.

Acknowledgments

This work was supported by the Advanced Research and Technology Development Program of the Office of Fossil Energy, part of which is managed by the Pittsburgh Energy Technology Center, U. S. Department of Energy under contract DE-AC05-84OR21400 with Martin Marietta Energy Systems, Inc.

References

1. Energy Information Administration, U.S. Department of Energy, Annual Energy Review (1989).

2. Haynes, H. W.; Borgialli, R. R.; Zhang, T., Energy & Fuels 1991, 5, 63.

3. Scott, C. D.; Lewis, S. N., in *Bioprocessing and Biotreatment of Coal* D. L. Wise, Eds. (Marcel Dekker, Inc., New York, 1991) pp. 275.

4. Ward, B., Biotechnology Techniques 1993, 7, 213.

5. Eligwe, C. A., Fuel 1988, 67, 451.

6. Davison, B. H.; Nicklaus, D. M.; Misra, A.; Lewis, S. N.; Faison, B. D., Applied Biochemistry and Biotechnology 1990, 24, 447.

7. Isbister, J. D.; Barik, S., in *Biotransformation of Low Rank Coals* D. C. Crawford, Eds. (CRC Press, Boca Raton, FL, 1993).

8. Laane, C.; Tramper, J.; Lilly, M. D., Eds., *Biocatalysis in Organic Media* (Elsevier, Amsterdam, 1987).

9. Scott, C. D.; Scott, T. C.; Woodward, C. A., Applied Biochemistry and Biotechnology 1993, 39/40, 279.

10. Scott, C. D.; Scott, T. C.; Woodward, C. A., Fuel 1993, 72, 1695.

11. Scott, C. D.; Lewis, S. N., Applied Biochemistry and Biotechnology 1988, 17/18, 403.

12. Scott, C. D.; Woodward, C. A.; Thompson, J. E.; Blankinship, S. L., Applied Biochemistry and Biotechnology 1990, 24/25, 799.

13. Takahashi, K.; Nishimura, H.; Yoshimoto, T.; Inada, Y., Biochem. Biophys. Res. Commun. 1985, 121, 261.

14. Asif, M.; Petersen, J. N.; Scott, T. C.; Cosgrove, J. M., Applied Biochemistry and Biotechnology 1993, 39/40, 535.

15. Asif, M.; Petersen, J. N., AIChE J. 1993, 39, 1465.

16. Wang, Y.; Petersen, J. N.; Kaufman, E. N., Appl. Biochem. Biotech. 1994, Submitted for Publication.

17. Asif, M.; Petersen, J. N.; Kaufman, E. N.; Cosgrove, J. M.; Scott, T. C., Ind. Eng. Chem. Res. 1994, In Press.

18. Wang, Y.; Petersen, J. N.; Kaufman, E. N., Chem. Eng. Sci. 1994, Submitted for Publication.

19. Kaufman, E. N.; Scott, T. C., Powder Technology 1994, 78, 239.

20. Kaufman, E. N.; Little, M. H.; Petersen, J. N., Chemical Engineering Science 1994, Submitted for Publication.

21. Przybyla, A. E.; Robbins, J.; Menon, N.; Peck, H. D., FEMS Microbiology Reviews 1992, 88, 109.

22. Laane, C.; Boeren, S.; Hihorst, R.; Veeger, C., in *Biocatalysis in Organic Media* C. Laane, J. Tramper, M. D. Lilly, Eds. (Elsevier, Amsterdam, 1987) pp. 65.

23. Scott, C. D.; Woodward, C. A.; Scott, T. C., Fuel Processing Technology 1994, In Press.

24. Farcasiu, M.; Smith, C., Energy and Fuels 1991, 5, 83.

25. Kennedy, S. C.; Bretton, R. H., AIChE. J. 1966, 12, 24.

VI.

MICROORGANISMS IN MINERAL PROCESSING OPERATIONS AND REMEDIATION

MICROORGANISMS IN MINERAL PROCESSING

R. W. Smith, M. Misra and A. Raichur

University of Nevada, Reno, NV 89557, USA

Abstract

The study of using microorganisms in place of mineral processing chemical reagents (flocculants, flotation modifiers and flotation collectors) is a recent development. Potential uses for microorganisms include their use in place of certain conventional flotation reagents and as flocculating agents in minerals engineering processes. In addition, in some studies microorganisms have been used to modify mineral surfaces and, thus, the conventional flotation or flocculation of the minerals. Some of the recent studies on such use of microorganisms and their derivatives are reviewed in this paper.

Introduction

Most microorganisms tend to have acidic isoelectric points (iep's) since with most organisms, anionic groups, particularly carboxylate groups, are more common on the surface than are cationic groups. Thus, most microorganisms, in the absence of adsorbing cations, will be negatively charged in aqueous solution except at quite acidic pH values [1,2]. There is, of course, some variation in the pH of organisms' iep values and there even are a few alkylophilic organisms such as *Bacillus firmus* that may possess neutral or even basic iep values. If the microorganisms are ruptured the derived cell fragments possess a more positive surface. Figure 1 illustrates the typical variation of the zeta potential as a function of pH for the hydrophobic microorganism *Mycobacterium phlei* and cell fragments derived from rupture of the organism. Note that the organism is highly negatively charged at mildly acidic to basic pH values. If heavy metal ions are present, adsorption of the metal ions onto a microorganism can reduce the charge on the organism or even, in some cases, the charge can be reversed from negative to positive. For example, the presence of 0.05 mM Cr(III) reverses the charge on *Bacillus subtilis* between pH 5 and pH 7.5 and 0.1 mM Cu(II) reverses the charge on the organism between pH 6 and pH 9 [3].

The hydrophobicity of the surfaces of microorganisms varies wildly depending on the proportion of fatty acid groups on the organism's surface to the various other functional groups and hydrophilic sections of the surface. Concerning planktonic bacteria, measured contact angles can vary at least from 15 to 70 degrees [4,5]. Unicellular algae, fungi and yeasts may

Mineral Bioprocessing II
Edited by David S. Holmes and Ross W. Smith
The Minerals, Metals & Materials Society, 1995

exhibit contact angles of similar magnitude. In general, it is likely that highly hydrophobic microorganisms such as *M. phlei* will only remain free floating in water if they are also highly charged.

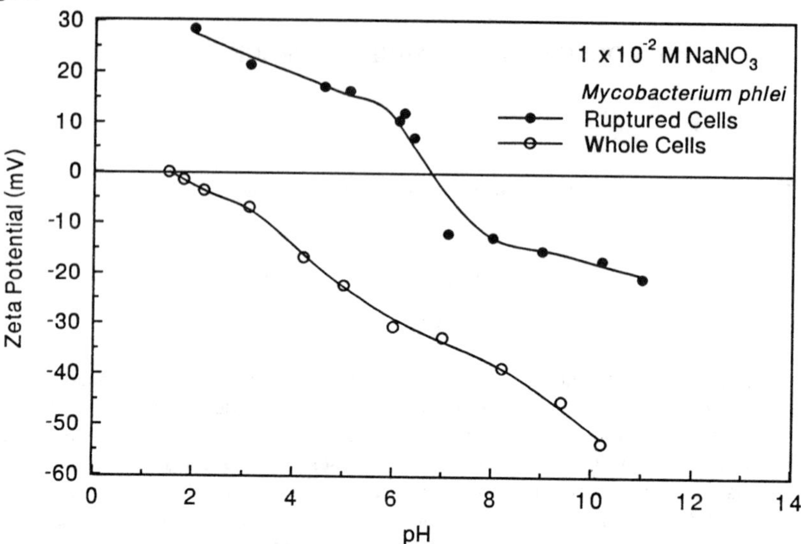

Figure 1 Zeta potential of whole and ruptured *Mycobacterium phlei* cells as a function of pH.

For microorganisms or products derived from them (cell fragments, soluble cell material) to function as flocculating or flotation reagents for minerals there must be adhesion of the organism or its derivatives to the mineral surfaces [6,7]. Even if the function may be simply to oxidize sulfide surfaces, some contact organism - sulfide mineral is likely.

Microorganisms will adhere to solid surfaces if the charge and hydrophobic interactions between the organism and the solid surface are conducive to adhesion [4,5]. In addition some microorganisms are likely to bind to solid surfaces if there is some specific binding between sites on the microorganism and the solid such as binding between an organism and iron sites on the organism [2] or, in the case of gram negative bacteria, there is a environmental gain for the bacteria to colonize on a surface [8]. In the latter case the organisms may colonize on the surface and form a biofilm that is held together by polysachharides produced by the bacteria.

Flocculation of Mineral Suspensions by Microorganisms

The possibility of using microorganisms (bacteria and fungi) as flocculants for Florida phosphatic clays was investigated as early as 1963 [9]. Subsequent work by various researchers through the mid 1980's, investigated the use of other bacteria, fungi, cyanobacteria as flocculants for various mineral suspensions in addition to phosphatic clays. For example, Bernstein [10] reported on the flocculation of organic and inorganic solid wastes using the products of lysed bacteria cells.

Recent studies [11-12] have included studies on the flocculation of fine hematite, calcite, and kaolinite suspensions using the yeast *Candida parapsilosus* and derivatives of the organism as flocculants. Figure 2 shows, for calcite, some of the results obtained. In the figures the flocculation efficiencies were calculated from the relationship:

$$E(\%) = 100 \frac{C_i - C_f}{C_i}$$

Where E(%) is flocculation efficiency, C_i is the initial concentration of solids at a depth of 15 cm down from the top of the water surface and C_f is the concentration of solids at the same depth at a set time.

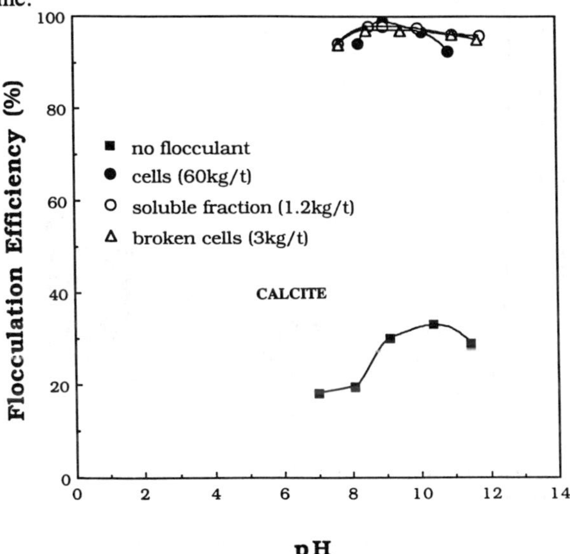

Figure 2 Flocculation efficiency for calcite suspensions by *Candida parapsilosis* and its derivatives as a function of pH for whole cells, cell fragments and the soluble cell fraction; 2 wt % solids, 0.75 min settling time, height = 15 cm down from the water surface.

It can be noted that with the cells and their derivatives excellent flocculation of the calcite can be obtained. However, a much greater weight of whole cells is required for equivalent flocculation as compared to the cell fragments and the soluble fraction of the cells. It is assumed that the soluble fraction consists of proteins, fatty acids and amphoteric surfactants.

Additional flocculation work was performed on the flocculation of fine hematite suspensions using *M. phlei* as the flocculant [13,14]. The microorganism is an excellent flocculant for hematite. This work was extended to studying the effect of the organism on the filterability of the settled suspension. When the organism was present filtration rate increased significantly.

Experimental work on mineral suspension flocculation was extended to the selective flocculation of coal from pyrite and ash using whole cells and cell fragments of *Mycobacterium phlei* as flocculants. Figure 3 shows some preliminary results on the specific flocculation of Illinois no. 6 coal.

Further work to date on this system has included contact angle and adsorption work on several coals and pyrite. Figure 4 illustrates the adhesion of *M. phlei* cells onto Illinois no. 6 coal and the effect of *M. phlei* cells and cell fragments on the contact angle on this coal, all as a function of pH. Figure 5 shows similar data on the adhesion of *M. phlei* onto Illinois coal pyrite and the resulting effect on contact angle on the mineral.

The two figures illustrate the substantial increase in hydrophobicity of this coal near pH 4 and the much less enhancement of the hydrophobicity of pyrite at the same pH value.

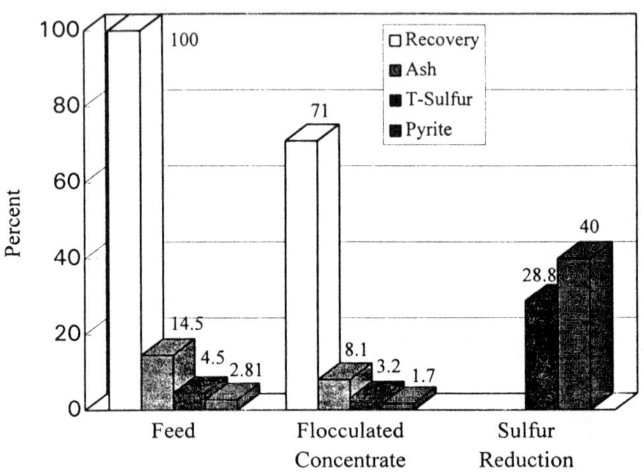

Figure 3 Selective flocculation of Illinois #6 coal with *Mycobacterium phlei*; pH 4.5; 100 ppm *M. phlei*; T-sulfur = total sulfur.

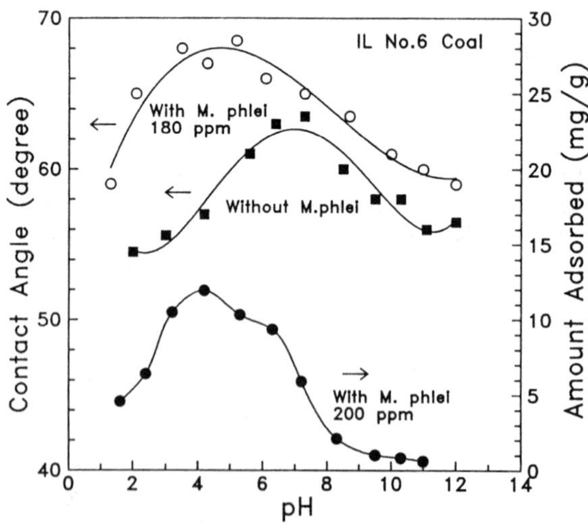

Figure 4 The effect of adsorption of *Mycobacterium phlei* cells on the contact angle on Illinois no. 6 coal; both shown as a function of pH.

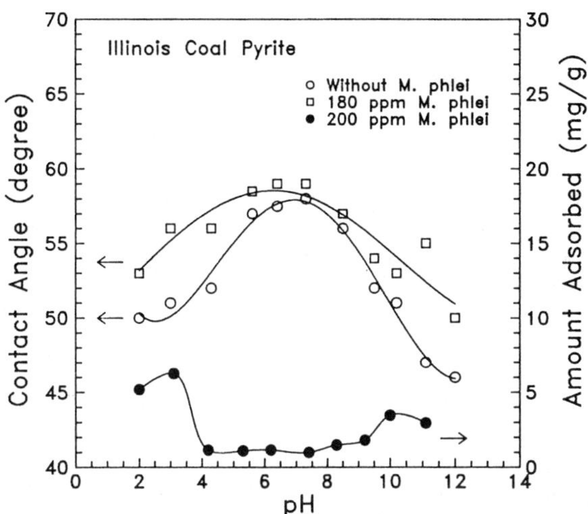

Figure 5 The effect of adsorption of *Mycobacterium phlei* cells on the contact angle on Illinois coal pyrite; both shown as a function of pH.

Yet other work was performed on the flocculation of Florida phosphatic clays using *M. phlei* as the flocculant [7]. It was found that the organism functioned as an excellent flocculant for dolomitic phosphatic clays but did not function very well as such for non dolomitic phosphatic clays. It was concluded that if the organism's cells were smaller than most of the mineral particles the organism would work well but not if they were larger than the mineral particles.

Mineral Flotation Using Bacterial Cells as Collectors

Recent work has shown that *M. phlei* will function as a flotation collector for hematite [13-15]. Figure 6 illustrates the contact angle on hematite as a function of pH in the presence of *M. phlei*. Figure 7 illustrates flotation of the mineral as a function of pH in the presence of the microorganism. These preliminary findings show that the organism can readily increase the hydrophobicity of a mineral and also demonstrate the effectiveness of the microorganism as a flotation collector. As with most all collectors, the effectiveness is a strong function of pH.

Modulation of Flotation by Microorganisms

Preconditioning with various microorganisms such as *Thiobacillus ferrooxidans* is known to cause depression of pyrite in coal flotation [16-18]. Also, contact angle on pyrite at the air water interface has been found to be influenced by bacterial conditioning with sulfide oxidizing bacteria [19]. In addition to *T. ferrooxidans*, a sulfate reducing bacterium, desulfovibro, is known to influence the floatability of sulfide minerals [20]. Furthermore, secreted metabolites from microorganisms could also influence the floatability of minerals.

Function of the microorganisms can be via surface oxidation of the pyrite leading to a decreased floatability or by adhering to the mineral and creating a less hydrophobic surface if the organism's surface is less hydrophobic than the mineral's. Evidence for the latter is to be found in the paper by Kawatra et al. [18].

A recent interesting study [19] has shown how pretreatment of sphalerite and galena with *T. ferrooxidans* can enhance the selective flotation separation of the two minerals.

Summary

It is evident from the data presented that microorganisms and products derived from the rupture of the organisms can often function in the place of conventional flotation and flocculation chemicals. The organisms cells and their derivatives function by adhering to a mineral surface creating a more hydrophobic or a more hydrophilic surface, or by altering the surface hydrophobicity through partly oxidizing the surface or by altering the surface in such a manner such that subsequent adsorption of reagents is altered.

The economics of such uses of microorganisms is not known. Although the organisms can often readily be cultured there will be a cost associated with this culturing, especially nutrient costs. Thus, it must be possible to easily culture the organisms using low cost carbon sources, etc. In some cases there may be costs associated with rupturing or otherwise manipulating the biomass. It may sometimes be difficult to obtain a sufficient weight of biomass for a large scale process. One attractive possibility would be to obtain a waste biomass from some other industrial process. At any rate the ultimate industrial use of microorganisms in the place of conventional reagents will depend on finding microorganisms or their products that are superior in their function and inexpensive. Considering the very little work that has been performed to date in such use of microorganisms and some of the interesting preliminary findings, there exists some optimism about ultimate commercialization of microorganism assisted flotation and flocculation.

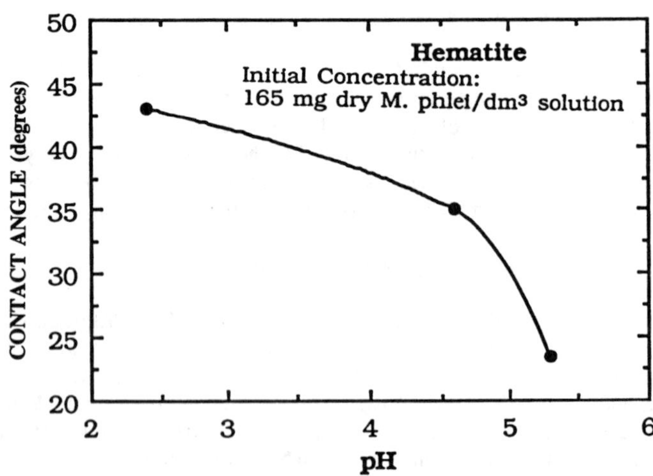

Figure 6 Contact angle on hematite in the presence of *Mycobacterium phlei* as a function of pH (from Smith et al. [15])

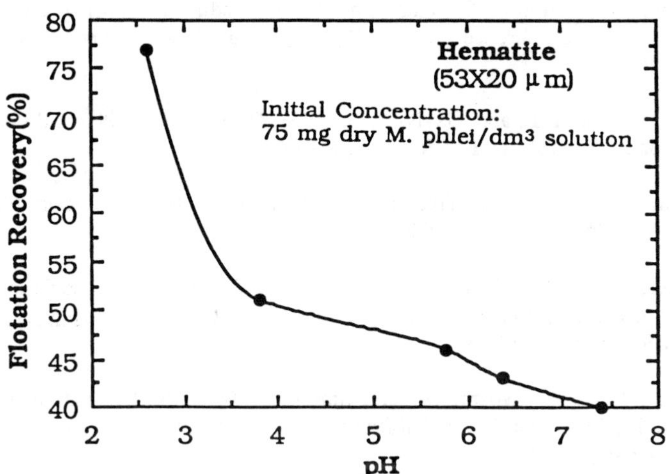

Figure 7 The effect of *Mycobacterium phlei* on the flotation of hematite as a function of pH (from Smith, et al. [15]).

References

1. Daniels, S. L., "Separation of Bacteria by Adsorption onto Ion Exchange Resins", (PhD Dissertation, University of Michigan, Ann Arbor, 1967).

2. Smith, R. W. and Misra, M., " Recent Developments in the Bioprocessing of Minerals", Mineral Processing and Extractive Metallurgy Reviews, 12, (1993), 37-60.

3. Collins, Y. E. and Stotzky, "Heavy Metals Alter the Electrokinetic Properties of Bacteria, Yeasts and Clay Minerals", Applied and Environmental Microbiology, 58, (1992), 1592-1600.

4. van Loosdrecht, M. C. M., Lyklema, J., Norde, W., Shraa, G. and Zehnder, "The Role of Bacterial Cell Wall Hydrophobicity in Adhesion", Applied and Environmental Microbiology, 53, (1987), 1893-1897.

5. van Loosdrecht, M. C. M., Lyklema, J., Norde, W., Shraa, G. and Zehnder, "Electrophoretic Mobility and Hydrophobicity as a Measure to Predict the Initial Step of Bacterial Adhesion", Applied and Environmental Microbiology, 53, (1987), 1898-1901.

6. Schneider, I. A. H., Misra, M. and Smith, R. W., "Flocculation of Mineral Fines with Ruptured Organisms", Biohydrometallurgy Technologies, TMS, (1993), 197-207.

7. Smith, R. W., Misra, M. and Chen, S., "Hydrophobic Bacteria as Flocculating Agents for Mineral Suspensions", Dispersion and Aggregation: Fundamentals and Applications, Engineering Foundation, (1994), 499-506.

8. Costerton, W., "Biofilm Structure and Metal Binding", (Paper presented at the Engineering Foundation Conference Minerals Bioprocessing (II) held at Snowbird, Utah, July 10-15, 1994).

9. Gary, J. H., Feld, I. L. and Davis, E. G., 1963, "Chemical and Physical Beneficiation of Florida Phosphate Slimes", USBM RI 6163, (1963), 33-34.

10. Bernstein, R. A., "Waste Treatment with Microbial Nucleo-Protein Flocculating Agent", US Patent No. 3,684,706 (1972).

11. Schneider, I. A. H., Misra, M. and Smith, R. W., "Bioflocculation of Fine Suspensions by *Candida parapsilosis* and Its Sonication Products", Reagents for Better Metallurgy, TMS, (1994), 197-208.

12. Schneider, I. A. H., Misra, M. and Smith, R. W., "Bioflocculation of Hematite Suspensions with Products from Yeast Cell Rupture", to be published in Chemical Engineering and Mineral Processing, (1994).

13. Smith, R. W. and Misra, M., "Mineral Bioprocessing: An Overview", Mineral Bioprocessing, TMS, (1991), 3-26.
293-301.

14. Dubel, J., Smith, R.W., Misra, M. and Chen, S., "Microorganisms as Chemical Reagents: the Hematite System", Minerals Engineering, 5, (1992), 547-556

15. Smith, R. W., Misra, M. and Chen, S., "Adsorption of a Hydrophobic Bacterium onto Hematite: Implications in the Froth Flotation of Hematite", Journal of the Society for Industrial Microbiology, 11, (1993), 63-67.

16. Attia, Y. A. and ElZeky, "Biosurface Modification in the Separation of Pyrite from Coal by Froth Flotation", Processing and Utilization of High Sulfur Coals, (1985), 673-682.

17. ElZeky, M. A. and Attia, Y. A., "Coal Slurries Desulfurization by Flotation Using Thiophilic Bacteria", Coal Preparation, 5, (1987), 15-37.

18. Kawatra, S. K., Eisele, T. C. and Bagley, S. T., "Studies of Pyrite Depression Dissolution in Pachuca Tanks and Depression of Pyrite Flotation by Bacteria", Biotechnology in Minerals and Metal Processing, SME, (1989), 55-61

19. Yelloji Rao, M. K., Natarajan, K. A. and Somasundaran, P., "Effect of Bacterial Conditioning of Sphalerite and Galena with *Thiobacillus ferrooxidans* on Their Floatability", Mineral Bioprocessing, TMS, (1991), 105-120.

20. Chen, C-Y. and Skidmore, D. R., "Adsorption of *Sulfolobus acidocaldarius* on Coal Particles", Flocculation in Biotechnology and Separation Systems, Elsevier Science, Publ., (1987), 415-426.

THE EFFECT OF SURFACTANT AND METALLIC IONS ON THE ADHESION OF *Streptomyces pilosus* CELLS ON THE MINERAL SURFACES

Zygmunt SADOWSKI

Technical University of Wroclaw, Institute of Inorganic Chemistry and Metallurgy of Rare Elements, 50-370 Wroclaw, POLAND

Abstract

The present study was undertaken for the investigation of the adhesion of *Streptomyces pilosus* cells to the mineral-liquid interface. Also, the effect of the accumulation of lead ions on the microbial cell adhesion was determined. A thin-layer wicking technique was used to measure the penetration rates of liquids (n-heptane, formamide and water). The polar components of the surface free energy of the microorganism cells were determined. The values of the free energy components of the *Streptomyces pilosus* cell surface show that the adsorption of lead ions causes a decrease of the free energy. The adsorption data show that an increase in the surface hydrophobicity of the calcite and barite particles caused the enhancement of the Streptomyces pilosus cells attachment. Accumulation of the lead ions by microbial cells causes an increase of the electrostatic interaction between the mineral surface and *Streptomyces pilosus* cells.

Mineral Bioprocessing II
Edited by David S. Holmes and Ross W. Smith
The Minerals, Metals & Materials Society, 1995

Introduction

The environmental problems associated with toxic metallic ions can be solved by the application of some microorganisms which have the ability to accumulate these ions. Initially, most work has been concentrated on the strain of *Streptomyces pilosus* as a candidate for the accumulation of ions [1-4]. Results show that these microbial cells can be adapted for the accumulation of toxic metal ions from solutions.

Another factor in determining an application for this microbial cell technique is the separation of cells which have accumulated ions from the natural environment. Several separation methods have been tested. For example:

i. flotation of cells with sodium oleate as a collector [5]
ii. selective flocculation by anionic flocculants [6]

However, the results obtained by using these separation methods indicated that both techniques, although available, need improvement.

The carrier separation method seems to be the method which allows better control of the conditions of separation [7]. The studies examining this new method should concentrate on the adhesion of the microorganism cells to the mineral surfaces which will act as carrier for these cells. The attachment of microorganism cells to the mineral-liquid interface should depend on the physico-chemical properties of both the microorganism cell-solution and the mineral-solution interfaces.

The thermodynamic model of the cell adhesion to the solid surface has been established [8-10]. According to this model, the cell adhesion energy can be calculated from the balance of interfacial energies.

$$G_{adh} = G_{C/S} - G_{C/L} - G_{S/L}$$

where: $G_{C/S}$ is the excess free energy per unit of the interface area (cell/solid),
$G_{C/L}$ is the excess free energy per unit of the interface area (cell/liquid),
$G_{S/L}$ is the excess free energy per unit of the interface area (solid/liquid).

The adhesion of the microorganism cells is favored if the adhesion energy has a negative value. The present study examined the use of the colloid approach for the establishment of a correlation between the cell adhesion and the mineral surface properties. The experimental data was obtained by using two colloidal techniques. The adhesion behavior of the microorganism cells to both the barite and calcite surfaces has been tested by the application of equipment similar to to that used by Clayfield and Lumb [11] for the fine particle adhesion test.

The thin-layer wicking technique has been developed by Chibowski [12,13] for the free energy of the mineral surface investigation.

Materials and Methods

Streptomyces pilosus strain RIA 1078 was obtained from the Institute of Immunology and Experimental Therapy (PAN) in Wroclaw, (Poland). The microorganisms were grown in a culture medium containing 10 g of peptone, 2 g of hydrated casein, 2 g of extracted yeast, 6 g of NaCl and 1000 ml of distilled water. The cells were always harvested after 48 hours of incubation at a temperature equal to 28 C. The pH of the solution was 7.0. After harvesting, the microorganism cells were separated by using a centrifuge at 4000xG for 10 minutes. One part of these microorganism cells was used for the biosorption of the metal ions experiment. The bioaccumulation procedure of metal ions has been described in the following papers [1-4].

The mineral samples (barite and calcite) used in the adhesion studies were highly pure samples supplied by WARD'S Natural Science Co. These mineral samples were dry ground in an agate mortar and screened. A fraction of -100 + 80 μm of both minerals was used for the adhesion studies.

The thin layer wicking experiment

The thin-layer wicking experiment provides a rapid and simple way to estimate the components of the free energy of the solid-liquid interface [12,13]. A Plexiglass chamber with a rectangular cross-section design was constructed. The test plate was covered with the microorganism film and was placed in a horizontal position in the chamber (Fig. 1).

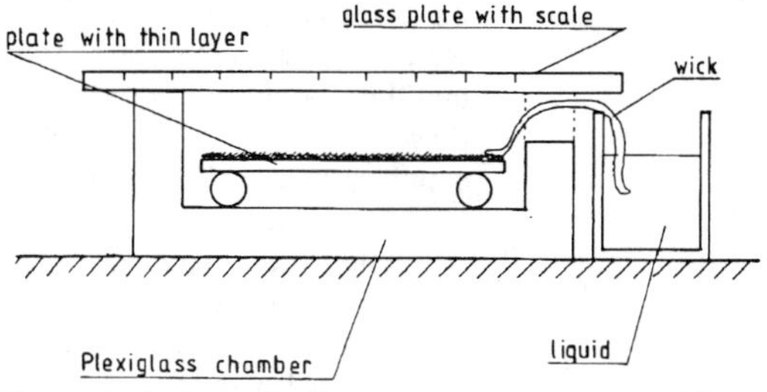

Figure 1 - Schematic diagram of equipment used for the wick experiments

The lid was a glass plate, so that the moving boundary of the liquid could be observed. On the glass plate the centimeter scale was marked. The tested liquid was transported with a vessel to the porous layer by a cotton wick 3 cm long. The surface of the liquid was at the same level as the plate. One end of the cotton wick was dipped into the liquid (n-heptane) and the other was touched to the edge of the porous layer. The viscosity of n-heptane was taken from the literature [13]; it was 0.409 cP at 20 C. Some of the plates were precontacted with the saturated vapor using n-heptane. For the polar components of the surface free energy determination distilled water and formamide (Reachim -Germany) were used.

Adhesion measurement

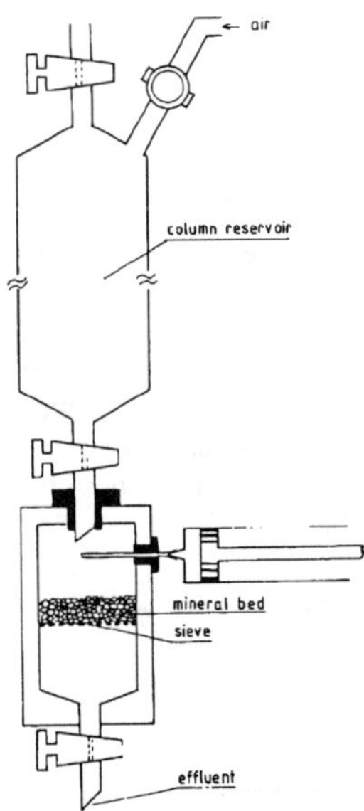

Figure 2 - Experimental equipment employed by Clayfield and Lumb [11].

Adhesion measurements were carried out according to the method which has been previously used by Clayfiel and Lumb [11] for a study of the adhesion of fine particles to the carrier surface. The equipment used in this study is shown in Fig. 2.

It consists of two main sections: a reservoir which contained a supporting electrolyte solution (0.9 % of NaCl) and a Plexiglass chamber. This chamber was equipped with a rubber inlet for the microorganism suspension's introduction. Inside, the chamber was divided into two parts. The plastic porous screen (40 μm) was used as a wall separating these two parts of the chamber. The mineral particles were placed on the screen. These particles were conditioned with the surfactant solution over 24 hours.

The electrolyte (250 ml) was placed in to the top reservoir. The flow of this solution was controlled and stabilized through the introduction of pressurized air to the reservoir. When the flow of the electrolyte was stabilized, the suspension of microorganism cells was introduced to the chamber. The time filtration of the electrolyte was noted. The concentration of the microorganism cells in the obtained solution after filtration through the particles cake was measured using a spectrophotometer. These spectrometer measurements were done in the range of 500 nm. The adhesion experiments were carried out at room temperature.

Results

The experiments using a thin-layer of microorganism cells were conducted on the cells without adsorption of Pb^{+2} ions. The first tests were done to determine the mean size (R) of the porous layer. The penetration rates of n-heptane on both the precontacted places with the n-heptane vapor and the bare surface of microbial cells are presented in Fig. 3.

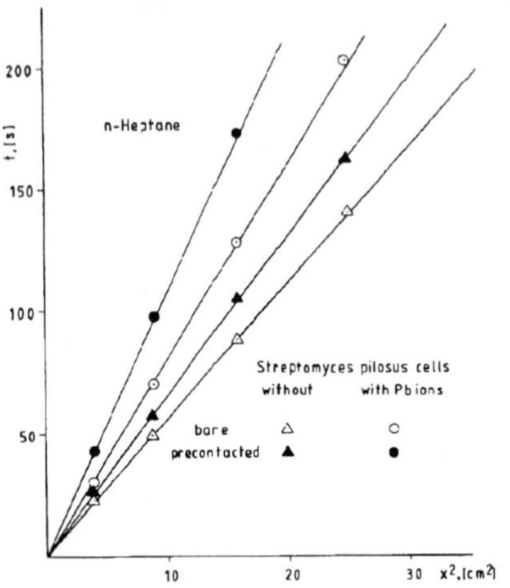

Figure 3 - A relationship between the time of n-Heptane wicking and the distance squred for Streptomyces p. cells, covered glaas plates.

As can be seen, straight-line relationships were found in these studies. This means that the Washborn's equation holds for these systems. The radius (r) of the capillary in the Washburn's equation must be replaced by an effective radius (R) of the porous material. The R values were determined from n-heptane penetration experiments using a plate precontacted with the n-heptane vapor to cover its surface with a duplex film. In this case the ΔG is equal to the surface tension of n-heptane.

Then, using the results from Fig 3, the apolar component (γLW_S) of the *Streptomyces pilosus* cell surface can be calculated. For these calculations the bare surface slide data (open points) were used. The average values of both the parameters R and (γLW_S) are given in Table I.

Table I Calculated Values of the Free Energy Components (in mJ/m²) of *Streptomyces pilosus* Cells Surface

Cell	R 10^{-5} cm	γ_s^{TOT}	γ_s^{LW}	γ_s^+	γ_s^-
Without lead	6.17	80.6	52.55	8.62	22.82
With lead	3.87	57.22	55.95	0.03	53.78

Next, using the same thin-layer of the microorganism's slides the experiments were carried on to find both the electron donor and the electron acceptor components. For these reasons water and formamid were used as testing liquids. The literature data including both the surface tension and the free energy components using liquids are collected in Table II.

Table II Surface Tension Components (mN/m) of the Liquids Used for Determination of *Streptomyces pilosus* Free Energy Components

Liquid	γ_s^{TOT}	γ_s^{LW}	γ_s^+	γ_s^-
Heptane	20.3	20.3	-	-
Formamid	58.8	39.0	2.28	39.6
Water	72.8	21.8	25.5	25.5

To determine the γ_s^+ and γ_s^- components for the microbial cells the wicking experiments were carried out on both the bare and precontacted surface plates with both polar liquids. Fig. 4 shows the results obtained for water and formamide tests on the plate which contains the *Streptomyces pilosus* cells which had been contacted with Pb ions. In Fig. 5 the wicking results obtained from both these liquids are presented. The plate which was used has been covered by the microbial cells which had not adsorbed the lead ions.

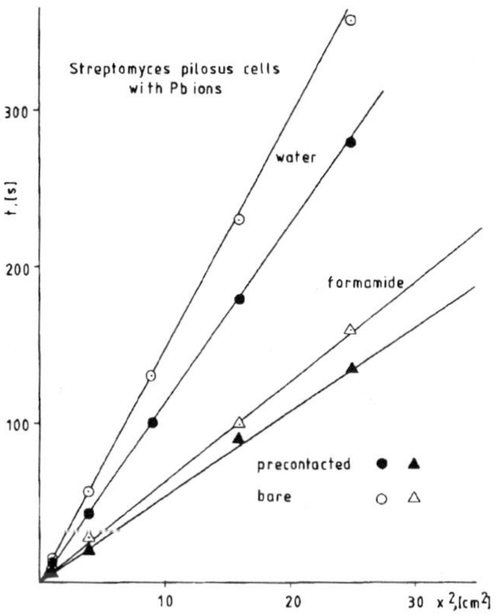

Figure 4 - Time of both water and formamide wick-wetting versus the distane squared for *Streptomyces pillosus* cells with lead ions

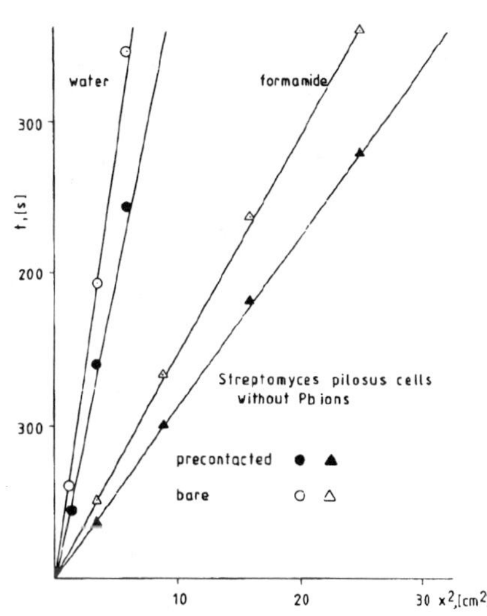

Figure 5 - Time of both water and formamide wick-wetting versus the distance squared for *Streptomyces pillosus* cells without lead ions

The polar components of the surface free energy of the microorganism cells were determined from the simultaneous solution of two equations for a pair of used polar liquids

$$x^2 = Rt/2\eta_L\ (W_A - W_c)$$

$$W_A = 2(\gamma_S^{LW} \cdot \gamma_L^{LW})^{0.5} + 2(\gamma_S^+ \cdot \gamma_L^-)^{0.5} + 2(\gamma_S^- \cdot \gamma_L^+)^{0.5}$$

$$Wc = 2\gamma_L$$

where:
x - the penetration distance,
R - radius of cappilary,
η_L - liquid viscosity,
W_A - the work of liquid adhesion,
W_C - the work of liquid cohesion,
γ_L - liquid surface tension,
γ^{LW}, γ^+,γ^- - surface free energy components.

The results obtained in this manner are shown in Table I. As can be seen, the surface of the microorganism cells shows strong electron donor properties while the electron acceptor properties are rather weak. The presence of lead ions on the microbial cell surface causes an increase of the value of the γ_s^- component and the value of the γ_s^+ component decreased to be very small. Generally in this study, the values of the free energy components of the *Streptomyces pilosus* cell surface show that this surface is weakly polar and the adsorption of lead ions causes a decrease of the total cell surface free energy.

The adhesion experiments were started from the correlation study between the amount of mineral bed used in a separate test and the adhesion of microorganism cells. For the next experiments the 1g of mineral samples has been designated.

An effect of the preliminary hydrophobed mineral surface on the adhesion of microbial cells was illustrated by the plotting of the concentration of cells as a function of the sodium oleate concentration of the solution which was contacted with the mineral particles (not with the microbial cells). The resulting graph for the barite and calcite particles is presented in Fig 6. This figure shows that an increase in the surface hydrophobisity of both minerals caused the enhancement of the *Streptomyces pilosus* cells adhesion. This behavior of microbial cells is a result of hydrophobic interaction between the mineral surface and the cell surface. The same role of hydrophobicity in the adhesion of bacterial cells to the hydrophobic sulfide minerals was noted by Solari et al. [14]. This hydrophobic interaction is almost independent of the electrostatic interactions. Phenomenon like this, has been observed for a variety of microorganism cells adhesion to a negatively charged polystyrene surface [8,15].

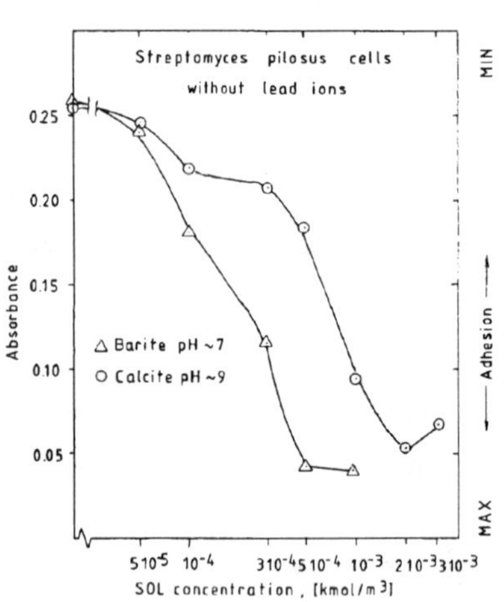

Figure 6 - Adhesion of Streptomyces p. cells without lead ions to various minerals

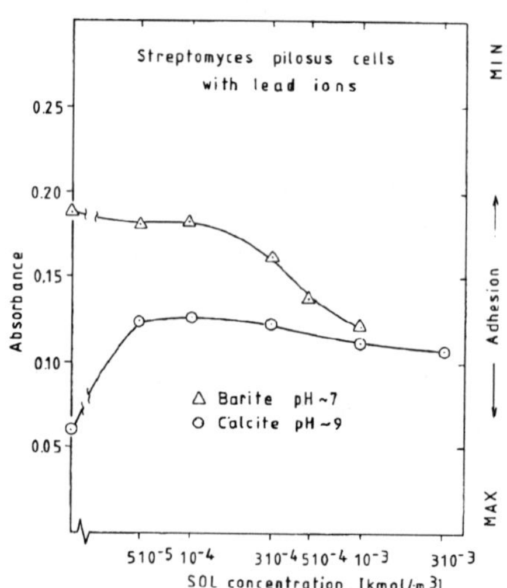

Figure 7 - Adhesion of *Streptomyces p.* cells with lead ionsto various minerals

The presence of accumulated metal ions on the microbial surface causes a change of the attachment mechanism. Fig. 7 presents the results of adhesion tests carried out with *Streptomyces pilosus* cells which accumulated lead ions. In this case the effect of electrostatic interaction seems to be stronger. The largest adhesion has been observed on the hydrophilic surface of the calcite particles. The accumulation of metal ions caused a change of the electrical potential on the microorganismsolution interface (the zeta potential is positive) [16]. This fact can be an explanation of the observed behavior of microbial cells.

Conclusion

1. The presented results showed that the microorganism cells can be treated as colloidal particles and the physicochemical model of interaction should be applied.
2. Free surface energy components calculated from thin-layer wicking experiments showed that the surface of *Streptomyces pilosus* cells is weakly polar and the accumulation of lead ions caused a decrease of the total free surface energy.
3. Adhesion of microbial cells to the mineral surface is a result of hydrophobic interactions which are important in the adhesion of *Streptomyces pilosus* cells without lead ions.
4. Accumulation of metal ions by microbial cells causes an enhancement of electrostatic interaction especially when the hydrophobic interaction is of low effect.

References

1. Z. Golab, B. Orlowska, M. Glubiak, K. Olejnik, **Acta Microbil. Polonica, 39**, (1990), 177-188.
2. Z. Golab, B. Orlowska, R.W. Smith, **Water Air Soil Pollution, 60**, (1991), 99-106.
3. Z. Golab, M. Glubiak, **Pollutants in Environment, 1**, (1991), 125-128.
4. Z. Golab, M. Breitenbach, A. Jezierski, **Water Air Soil Pollution**, in press.
5. Z.Sadowski, Z. Golab, R.W. Smith, **Biotech. Bioeng. 37**, (1991), 955-959.
6. Z. Sadowski, **Pollutants in Environment, 2**, (1992), 93-96.
7. Z. Sadowski, **Pollutant in Environment, 3**, (1993), 58-62.
8. W.Norde, J.Lyklema, **Colloids Surfaces, 38** (1989), 1-14.
9. M.C.M. van Loosdrecht, J. Lyklema, W. Norde, G. Schraa, A.J.B. Zehnder, **Appl. Environ. Microbiol., 53**, (1987), 1893-1897.
10. R.G. Reddy, in **Mineral Bioprocessing**, R.W.Smith, N.Misra (Eds.), The Mineral, Metal, Material Society, 1991, 453-467.
11. E.J. Clayfield, E.C.Lumb, Discuss. **Faraday Soc., 42**, (1972), 413-416.
12. E.Chibowski, **J.Adhesion Sci Technol., 6**, (1992), 10691090.
13. L.Holysz, E.Chibowski, **Langmuir, 8**, (1992), 717-721.
14. J.A. Solari, G. Huerta, B. Escobar, T. Vargas, R. Badilla-Ohlbaum, J. Rubio, **Collids Surfaces**, (1992), 159-166.
15. van Loosdrecht, J. Lyklema, W. Norde, A.J.B. Zehnder, **Microbiol. Rev., 54**, (1990), 75-93.
16. R.W. Smith, **Mineral Process. Extract. Metall. Rev., 4**, (1989), 277-304.

EVALUATION OF SULPHATE REDUCING BACTERIA AND RELATED PROCESS PARAMETERS FOR DEVELOPING A PASSIVE TREATMENT METHOD

Nural Kuyucak and Pascale St-Germain
Noranda Technology Centre, 240 Hymus Blvd., Pointe Claire, Quebec, Canada, H9R 1G5

Abstract

The use of sulphate reducing bacteria (SRB) for treating acid mine drainage (AMD) has recently received great attention. Several institutions, including mining and metallurgical industries, have investigated SRB-based processes as an alternative to conventional treatment methods for removing metals and for increasing alkalinity. Some research efforts have been devoted to developing active processes where reactors containing SRB are operated under controlled conditions. Significant work has been done on developing *in situ* passive processes such as engineered wetlands, biotrenches, and in pit and mine shaft applications.

Noranda Technology Centre (NTC) has assessed the capabilities and limitations of the SRB process with the aim of developing a passive process for treating AMD *in situ* in open pits. NTC took a step approach and conducted very intensive research, in order to understand the SRB-based passive treatment and the related process parameters. The work performed included initial bench-scale studies to gain insights into process fundamentals where pure chemical nutrients were used, selection of inexpensive/ alternative nutrient sources, and investigations of process start-up and scale-up parameters, using large-scale continuous reactors. In addition, the technical and economic feasibility of the process toward implementation of the technology for a given mine site was assessed. It was found that although the concept was viable and the process had great potential for treating AMD, the SRB-based technology would be most successful in treating low load situations. The limitations identified and research needed are discussed.

1. Introduction

In general, sulphide precipitation offers some advantages over conventional hydroxide precipitation which is a commonly used method for treating acid mine drainage in mining industries (Bhattacharryya *et al.* 1981; Kuyucak *et al.* 1991; Hammack and Edenborn 1992). These advantages include a high degree of metal removal (e.g. Cu, Cd, Zn, Pb, As, Hg, Ni and Fe) at low pH (pH 3-6) and improved sludge characteristics, i.e. chemically more stable and denser/less voluminous sludges (Peters and Ferg 1987). However, the high cost of sulphides discourages their use in treating AMD. The use of sulphate reducing bacteria has been proposed by several researchers to obtain alternative inexpensive sulphide reagents, due to their ability to convert sulphate contained in AMD to H_2S. During the process, SRB utilize a carbon source and generate alkalinity as well (Kuyucak et al. 1991). The hydrogen sulphide generated reacts with metals to form

Presented at the Biomineral Bioprocessing II Conference, July 10-15,1994, Snowbird, Utah

Mineral Bioprocessing II
Edited by David S. Holmes and Ross W. Smith
The Minerals, Metals & Materials Society, 1995

insoluble metal complexes. The organic carbon is converted to bicarbonates. The reactions are shown below:

$$2\ H^+ + SO_4^{2-} + 2CH_2O \xrightarrow{SRB} H_2S + 2\ HCO_3^- \qquad \text{Eq. 1}$$

$$H_2S + M^{2+} \text{----->} MS\downarrow + 2\ H^+ \qquad \text{Eq. 2}$$

Where: CH_2O and M refer to the organic compounds used as nutrients and metal ions, respectively.

Several investigators have been exploring SRB processes as a possible basis for the development of cost-effective treatment for wastewaters including mining effluents, which have high acidity and contain sulphate and metal ions. Possibilities include passive systems for treating mine effluents (e.g. constructed wetlands, biotrenches and open pits) and reactor applications for treating metallurgical process effluents (Hedin and Nairn 1992; Kalin 1992; Wildeman 1992; Barnes *et al.* 1991; Béchard *et al.* 1991; Kuyucak *et al.* 1991; Maree and Strydom 1987).

The work carried out at NTC aimed to evaluate the potential use of sulphate reducing bacteria (SRB) for treating AMD *in situ* in open pits to remove metals and increase alkalinity. The results of the earlier work (Phase I), in which pure chemicals were used as nutrients, indicated that open pit application of the process would technically be feasible if the open pit could be well manipulated to be used as a big, deep natural reactor. The process economics would be basically dependent on the cost of nutrients and the degree of achieved treatment.

Further research (Phase II) has, therefore, been undertaken to investigate alternative/inexpensive nutrient sources for SRB, and the parameters required for a successful process start-up and scale-up for field demonstration of the process. In addition, technical and economical feasibility of the process were assessed and the key questions to be addressed were defined. The results of Phase II studies are discussed below.

2. Materials & Methods

2.1 Materials

Nutrients. Selection of alternative nutrient sources was performed on two main groups of waste materials: cellulosic wastes and organic wastes. The cellulosic wastes consisted of: wood pulp, sawdust, bark, maple leaves, oat straw, fuel peat and horticultural peat. The organic wastes included: cow manure, distillers' dried grains, brewers' dried grains, dehydrated whey, molasses and starch. Combination of wood pulp, manure and brewers' dried grains was retained for all other tests. In control tests to assess the ability of the nutrient types to support SRB activity, chemical nutrients such as a combination of lactic acid (15 g/L), $(NH_4)_2SO_4$ (1.8 g/L) and K_2HPO_4 (0.35 g/L) was used.

Substrate. In nutrient selection experiments, the substrate consisted of limestone and sand, in a mixture of 1:1. In scale-up parameters tests, the substrate was simply limestone. Materials examined as substrate alternatives in 2-L batch tests were: limestone, phosphate, limestone and sulphidic rocks in a ratio of 1:1, sulphidic rocks and wood chips. In another batch test, the position of nutrients vs substrate and its effect on the treatment efficiency was addressed. Mainly, three ways were looked at: nutrients on top of substrate, mixing both, sandwich nutrients between 2 layers of substrate. For this experiment, the substrate was gravel.

2.2 Water Types

Although all tests were conducted with moderate strength AMD (F-Group) in nutrient selection studies, low strength AMD (Waite-Amulet) and tap water were also used. Water analyses are presented below in Table I.

TABLE I: Analysis of Tap Water and AMD Used in Nutrient Selection Tests.

Water Source	Strength	Parameters (mg/L)					
		pH	[Al]	[Cu]	[Fe]	[Zn]	[SO_4^{2-}]
Tap water		5.0	1.1	0.93	0.22	n.d.	50.8
Waite-Amulet AMD (WA)	low	3.0	26.8	1.3	179	6.5	585
F-group (Mattabi) AMD*	medium	2.5	131	45.2	209	349	3972

* Concentrations showed a gradual increase over the time (e.g. 450, 4500 mg/L of Zn and SO_4, respectively).

2.3 SRB Inoculum

SRB enriched from mine sediments, as described earlier (Kuyucak *et al.* 1991), were used in all studies. The inoculum was 1% of the working water volume. In one batch test which aimed at studying the effect of size inoculum on treatment efficiency, different inoculum proportions were added: 0.1, 1, 3, 5 and 10%. In parallel, two different inoculation techniques, i.e. pouring vs injection, were looked at. In this experiment only 0.1, 1 and 10% inoculum were studied.

2.4 Temperature

Effect of temperature on the process was addressed in two tests: at start-up, in 2-L batch tests where reactors were put at either 20, 10 or 4°C, and 5-L continuous kinetic tests where reactors were put at room temperature and 10°C.

2.5 Procedures

Different types of reactors and experimental modes were used in each study. All batch studies, including nutrient experiments and three key start-up parameters (solid substrate, inoculation technique and start-up temperature), were performed in 2-L reactors (glass bottles) for a 35-day period, usually. Larger size reactors (e.g. 5-L, 280-L drum and a 5-m high, 160-L volume column) in continuous mode were used to investigate process kinetics and major parameters for process scale-up (e.g. temperature, nutrient consumption, effect of volume to area ratio, effluent quality). Except for the drum, all reactors were made of plexiglass and equipped with sampling ports at the side. The drum was a standard storage container of 280-L capacity, opened at the top. As the bottles, drum and tall column were kept open, the 5-L reactors, where kinetics and off-gases analyses were determined, were closed and tightly sealed. In nutrient studies, control tests for the presence of SRB were also included. All tests were run in duplicates. pH adjustments were done using $Ca(OH)_2$.

Usually, a gravel sediment of varying thickness, depending on reactor type, was first layered at the bottom of each reactor. SRB inoculum (1%) was then directly added to the sediment, the nutrient (wood waste, manure and brewer's dried grain) was placed on top of the sediment, giving a 2-4 in thickness (30" in tall column) and water (AMD) with pH of 4.5-5 was poured over the materials until the reactor became full. Then the reactors were left until the onset of SRB activities was confirmed with the increase in pH, decrease in ORP, dark color and distinct odor of the system. Once the system was established, fresh AMD was added in a semi-continuous or continuous mode as required. During the initial batch mode in 5-L reactors and tall column, the water run through those reactors was recirculated with slow flow rate to accelerate SRB establishment.

In tall column tests, since signs of activity were seen only in the organic bed where channelling was observed, nutrients were resuspended throughout the entire column. The aims of this resuspension were to speed up the treatment of the water, to eliminate channeling and to obtain an even distribution of SRB in the nutrient bed. The nutrients had to be resuspended three times before suitable conditions could be established for SRB growth, i.e. ORP < -150 mV.

2.6 Analytical Methods

Carbon (C) and phosphorus (P) content of nutrients were analyzed by a LECO total carbon analyzer and an inductively-coupled plasma (ICP) after nitric and perchloric acid digestion, respectively. Total nitrogen (N) was determined using the Kjeldhal method. Oxidation-reduction potential (ORP) and pH were measured using appropriate electrodes connected to a Beckman pl12 meter. An ICP was used for analysis of metal ions in solution samples, which were filtered through a 0.45 μm Millipore filter and acidified with 3% HCl. Dissolved oxygen (DO) was determined using a polarographic electrode connected to a laboratory dissolved oxygen meter. In tall column, pH, ORP, DO and temperature were monitored *in situ* using a H20 Multiparameter analyzer which was connected to a data logger.

Except for total organic carbon (TOC), which was preserved by acidification to pH 2, volatile suspended solids (VSS), total suspended solids (TSS), biological oxygen demand (BOD), total organic carbon (TOC) and chemical oxygen demand (COD) analyses were performed on unfiltered 500 mL samples, preserved at $4^{0}C$. They were analyzed respectively as described in Standard Methods (1989), 5310B, 2540E, 2540D, 5210B and 5220C. The fraction of VSS over TSS indicates the fraction corresponding to microorganisms in TSS. For the gas experiments, the gas sample was taken every week and analyzed for H_2S, CO_2, CH_4 and N_2 by gas chromatography.

S^{2-} concentration was determined through titration of samples with $Pb(ClO_4)_2$. Samples were previously mixed with a sulphide antioxidant buffer. The end point was determined from the ORP curve as a function of volume added. Alkalinity was determined as $CaCO_3$ equivalents through titration of a 25 mL sample with 1N H_2SO_4. The alkalinity concentration was determined from the volume of acid necessary to decrease the pH to 4.5.

3. Results

3.1 Nutrient Selection

Initial experiments involved investigations of cellulosic wastes as sole nutrients for SRB. Cellulosic wastes tested included straw, maple leaves, wood pulp, sawdust, bark, fuel and horticultural peat where materials showed greatest potential, in decreasing order. Results indicated that cellulosic wastes alone would sustain SRB growth. Particularly when moderate strength AMD is used, the changes in pH, ORP, and SO_4 removal were significantly suppressed. Therefore, use of a mixture of wood and organic wastes was considered. Subsequently, organic and cellulosic wastes were analyzed for their C, N, and P content and were mixed with each other in order to obtain optimum C:N:P ratio of 110:7:1, which would support high SRB activity (Gerhardt 1981). The analysis of each material tested is tabulated below (Table II). Several combinations of cellulosic and organic waste materials were examined. Organic wastes considered for these tests were manure, whey, brewers' yeast and brewers' dried grain.

TABLE II: C, N, P Ratios in Waste Materials.

Substrate	Organic Carbon	Organic Nitrogen	Organic Phosphorus
Raw wastewater sludge*	26.2	2.8	1
Brewers' yeast	31.5	1.06	1
Digested wastewater sludge*	32.3	2.4	1
Distillers' grains	58.4	3.84	1
Whey	60.3	2.38	1
Brewers' grains	85.6	4.65	1
Cow manure*	100	3.16	1
Sugar maple leaves*	191	7.18	1
Oat straw	209.5	3.85	1
Bark (3 years)**	519.2	14.6	1
Fuel peat	743.8	61.3	1
Straw*	915	5.97	1
Cane molasses	1481.8	13.6	1
Wood pulp	1595	56.7	1
Bark (30 years)	1610	64	1
Corn starch	2322	18.3	1
Oak leaves*	1660	11.4	1
Newspaper*	2760	0.42	1
Sawdust	11300	65	1
Pine sawdust*	30200	26.9	1

* ref.: N. Kuyucak, 1991.

** stored outdoors for 3 or 30 years.

All the mixtures examined indicated changes in ORP, pH, metals removal and of BOD release with varying degrees. The mixture of wood pulp, manure and brewers' dried grain exhibited the closest results to those obtained from the control tests which consisted of lactic acid, ammonium sulphate, and potassium biphosphate salts as nutrients. The patterns observed for changes in pH, ORP and SO_4 levels in both, chemical and natural nutrient tests, closely followed each other (Figure 1) where low-strength AMD (e.g. Waite-Amulet AMD; Table I) was used.

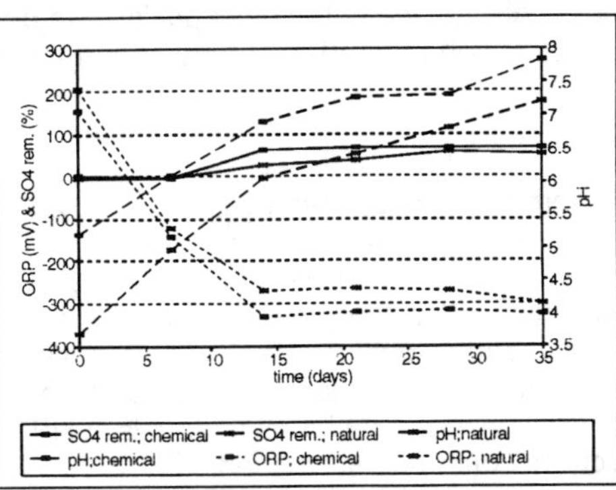

Figure 1 SO_4 and Zn removal and pH increase with the use of the mixture of woodpulp, brewers' dried grain and manure vs. lactic acid, ammonium sulphate and potassium biphosphate.

3.2 Process Parameters Investigated for Start-up

The optimal conditions required for a successful process start-up in the pit were investigated in conjunction with the use of natural nutrients (wood pulp, manure and brewers' dried grains). The parameters examined were: type of materials to be used and ways of preparing the sediment layer; need for SRB inoculum and the way of addition; and effect of temperature and the possibility of acclimatization, as described by Ahonen and Tuovinen (1989).

In the earlier study, it was found that presence of a substrate (sand and gravel) greatly enhanced SRB performance in AMD (Kuyucak *et al.* 1991). In order to develop a generic process, rather than a site specific one, various types of materials were investigated, along with different fashions for mixing such materials with nutrients, in the present study.

Type of Sediment and Sediment Layer Preparation. The results revealed that phosphate rocks and limestone (gravel) would be preferable to wood chips or sulphidic rocks, but the latter can be used. When limestone was mixed with sulphidic rocks the process was significantly improved as indicated by the increase in pH. When wood chips and sulphidic rocks were used alone, the increase in pH was delayed by 16-18 days. However, as time went by, their performance was similar to that of other materials, indicating that a wide range of materials (particularly those which have a large surface area, proper permeability, and buffering capacity) can be used to form the sediment layer; the process is not limited by the type of material (Figure 2).

In terms of metals removal, Cu and Zn were readily extracted from the water. However, significant Fe removal (> 90%) was delayed in the presence of sulphidic rocks and wood chips. Al removal seemed to be dependent on the sediment material used but with all materials, SRB activity could be established.

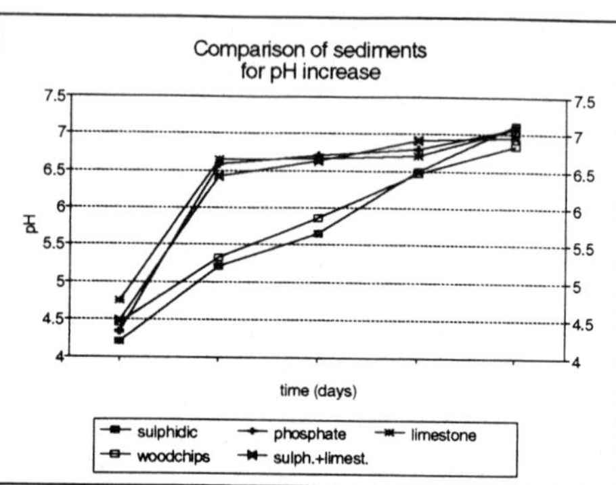

Figure 2 Use of different materials as a sediment layer.

Effect of Inoculum Proportions and Inoculation Method. Two inoculation methods (direct pouring and injection) were compared for inoculum proportions of 10%, 1%, 0.1%. The results indicate that SRB inoculation would be essential to obtain positive measurable differences in process parameters in a reasonable time frame. Based on SO_4 removal and ORP results, the optimum amount of SRB inoculum is 1% and injection of the SRB inoculum is better than pouring it. Use of 0.1% inoculum resulted in a lag time of 7 days as compared with other inoculum portions. The final SO_4 removal was between 58 and 63%.

Effect of Start-up Temperature. Only the tests initiated and performed at room temperature showed SRB activity after 42 days, as observed by relatively high SO_4 removal, as well as significant metal removal, as compared with results from 10 and 4°C incubation. Experimental results showed that start-up temperature was very important, as well as the incubation temperature of the inoculum, in facilitating the process start-up.

3.3 Investigations of Parameters for Process Scale-up (5-L Reactors, Drum and Tall Column Tests)

The process scale-up parameters investigated were: kinetics at 20 and 10°C, composition of off-gas, influent entrance to the system (e.g. from the surface, over the sediment or in/through the sediment), nutrient consumption, effect of surface to volume ratio of the reactor, SO_4 reduction and metals removal rates, reaction zone, effect of turnovers, comparison of start-up methodologies.

Rate of Sulfate Reduction and Effect of Temperature. The results showed that there is a relationship between rates of sulphate reduction, metal removal and dissolved organic carbon present in the system. Alkalinity generation could not directly be coupled to SO_4-reducing activity, since in some cases no increase in pH was observed with the increase in net SO_4 removal. At the steady state conditions, SO_4 removal rates reached around 300 mmol/m^3/d for a retention time of 280 days in the drum or 61 days in the nutrient bed. Only under reduced conditions were all metals, except Mn, removed to very low levels indicating S^{2-} precipitation. In this case, the total amount (mol) of SO_4 reduced was in excess of total amount (mol) of metals removed.

There was a decrease in SO_4 reduction rate from 700 to 400 mmol/m^3/d corresponding to 57% reduction by decreasing the temperature from 20 to 10°C. The total amount of TOC concentration was shown to be dependent on temperature. The TOC concentration was reduced by half with a decrease in temperature from 20 to 10°C. The results obtained from 5-L reactors and drum tests with regard to sulphate reduction rates were in agreement. However, the effect of limestone on the overall process performance was more pronounced in 5-L reactors since AMD relatively uniformly passed through the sediment layer consisting of gravel in those reactors.

Effect of Treatment on Water Quality. The effluent quality obtained from the SRB process for 14 months is summarized in Table III. The concentration of Al, Fe and Zn fluctuated between non-detectable levels by ICP to 10-20 mg/L, depending on the conditions. Under oxidized conditions, the system would release Zn, Fe and Al whereas, As, Cd and Cu were removed to undetectable levels under all circumstances indicating that other metal removal phenomena were involved in the process (e.g. bio-complexation, bio-adsorption and precipitation). Particularly, the Zn concentration follows very closely the SO_4 reducing activity. Mn removal was always poor.

Basically, the concentration in BOD, COD, and TOC significantly increased with the addition of fresh nutrients indicating that minimal treatment to decrease BOD to acceptable levels would be required (e.g. aerated lagoon). The addition of fresh nutrients might help in reestablishing the system by replenishing phosphorous (TP) and carbon (TOC), as well as providing bacterial sources.

Organic indices such as TSS, VSS, BOD, COD, TOC, TKN and TP for the final effluent quality were also analyzed and represented in Table IV.The comparison of results of the filtered vs, unfiltered samples for a few heavy metal ions showed that the difference was not significant for all metals except Fe. These results imply that metals are not in the colloidal form in the effluent and any additional filtration system will not be required in full-scale applications.

Table III: Concentration of Metal Ions and Sulfate in the Effluent Taken From the Drum as Analyzed at Each Consecutive Month (for 14 months)

Parameter (mg/L)	Initial AMD (average)	Quality of effluent at each month													
		35	66	96	125	159	188	226	245	259	300	346	370	398	431
pH	2.5	6.8	6.6	6.6	6.4	6.2	6.0	5.2	4.8	5.4	6.3	6.8	6.8	6.5	6.4
Al	173	0.44	0.188	0.3	0.36	0.37	0.39	1.1	1.8	nd	nd	0.28	0.45	0.55	0.51
As	1.0	nd	0.09	nd	nd	nd	nd	nd	nd	nd	nd	nd	nd	nd	nd
Cd	1.5	nd	nd	nd	nd	nd	nd	0.0555	0.04	0.04	0.044	nd	nd	nd	nd
Cu	47	nd	nd	nd	0.044	0.04	0.02	0.17	0.072	0.05	nd	nd	nd	nd	nd
Fe	160	0.74	0.19	0.69	0.76	4.77	4.66	2.66	8.99	1.12	0.19	0.21	0.54	0.37	0.13
SO_4	4000	1682	1721	2837	3720	3360	3300	3450	2989	2989	2490	2297	2524	3210	2460
Mn	38	1.11	6.99	13.5	21.5	23.6	23.4	21.1	25.3	26.0	24.3	21.5	21.8	20.2	19.5
Zn	350	0.38	nd	11.4	1.44	1.33	1	12.8	0.01	18	nd	nd	0.27	0.11	nd

Table IV: Effluent Quality in Terms of Organic Parameters

Organic Parameter	Quality of water before the addition of fresh nutrients (mg/L)	Quality of the water after the addition of fresh nutrients (mg/L)
Time of sampling (days)	257	319
TSS	156	80
VSS	35	47
BOD	30	84
COD	79	260
TOC	27.2	123
TKN	28.1	26
TP	<0.05	0.12

Reaction Zone Determination. From the tall-column tests it was found that the reaction zone was situated in the organic layer. To maintain a good activity in the reactor, contact between the organic nutrient and water to be treated would be required. Over time establishment of SRB in the sediment took place. Resuspension of the nutrient bed, which also simulates turnover, did not deteriorate the system perfomance; on the contrary, it improved metal removal and homogeneous distribution of bacteria in the nutrient bed.

Importance of Surface Area to Volume Ratio. The results showed that to get a true representation of the SO_4 reduction rate, retention time in the bed volume must be taken into consideration. However, when comparing drum with tall column tests, the maximal use of the nutrients can only be ensured by the presence of a large surface area, which provides better contact for nutrients, bacteria and AMD. In a small area and high depth case, uniform flow distribution within the nutrient bed cannot be maintained and the process efficiency declines.

Effect of Influent Addition Method. The use of a baffle to introduce AMD over the sediment resulted in constant fluctuations in pH, with the general trend decreasing in the long-term. Adding influent through a tube (subsurface flow) in the sediment seemed to favor a more constant trend from day to day implying that better contact was provided between the influent, the nutrients and SRB. However, the baffle concept can also be used, particularly on a pulse base.

Composition of Off-Gases. The qualitative analysis of gases produced at 20^{0}C showed that the gas was composed of average of 1.5% CO_2 and 98.5% N_2/O_2. Some CH_4 (e.g. 0.023%) and no H_2S could be detected. When the reactor was transferred at 10^{0}C, CH_4 stopped being generated, indicating that activity of methanogens was disturbed. The switch from batch to continuous mode resulted in an increasing trend in CO_2 and CH_4 concentrations and an inversely decreasing trend for N_2/O_2 content.

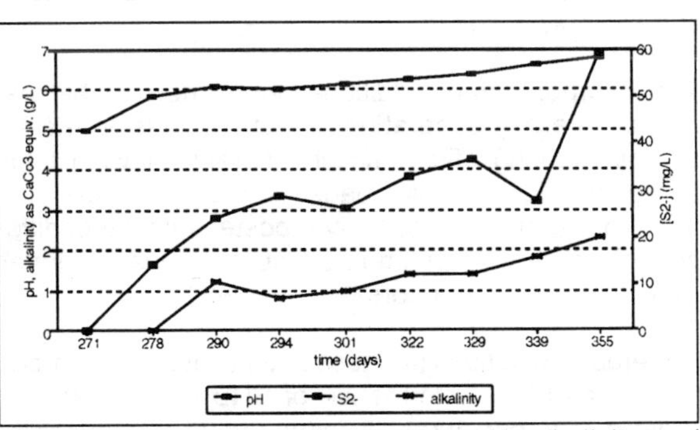

Figure 3 The relationship between pH, alkalinity and S^{2-} generation

Relationship Between S^{2-} and Alkalinity Generation and pH Behaviour. Alkalinity generation might not be directly coupled to SO_4 reduction. pH, S^{2-} and HCO_3^- concentrations were closely monitored throughout the tests, particularly at steady state conditions, and the results are illustrated in Figure 3. The trend in the pH curve was closer to the trend in the alkalinity generation curve as opposed to the S^{2-} curve, implying that the influence of alkalinity generation on pH is predominant over SO_4 reduction.

Nutrient Consumption Rate. The carbon requirement was computed based on Equation 1 where 2 C are required to reduce 1 SO_4. The amount of C in mol (or mol/m^2) present in the nutrient bed was calculated using Table II. Then the C (nutrient) utilized for the amount of sulphate reduced was correlated per day. The nutrient consumption rate for the first 215 days at room temperature was 0.07%/d. Later, it decreased and stabilized around 0.05%/d.

4. Discussion

4.1 General Concept

The increase in alkalinity and removal of metals from AMD with the SRB process are the result of several chemical and biological reactions. The efficiency of the system depends on several process parameters such as: characteristics of the water being treated (e.g. acidity, type and concentration of metals), environmental conditions (e.g. temperature, redox potential), percentages of gases generated from the reactions (e.g. CO_2, H_2S, CH_4) and the conversion of CO_2 to HCO_3^- in the effluent. The effluent pH is usually maintained by the adjustment of the carbonic acid/bicarbonate system to an optimum level at which pH/alkalinity is increased and buffering is obtained.

Synergism between three groups of microorganisms (acidogens, methanogens and sulphate reducers) is essential for the success of the process, particularly when natural organic materials are used as nutrients (Tuttle *et al.* 1969). Organic matter, which is biochemically converted by acidogens to several acidic and basic products, serve as nutrients to SRB and contribute to the pH within an anaerobic environment (Westrich and Berner 1984; Holder *et al.* 1984; Maree and Strydom 1987; Kuhl and Jorgensen 1992; Scheeren *et al.* 1992).

The incorporation of additional limestone into the reactors was suggested to boost the rate of alkalinity generation by both increasing the overall rate of limestone dissolution and by buffering influent acidity, thereby allowing sulphate reduction to occur uninhibited at higher influent acidity loads (Hedin *et al.* 1991). Dvorak et al., 1991 also pointed out that the use of a previously composted mixture of different organic materials, which were conditioned with gypsum and limestone, was important to increase the rate of sulphate reduction in their reactors.

Anaerobic conditions are necessary to assure the production of suitable types of nutrients, as well as high performance of SRB and limestone. Anaerobic degradation of simple organic compounds results in the formation of organic acids. However, sometimes oxical degradation may also be required to oxidize organics whose complex structure is difficult to break down anaerobically and oxidize CH_4 formed in the system (Morel 1983).

At low temperatures, the rates of chemical and biochemical reactions are reduced, resulting in less alkalinity and H_2S generation. The cellulose decomposers are mainly mesophillic microorganisms and, consequently, are more sensitive to low temperatures than SRB.

4.2 Critical Parameters at Process Start-up and During Operation

At process start-up, it is necessary to establish a correct start-up procedure (Scheeren *et al.* 1992; Wildeman 1992; Dvorak *et al.* 1991). The parameters considered to be important at start-up are: pH and reducing conditions, flow rate and composition of the influent, sufficient nutrients (e.g. C, N, P and SO_4), good contact of AMD with substrate and temperature. Scheeren et al., 1992 stated that high pH (>6) and sufficient nutrients have to be maintained at start-up; however, as soon as methanogenic activity is detected (via methane gas production), the pH and dosage of phosphorous can be reduced. The concentration of an active microcosm is another critical factor. Since the sulphate reduction and $CaCO_3$ dissolution and sulphide movement processes are mainly diffusion controlled, the system should be designed to overcome mass transfer limitations (Morel 1983; Kuhl and Jorgensen 1992).

Although all chemical and biochemical reactions including SO_4 reduction, metal sulphide formation and limestone dissolution described above are exothermic (Kuyucak *et al.* 1990), in a large water reservoir (e.g. pit), which continuously receives ground and surface water, under natural environmental conditions the chance of producing a heat gradient in the *in situ* pit process seems to be unlikely. Loading rates of acidity and metals are also critical. Although the decrease in temperature lowers the bacterial activity, SO_4 reduction does not stop. Therefore, the actual temperature should be taken into account while designing the process.

The need for an optimum amount of substrate and surface area in the wetland application of the SRB process was substantiated by several investigators (Hedin and Nairn 1991; Nairn and Hedin 1992; Wildeman 1992). The actual reaction zone was found to be mainly in the substrate layer (Holder *et al.* 1984). An optimum 12-18 in. thickness and certain properties (e.g. permeability) for the substrate to be used in constructing wetlands was suggested to eliminate channelling and shortcircuiting. Even gentle mixing is recommended to eliminate mass transfer limitations and to remove trapped H_2S in the substrate (Scheeren *et al.* 1992).

4.3 Metals Removal

According to the literature (Hammack and Edenborn, 1992; Tarutis, *et al.*, 1992; Wildeman, 1992), copper, zinc, cadmium, iron, nickel and lead are retained within the reactors or wetlands as insoluble monosulphides, following their reaction with bacterially generated H_2S. Aluminum and manganese are hydrolyzed within the reactors and are retained as insoluble hydroxides or carbonates as a result of being pH dependent. Zinc may be retained in both carbonate and sulphide phases depending on pH and availability of S^{2-} species. Removal of some metals (e.g. Cu, Cd, As) to low levels under all circumstances was attributed to binding with organic matter and/or other oxide minerals as well.

4.4 Economics, Longevity and Conclusions

The economics of the process, which depend on the amount of nutrients required for the pit considered, have been calculated and compared to the cost of conventional lime neutralization. The SO_4 reduction rate found (e.g. 0.3 mol/m^3/d H_2S at 20^{0}C or 0.15 mol/m^3/d H_2S at 10^{0}C) is in well-agreement with the rates reported in the literature (Bolis *et al.*, 1991; Devorak, *et al.*,1992; CD&M, 1991). Assuming that AMD contained 20 mol/d metals, a ton of substrate material costs $40. The **conservative rate** (0.15 mol/m^3/d H_2S), the amount of substrate required and the total initial cost of the process have been estimated to be 40 000 tonnes and $ 1.6 million, respectively. Since the surface area in the pit considered would be only 5000 m^2, the substrate layer would be 8-m deep. Although the life expectancy of such a substrate layer is expected to be 120 y, as discussed above, efficiency of the system becomes uncertain due to its thickness. In addition to the mass transfer limitations expected, temperature at the bottom of the pit all year round is $<10^0$C.

In conclusion, loadings in open pits such as F-Group is too high for passive biological systems but the concept is viable for treating low load situations. The research needs to search easily degradable, yet inexpensive, nutrients and to overcome low temperature and mass transfer drawbacks.

Acknowledgements

The authors thank Mr. K.G. Wheeland for his support of this project and for his intensive review and comments provided during the course of the study.

References

Ahonen, L. and O. H. Tuovinen. 1989. Microbiological Oxidation of Ferrous Iron at Low Temperature. *Appl. Environ. Microbiol.*, 55(2): 312-316.

Barnes, L.J., Janssen, P.J.H. Scheeren, J.H. Versteegh and R.O. Koch, 1991. Simultaneous Microbial Removal of Sulphate and Heavy Metals from Waste Water. In Proceedings of EMC'91 Conference: Non-Ferrous Metallurgy-Present and Future. p. 391-403. (Brussels, Belgium, September, 1991).

Béchard, G., S. Rajan and R. G. L. McCready. 1990. Microbiological Treatment of AMD; Laboratory and Field Study Uptake. In Biominet Proceedings. p. 130-151. (CANMET, Ottawa, Ontario, November 1990).

Bhattacharryya, D., A. B. Jumawan, G. Sun, C. Sund-Hagelberg and K. Schwitzgebel. 1981. Precipitation of Heavy Metals with Sodium Sulphide. *A.I.Ch.Eng. Symposium Series* 77:31-38.

Bolis, J.L., T.R. Wildeman and R.R. Cohen, 1991. The use of Bench Scale Permeaters for Preliminary Metal Removal from AMD by wetlands. National Meeting of American Society for Surface Mining and Reclamation, Durango, CO. May 14-17, 1991.

Camp Dresser and McKee, Inc. Denver, Colorado, 1991. Passive Treatment System, Pilot Plant Interim, 2nd progress report, submitted to Noranda Grey Eagle Mine.

Dvorak, D. H., R. S. Hedin, H. M. Edenborn and P. E. McIntire, 1992. Treatment of Metal Contaminated Water using Bacterial Sulphate Reduction: Results from Pilot Scale Reactors. *Biotechnol. Bioeng.* 40:609-616.

Gerhardt, P. 1981. Manual of Methods for General Microbiology, Washington D.C.: ASM Publ.

Hammack, R. W. and H. M. Edenborn. 1992. The Removal of Nickel from Mine Waters Using Bacterial Sulfate Reduction. *Appl. Microbiol. Biotechnol.* 37:674-678.

Hedin, R. S., D. H. Dvorak, S. L. Gustafson, D. M. Hyman, P. E. McIntire, R. W. Nairn, R. C. Neupert, A. C. Woods and H. M. Edenborn. 1991. Use of Constructed Wetland for the Treatment of Acid Mine Drainage at the Friendship Hill National Historic Site, Fayette County, PA. Final Report by USBM, Pittsburgh, PA.

Hedin, R. S. and R. W. Nairn. 1992. Designing and Sizing Passive Mine Drainage Treatment Systems. In Proceedings of the 13th Annual West Virginia Surface Mine Drainage Task Force Symposium. (Morgantown, WV, April, 1992).

Holder, G. A., G. Vaughan and W. Drew 1984. Kinetic Studies of the Microbiological Conversion of Sulphate to Hydrogen Sulphide. *Wat. Sci. Tech.* 17:183-196.

Kalin, M. 1992. Decommissioning Pits with Ecological Engineering. In Proceedings of 24th CMP Conference. (Ottawa, Ontario, January, 1992). Paper no 27.

Kalin, M. 1992. Decommissioning Open Pits with Ecological Engineering. In Proceedings of CIM Conference. Chap. 32. (Ottawa, Ontario, February, 1992).

Kuhl, M. and B. B. Jorgensen. 1992. Sulphate Reduction and Sulphide Oxidation in Compact Microbial Communities of Aerobic Films. *Appl. Environ. Microbiol.* 58(4): 1164-1174.

Kuyucak, N., D. Lyew, and K. G. Wheeland. 1990. Lower Cost Natural Processes for Treating AMD, Noranda Technology Centre Research Report, EN-9305-003 and LN1-9055: RR 90-1.

Kuyucak, N., D. Lyew, P. St-Germain, and K. G. Wheeland. 1991. *In Situ* Bacterial Treatment of AMD in Open Pits. In Proceedings of 2nd International Conference on the Abatement of Acidic Drainage. p. 336-353, vol. 1. (Montreal, Quebec, September, 1991).

Maree, J. P. and W. F. Strydom. 1987. Biological Sulphate Removal from Industrial Effluent in an Upflow Packed Bed Reactor. *Wat. Res.* 21(2): 141-146.

Morel, F. M. M. 1983. Principles of Aquatic Chemistry. NY: John Wiley & Sons Publ., pp. 336-338.

Nairn, R.W. and R.S. Hedin, 1992. Designing Wetlands for the Treatment of Polluted Coal Mine drainage.

Peters, R. W. and J. Ferg. 1987. Dissolution/Leaching Behavior of Metal Hydroxide/ Metal Sulphide Sludges from Plating Wastewaters. *Hazardous Waste and Hazardous Materials.* 4(4): 325-331. Mary Ann Liebert Inc. Publ.

Scheeren, P. J. H., R. O. Koch, C. J. N. Buisman, L. J. Barnes and J. H. Versteegh. 1992. New Biological Treatment Plant for Heavy-Metal Contaminated Groundwater. *Trans. Instu. Min. Metall.* 101: C190-C199.

Tuttle, J. H., P. R. Dugan, and C. I. Randles. 1969. Microbial Sulphate Reduction and its Potential Utility as an Acid Mine Water Pollution Abatement Procedure. *Appl.Microbiol.*, Feb., p. 297-302.

Westrich, J. T. and R. A. Berner. 1984. The Role of Sedimentary Organic Matter in Bacterial Sulphate Reduction. *Limnol. Oceanogr.* 29(2):236-249.

Wildeman, T. R. 1992. Constructed Wetlands That Emphasize Sulphate. In Proceedings of 24th CMP Conference. (Ottawa, Ontario, January, 1992). Paper no. 32.

Wildeman, T. 1992. Decommissioning Open Pits with Ecological Engineering. Chap. 32. In Proceedings of CIM Conf. (Ottawa, Ontario, February, 1992).

Degradation of the Metals Tailing Pond Waters Containing Cyanides by Alginate Immobilized Cells of *Pseudomonas putida*

Kirit D. Chapatwala , G.R.V. Babu, Eddie R. Armstead and James H. Wolfram[1].

Division of Natural Sciences, Selma University, Selma, Alabama 36701
[1] Biotechnology, INEL, EG & G, Idaho Inc., Idaho Falls, Idaho 83415.

ABSTRACT

Cyanide compounds are extensively used in our highly industrialized society. These compounds are generally ended into the environment via the industrial wastewater streams. Increasing accumulation of such compounds in the ecosystem may cause deleterious effects, as most of them are highly toxic and tend to destabilize the ecological balance by inhibiting the beneficial microbial growth. The bacterium capable of utilizing cyanide as a sole source of carbon (C) and nitrogen (N) was isolated and identified from environmentally contaminated sites. The species *Pseudomonas putida* was selected and immobilized in calcium alginate. The immobilized cells were used in the air-uplift-type fluidized bed batch bioreactor to study degradation and removal of cyanide present in the metals tailing pond water (contaminated wastewater). The immobilized cells of *P. putida* converted cyanide present in the wastewater into NH_3 and CO_2. The present study concludes that complete remediation of the metals tailing pond water containing cyanides can be achieved with the alginate immobilized cells of *P. putida* using the air-uplift-type fluidized bed bioreactor.

Mineral Bioprocessing II
Edited by David S. Holmes and Ross W. Smith
The Minerals, Metals & Materials Society, 1995

INTRODUCTION

Bacteria can degrade and thereby detoxify a wide variety of cyanides (1-3). Cyanide is a highly toxic and metabolic inhibitor (4). Considerable amount of cyanides are generally found in the industrial wastewater due to industrial production and use in industries like metal plating and ore leaching (5). Since higher concentration of cyanides are highly toxic for biodegradation, they are generally removed by chemical means prior to biological treatment (6). However, the application of the biodegradation of cyanides has been reported earlier (7,8). The immobilized cells of *Pseudomonas putida* have been shown to degrade higher concentration of cyanides into non-toxic compounds - NH_3 and CO_2 when compared to free cells (3). Therefore, in the present investigation, an attempt was made to study the degradation of metals tailing pond waters containing cyanides by the calcium alginate-immobilized cells of *P. putida* in an air-uplift-type fluidized bed batch reactor.

MATERIALS AND METHODS

Chemicals:

All chemicals (99% purity) including sodium alginate (Type VII) were purchased from Sigma Chemical Co., St. Louis, MO.

Isolation of Pseudomonas putida:

The bacterium selected was isolated from soil samples collected from industrial waste sites. The sample was diluted (1:10) with sterile minimal medium and the suspension was incubated at 25°C for 1 h. One-milliliter of the suspension was inoculated into 9-ml of medium supplemented with different (50-400 ppm) concentrations of NaCN. The samples were incubated at 25°C for 7 days and the tubes were examined for turbidity. After five to seven serial transfers, turbid samples were streaked onto plates containing NaCN as the sole source of C and N. Successive streaking of the colonies onto NaCN agar allowed isolation of a pure culture that could utilize NaCN as its sole source of C and N.

Identification Pseudomonas putida:

The isolates were characterized by using the rapid identification kits (Flow Laboratory Kits for Gram-negative microorganisms. Flow Laboratories Inc., McLean, VA and the API 20E Test Kit, Analytab Products, Plainview, NY). The results were also confirmed by a computer survey available from these suppliers. Subsequently, the specific substrate utilization tests were performed as described in the *Bergey's Manual of systematic Bacteriology* (9). and as described by Smibert and Krieg (10).

Immobilization of Pseudomonas putida in Alginate:

The cell-paste was suspended in 100 ml of 0.85% sterile saline and mixed with 100 ml of 4% sterile sodium alginate. The alginate-cell mixture was extruded dropwise through a 25-gauge needle from a height of about 20 cm into cold 0.2 M $CaCl_2$. Each drop was hardened into a bead containing entrapped cells of *P. putida*. These beads were further hardened by stirring them in $CaCl_2$ for 30 min and they were then stored at 5°C for 24 h.

Batch Bioreactor Experiment:

The batch bioreactor experiment was performed in an 800-ml air-uplift-type fluidized bed batch reactor (Fig. 1). The reactor was filled with 750 ml of metals tailing pond water containing 194.21 mg ^-CN (89.64 mg C and 104.57 mg N) and 50% bead volume. The reactor was aerated with CO_2 and NH_3 free air (200 ml/min) at 25°C. The degradation of ^-CN in bioreactor was monitored by determining pH, bacterial growth (outside the beads, if any), dissolved and gaseous NH_3 and gaseous CO_2 for 240 h. Gaseous NH_3 and CO_2 were trapped in 0.5 M boric acid and 0.5 M KOH respectively. The samples (5 ml) were collected at regular intervals for the analysis. The experiments were repeated three times with triplicates and the mean values are reported. The values presented in the investigation are significant at $P < 0.05$.

Analytical Methods:

Growth in liquid culture was estimated from the A_{546} using 10 mm cuvettes (Gilford Spectrophotometer, Ciba-Corning, USA). The Beer-Lambert Law was followed for an absorbancy of 1.0 containing 653 mg dry wt per ml (4). Samples (5 ml) were taken from effluent port and bacteria were removed by centrifuging at 15,000 x g for 10 min at 5°C (Beckman Instruments, USA). Dissolved NH_3 was determined colorimetrically by the Berthelot's procedure as described by Kaplan (11). The gaseous NH_3 was determined by back titration of boric acid (known concentration) with 0.5 M KOH. The gaseous CO_2 was first dissolved in known volume of 0.5 M KOH and back titrated for free KOH using 0.5 M HCl.

RESULTS AND DISCUSSION

The bacterial species, *P. putida* was isolated and identified from contaminated site samples supplemented with 100 ppm of NaCN as a sole source of C and N. The colonies on minimal mineral salt agar plate-NaCN as source of C and N, were aerobic, small (1-3 mm in diameter), circular, convex, with entire margin, beige and creamy in texture and fluorescent.

The cellular morphology of the *P. putida* was gram-negative, small, rod-shaped, motile, non-spore forming and non-capsulated. Oxidase, catalase, and arginine dihydrolase reactions were positive. Growth was observed on sodium benzoate, MacConkey, and glucose plates but not on xylose and maltose and failed to hydrolyze gelatin.

Optimal conditions for the growth of *P. putida* were pH 6.7 and 25°C. No growth was observed below 5.0 or above 8.5 pH, and at 10°C or 55°C.

The typical yield following the immobilization was about 0.6 g of beads per ml of cell-alginate suspension. Individual beads had a diameter of 1-2 mm with an average wet and dry weight of 15 and 0.8 mg respectively. At the time of immobilization, each bead contained approximately 1-2 x 10^8 viable cells, as determined by pour plate count method of disrupted beads (13).

The pH of medium was increased in a time-dependent manner from 8.2 to 9.4 during the degradation of ^-CN present in the metals tailing pond water (Fig. 2). The data in Fig. 3 & 4 indicate that the production of CO_2 and total NH_3 was directly proportional to the degradation of ^-CN. At around 192 h of incubation, 50% of CO_2 and NH_3 production was observed with the degradation of

50% total ^{-}CN. The NH_3 content in general was higher than CO_2 during the ^{-}CN degradation by the immobilized cells of *P. putida.*

The metabolism of NaCN by *P. putida* has been very well documented (1-3). However, the enzymatic pathways involved in the production of ammonia from cyanide are not clear (14, 15). The present study demonstrates that the cells of *P. putida* immobilized in alginate could degrade ^{-}CN of metals tailing pond water into CO_2 and NH_3. This suggests that the mechanism involved in the conversion of ^{-}CN of metals tailing pond water to CO_2 and NH_3 by the immobilized cells of *P. putida* is identical despite source or type of cyanide compounds.

The data on mass balance show that complete conversion of ^{-}CN was achieved with *P. putida* immobilized in the alginate matrix. This confirms that the immobilized cells of *P. putida* would utilize ^{-}CN present in the metal tailing pond water as the sole source of C and N and thereby suggesting that immobilized cell technology can be employed in the field of cyanides bioremediation.

ACKNOWLEDGEMENTS

We thank the Engineering Foundation - Conference on Mineral Bioprocessing II. This work was supported by Sub-Contract No.C85-110800 from Idaho National Engineering Laboratory, Idaho Falls, ID.

REFERENCES

1. White, J.M., D.D. Jones, D. Huang and J.J. Gauthier. 1988. Conversion of cyanide to formate and ammonia by *Pseudomonad* obtained from industrial wastewater. ***J. Ind. Microbiol.***. 3:262-272.

2. G.R.V. Babu, J.H. Wolfram and K.D. Chapatwala. 1992. Conversion of sodium cyanide to carbon dioxide and ammonia by immobilized cells of *Pseudomonas putida*. ***J. Ind. Microbiol.***. 9:235-238.

3. Chapatwala, K.D., G.R.V. Babu and J.H. Wolfram. 1993. Screening of encapsulated micorbial cells for the degradation of inorganic cyanides. ***J. Ind. Microbiol.*** 11:69-72.

4. Vennesland, B., E.E. Conn, C.J. Knowles, J. Westley and F. Wissing (eds), ***Cyanide in Biology,*** Academic Press, Inc. (London), Ltd., London.

5. Knowles, C.J. 1976. Microorganisms and Cyanide. ***Bacteriol. Rev***. 40:652-680.

6. Knowles, C.J. and A. W. Bunch. 1986. Microbial cyanide metabolism. ***Adv. Microb. Physiol.*** 27:73-111.

7. Mudder, I.I., and J.L. Whitlock. 1984. Biological treatment of cyanidation wastewaters. In: ***Proceedings of the 38th Annual Purdue Industrial Waste Conference*** (Bell, J.M. Ed.), Butterworth Publishers, Boston, 279-287.

8. Shivaraman, N. and N.M. Parhad. 1984. Biodegradation of cyanide in a continuously fed aerobic system. ***J. Environ. Biol.*** 5:273-284.

9 Palleroni, N.J. 1984. Gram-negative aerobic rods and cocci. In: ***Bergey's Manual of Systematic Bacteriology.*** Vol. II (Krieg, N.R. and J.G. Holt, eds.) 140-149. Williams and Wilkins, Baltimore.

10.Smibert, R.M. and N.R. Krieg. 1981. General characterization, In: ***Manual of Methods for General Bacteriology*** (Gerhardt, P., ed.). 409-433. American Society for Microbiology, Washington, D.C.

11.Harris, R. and C.J. Knowles. 1983. Isolation and growth of a *Pseudomonas* species that utilizes cyanide as a source of nitrogen. ***J. Gen. Microbiol.*** 129:1005-1011.

12.Kaplan, A. 1969. The determination of urea, ammonia and urease. ***Methods Biochem. Anal.*** 17:311-324.

13.O'Reilly, K.T. and R.L. Crawford. 1989. Kinetics of *p*-cresol degradation by an immobilized *Pseudomonas* sp. ***Appl. Environ. Microbiol.*** 55:866-870.

14.Ingvorsen, K., B.H. Pederson and S.E. Godtfredsen. 1991. Novel cyanide hydrolyzing enzyme from *Alcaligenes xylosoxidens* subsp. *denitrificans*. ***Appl. Environ. Microbiol.*** 57:1783-1789.

15.Kunz, D.A., O. Nagappon, J.S. Avalos and G.T. Delong. 1992. Utilization of cyanide as a nitrogenous substrate by *Pseudomonas fluorescens* NCIMB 11764 : Evidence for multiple pathways of metabolic conversion. ***Appl. Environ. Microbiol.*** 58:2022-2029.

Legend for Figures

Fig. 1. Schematic diagram of the air-uplift-type fluidized bed batch reactor.

BA	: 0.5 M Boric acid	: To trap NH_3 in the air
KOH	: 0.5 M Potassium hydroxide	: To trap CO_2 in the air
CO_2	: Carbon dioxide	: Produced during ^{-}CN degradation
NH_3	: Ammonia	: Produced during ^{-}CN degradation

Fig.2. Changes in pH [●——●] and per cent utilization of ^{-}CN [O——O] during the degradation of metals tailing pond water by the immobilized cells of *P. putida* at different incubation periods.

Fig. 3. Per cent changes in CO_2 [●——●] production and ^{-}CN [O——O] utilization during the degradation of metals tailing pond water by the immobilized cells of *P. putida* at different incubation periods.

Fig. 4. Per cent changes in NH_3 [●——●] production and ^{-}CN [O——O] utilization during the degradation of metals tailing pond water by the immobilized cells of *P. putida* at different incubation periods.

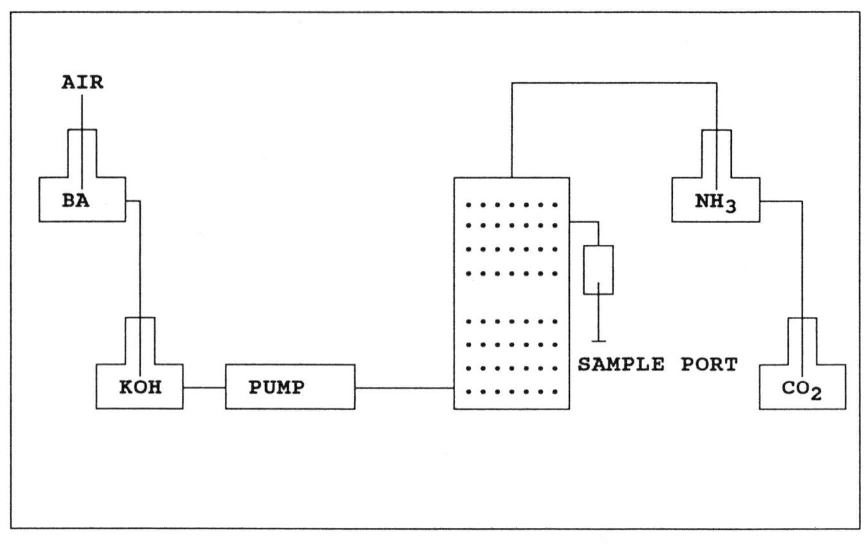
AIR
BA
KOH
PUMP
SAMPLE PORT
NH3
CO2

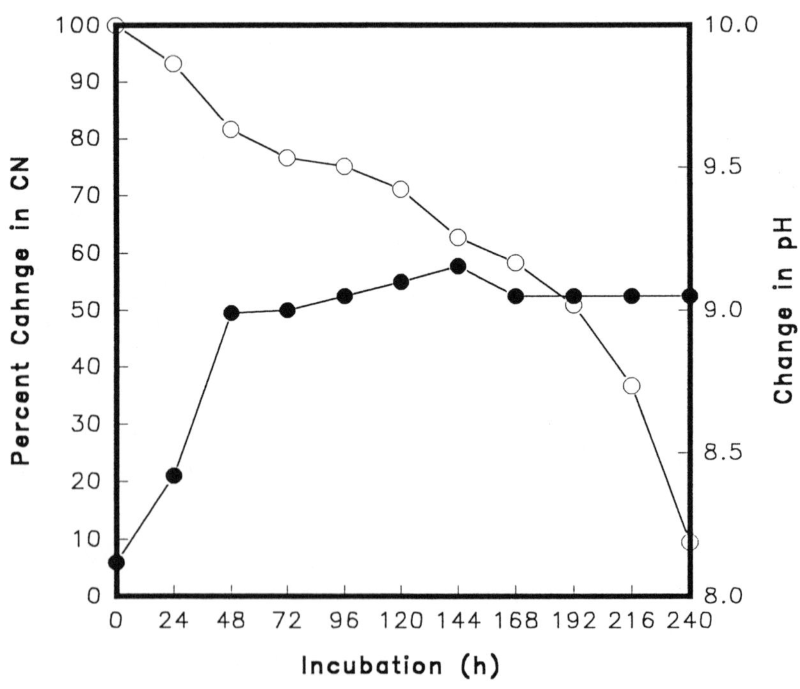
Percent Cahnge in CN
Change in pH
Incubation (h)
0
10
20
30
40
50
60
70
80
90
100
8.0
8.5
9.0
9.5
10.0
0 24 48 72 96 120 144 168 192 216 240

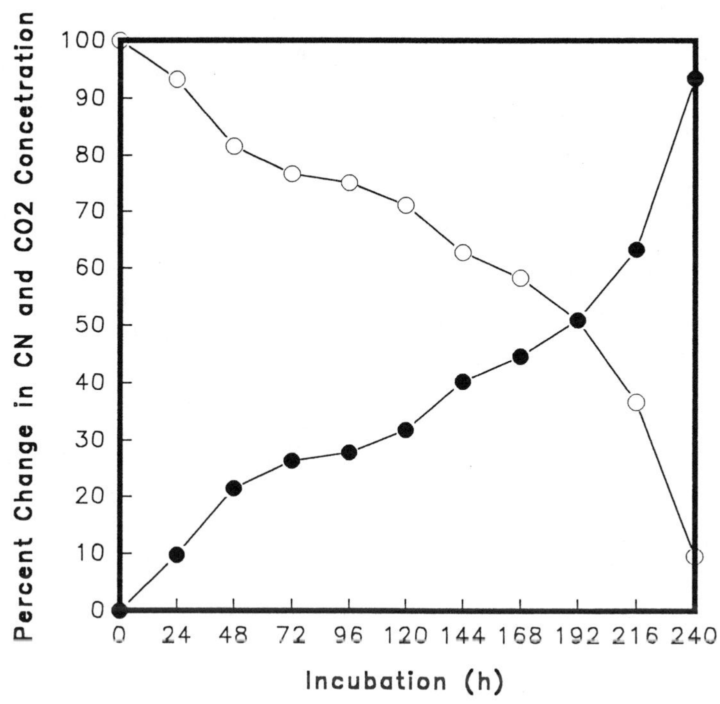

Percent Change in CN and CO2 Concetration
100
90
80
70
60
50
40
30
20
10
0
0 24 48 72 96 120 144 168 192 216 240
Incubation (h)

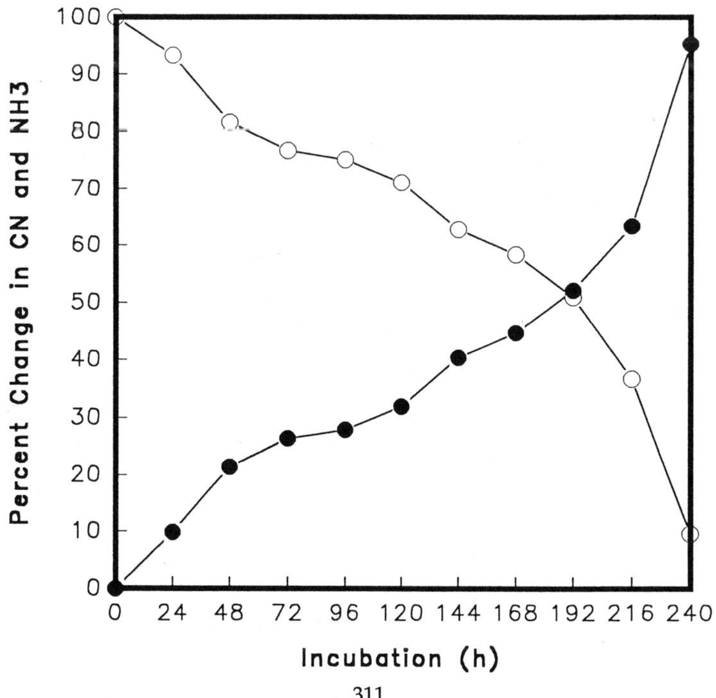

Percent Change in CN and NH3
100
90
80
70
60
50
40
30
20
10
0
0 24 48 72 96 120 144 168 192 216 240
Incubation (h)

The Effect of Growth Medium of *Thiobacillus ferrooxidans*

on Pyrite and Galena Flotation

M. Misra and S. Chen

Department of Chemical and Metallurgical Engineering

University of Nevada, Reno

Reno, NV 89557

ABSTRACT

The bacterium, *Thiobacillus ferrooxidans*, has been used for depression of pyrite during the flotation separation of coal particles from associated minerals. However, the utilization of *T. ferrooxidans* as a regulator in the differential flotation of sulfide minerals is yet to be established. The conventional reagent used for pyrite depression is cyanide which is toxic and in many cases nonselective. The objective of the present investigation is to determine if *T. ferrooxidans* can be used for the selective flotation of galena from pyrite. In view of this objective *T. ferrooxidans* was grown in a variety of culture media to render bacteria with different surface properties such as contact angles and zeta potentials. The influence of bacterial surface properties on adhesion and subsequent hydrophilic/hydrophobic balance of the pyrite mineral surface with regard to flotation recovery was examined. The flotation response of pyrite and galena was interpreted in terms of surface modifications induced by the adsorption of bacteria.

Mineral Bioprocessing II
Edited by David S. Holmes and Ross W. Smith
The Minerals, Metals & Materials Society, 1995

INTRODUCTION

The bacterium, *Thiobacillus ferrooxidans*, has traditionally been used for the biooxidation of refractory gold ores and bioleaching of selected sulfide minerals. In recent years modest amount of work has been directed to the selective depression of pyrite (FeS_2) from complex coal-pyrite matrix (Dogan et al., 1985, 1986; Harada and Kuniyoshi, 1985; Townsley et al., 1987). Initially, pyrite present in coal is oxidized by biotreatment so that hydrophobic coal can be separated from hydrophilic pyrite by a variety of flotation methods (Ohmura and Saiki, 1994). It has been postulated that superficial surface oxidation and formation of hydrophilic sites on pyrite is responsible for depression (Zeky and Attia, 1987). However, the utilization of such a processing strategy for coal-pyrite depression is yet to be commercialized.

In complex sulfide mineral flotation systems, the most often used depressant for pyrite is potassium and/or sodium cyanide. Cyanide is known to be toxic and environmental regulations are stringent for final disposal of cyanide bearing effluents. Due to the environmental regulations and constrictions, there is a need to develop an alternative pyrite depressant. One of the potentially promising approaches involves the use of biological reagents such as *Thiobacillus ferrooxidans*. The interaction of *T. ferrooxidans* with sulfide minerals has been the subject of numerous investigations (Andrews, 1988; Bennett and Tribush, 1978; Berry et al., 1978; Sugio, et al., 1978). The underlying phenomenon which determines the pyrite-bacterium interaction is the attachment of the microorganism to the mineral surface. This attachment or adhesion is influenced by the surface properties of the bacterium which in turn are dependent on culturing conditions and composition of the growth media. In this paper the effect of different culture media on selective attachment and subsequent depression of pyrite is discussed. Although preliminary in nature, the studies have shown that by adjusting the growth medium the adhesion of the organism to pyrite can be increased and selective separation of galena from pyrite can be achieved.

MATERIALS AND METHODS

Microorganism, medium, and conditions of cultivation The iron oxidizing bacterium, *Thiobacillus ferrooxidans*, was obtained from American Type Culture Collection. The nutrient composition of the three different culture media viz., 9K Standard Medium (SM), Phosphate Deficient Medium (PDM), and Pyrite Added Medium (PAM) used for the growth of bacteria is presented in Table 1. The cultivation of *T. ferrooxidans* was carried out in four 250 ml. flasks which were placed in the shaker bath at 30^0C for four days. Iron-free cell suspensions were used in all preparations. For this purpose 1 L of a 4-day old bacterial culture was first passed through Whatman No. 44 filter paper in order to remove the precipitates. The filtrate was then spinned in a refrigerated centrifuge (JZ-21 Beckman, SS-34 rotor) at 15,000 rpm for 30 minutes. The residual pellet was suspended in a solution of sulfuric acid at pH 1.8 and allowed to stand for 2 hours in a refrigerator to permit settling of any precipitate. The supernatant containing the cells was again centrifuged as described earlier. Washing with sulfuric acid was repeated until the cell suspensions were free from iron. The cell pellet was finally suspended in 150 ml sulfuric acid (pH=1.8) and stored in a refrigerator at 4^0C. The concentration of *T. ferrooxidans* was determined by a cell counting technique by using a polarizing optical (Nikon) microscope.

Minerals The samples of pyrite and galena were obtained from Ward's National Science Establishment. The purity of pyrite and galena was reported to be > 95%.

Table 1. Nutritional composition of the three different culture media used in the growth of *Thiobacillus ferrooxidans*

9K Standard Medium	Phosphate Deficient Medium	Pyrite Added Medium
3.0 g/L $(NH)_4SO_4$ 0.5 g/L $MgSO_4 \cdot 7H_2O$ 0.1 g/L KCl 0.5 g/L K_2HPO_4 142.2 g/L $FeSO_4 \cdot 7H_20$	3.0 g/L $(NH)_4SO_4$ 0.5 g/L $MgSO_4 \cdot 7H_2O$ 0.1 g/L KCl 42.2 g/L $FeSO_4 \cdot 7H_2O$	3.0 g/L $(NH)_4SO_4$ 0.5 g/L $MgSO_4 \cdot 7H_2O$ 50.0 g/L Pyrite

Contact Angle Measurement For measuring contact angles the mineral sample was cut into a 2 x 2x 2 cm cube and one of the surfaces was wet ground under a stream of deionized distilled water on grit papers #320 to #600. The sample was then wet polished on a broad cloth mounted on a plate using 1.0 μm and 0.5μm abrasive alumina paper. After careful polishing, the specimen was cleaned with a jet of deionized distilled water to remove dirt and alumina powder. Prior to using in the contact angle measurement the sample specimen was polished with a clean broad flannelette. Plastic surgical gloves were used at all times to handle the specimen in order to avoid contamination.

Contact angle measurements were made using a Rame-Hart Contact Angle Goniometer (Model No. 100). Freshly prepared *T. ferrooxidans* suspensions were pH adjusted with dilute sulfuric acid (pH=2.4) and conditioned together with pyrite sample on a rotary shaker at 25 ^{0}C for 30 minutes. The pyrite sample was separated from the suspension and dried at room temperature. The captive-bubble method was used for the contact angle measurements (Misra, 1984).

Electrokinetic Measurements Electrophoretic mobilities of the bacterium, obtained from three different culture media, were determined using the Lazer Zee Meter. The pH of the suspensions was measured using Fisher 925 pH meter in conjunction with Fisher glass electrode. The bacterial suspensions were prepared in an electrolyte solution (100 ml of 5 x 10^{-4} M NaCl), and the pH was adjusted using NaOH and HCl. The samples were further conditioned for 15 minutes and the zeta potential values were measured. The reported values are the average of 10 readings.

Adhesion of *T. ferrooxidans* on Pyrite Surface Adsorption of *T. ferrooxidans* on pyrite surface was carried out in a 100 ml glass stoppered flask. 100 mg Pyrite sample (74 x 53 μm) was washed with sulfuric acid (pH=1.0) for 10 minutes and mixed with 50 ml of *T. ferrooxidans* (10^8 cells/ml at pH=2.0). The suspension was stirred for a period of 30 minutes to allow for the adsorption to reach an equilibrium value. The stirring was stopped and the suspension was allowed to settle. The supernatant was withdrawn to determine the equilibrium concentration of the bacterium. The amount of *T. ferrooxidans* adsorbed on pyrite surface was determined by the difference in the concentration before and after adding the pyrite. The adsorption density was calculated using the following equation:

$$\Gamma=(C_1-C_2)V/W \qquad (1)$$

where Γ is the adsorption density expressed by the number of *T. ferrooxidans* cells adsorbed per unit weight of pyrite, C_1 and C_2 refer to the bacterial concentration (cells/ml) before and after adsorption, V is the volume of the *T. ferrooxidans* solution (ml), W is the weight (g) of the pyrite sample used in the experiment.

Flotation Experiments The mineral samples were ground to 100 x 270 mesh in a ceramic ball mill and dry sieved. Recovery of pyrite and galena was accomplished by a combination of biotreatment and collector addition prior to the flotation. Purified sodium isopropylxanthate was used as a collector molecule and argon gas with a flow rate 104 ml./min. was used in all flotation tests. In order to determine the optimum conditions for the recovery, the operating variables such as pre-washing time of the minerals with dilute sulfuric acid, time and pH of bioconditioning with *T. ferrooxfdans,* collector (xanthate) concentration, flotation time and pH were investigated. The preliminary results have lead to the establishment of optimum conditions which are as follows: (i) pre-washing time 10 min., (ii) biotreatment with *T. ferrooxidans* at pH 1.8 and conditioning for a period of 30 min., (iii) xanthate addition (1.6×10^{-4} M) at pH 9 and a conditioning time of 5 min., and (iv) flotation time 90 seconds. A Hallimond tube was used in all flotation tests. The schematic of the flotation test is shown in Figure 1. It must be emphasized that the bioconditioning was carried out at a lower pH 1.8 and collector addition and flotation was performed at higher pH 9.0. The reason being that the bacterial adhesion and its biogeochemical reactions are optimal at lower pH values, whereas xanthate as a collector is effective only at higher pH values. Further, xanthate is believed to decompose in the solutions of low pH values. Flotation experiments were conducted mainly using pyrite and the bacteria harvested from three different culture media. Flotation of galena was performed using the bacteria obtained from the PDM culture.

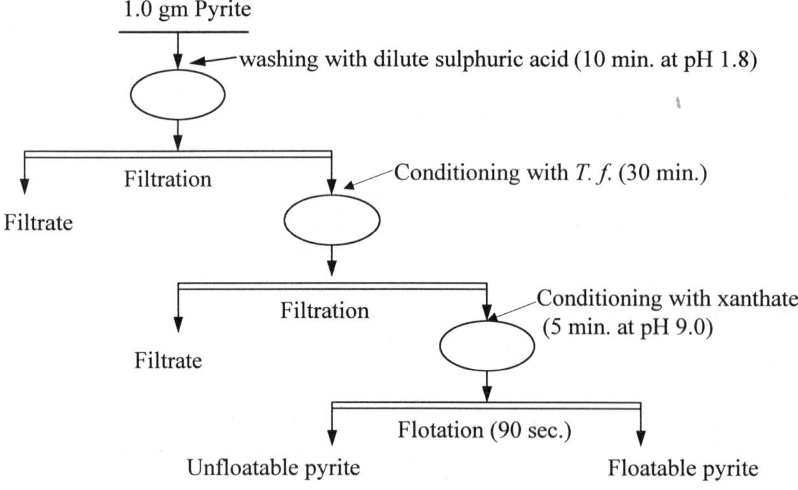

Figure 1. Flotation Flowsheet

RESULTS AND DISCUSSION

Contact Angles Contact angle of water bubble placed on polished pyrite and galena surfaces were measured as a function of *T. ferrooxidans* concentration, and the results are plotted in Figure 2. The contact angle of pyrite without *T. ferrooxidans* was noted to be 49^0. With the adsorption of *T. ferrooxidans* (at a concentration of 1.9×10^8 cells/ml), the contact angle of pyrite was decreased to 23^0 indicating the decreased hydrophobicity of mineral. As can be seen, *T. ferrooxidans* cultured in PDM can make pyrite surface significantly more hydrophilic than that grown in the SM. The contact angle of galena, without the adsorbed *T.*

ferrooxidans, was observed to be 50^0. However, with the adsorption of *T. ferrooxidans* the contact angle was increased to $57\text{-}59^0$ and remained at that value independent of the concentration of *T. ferrooxidans.* This observation indicates that as a result of *T. ferrooxidans* adsorption on galena, unlike in the case of pyrite, the mineral becomes slightly more hydrophobic.

Zeta Potentials Figure 3 shows the zeta potential of *T. ferroxidans* cultured in different media as a function of pH. It is evident that the zeta potential values of *T. ferroxidans* derived from different culture media are different. A greater variation among the bacteria was noted in the PZC values. *T. ferroxidans* grown in SM, PDM, and PAM were observed to have PZC values at pH=2.0, pH=3.2, and pH=3.7 respectively. Similar increases in the PZC values for the bacteria grown on several mineral surfaces have been reported (Devasia et al., 1993)

Adhesion of *T. ferroxidans* on Pyrite Surface Adhesion of bacteria, grown from three different culture media, was determined on the pyrite surface at pH 2.0. The particle size of the pyrite sample used in these experiments was in the range 75 x 53 μm. It is well known that the nutritional composition of the culture medium greatly influences the growth phase (Smith et al., 1992) and the hydrophobicity of the bacterium (Marshall, 1976). The results obtained in this study are shown in Figure 4. It is evident that the growth medium influences the surface properties of bacteria. The adsorption of the bacteria cultured in the standard 9K medium was found to be the lowest. In comparison to the standard 9K medium, the adsorption on the other two media viz., PDM and PAM was found to be higher by 80% and 140% respectively. In general, microorganisms are sensitive to the nutritional requirements in their growth phase and starvation brings about changes in the nature of bacteria cell surface and their hydrophobicity (Albertson et al., 1987; Doyle and Rosenberg, 1990; Wibawan et al., 1992). In particular, Jerez et al., (1992) have reported that *T. ferrooxidans* undergoes several changes upon phosphate starvation. Amaro et al., (1993) reported that phosphate starvation increased the capacity of *T. ferrooxidans* to attach itself to the sulfide minerals and elemental sulfur. Such an increase was attributed to the changes in the outer membrane proteins and a 25% increase in the lipopolysaccharides in comparison to the bacteria grown in the standard 9K medium. Phosphate starvation was shown to increase the proportion of high molecular weight proteins with a seemingly corresponding decrease in the low molecular weight fraction(Jerez et al., 1992).

Flotation Recovery of Pyrite and Galena The flotation results obtained for pyrite with *T. ferrooxidans* derived from three different culture media are plotted as a function of the bacteria concentration in Figure 5. The conditions of the experiment are described in Figure 1. In each case it was observed that the amount of pyrite reported to the froth decreased with an increase in the bacterium concentration. *T. ferrooxidans* harvested from different growth media showed large variations in the depression of pyrite and that cultured in the PAM depressed the pyrite to the greatest extent in the bacterium concentration range studied.

Figure 6 shows the results of pyrite flotation, using *T. ferrooxidans* grown in PAM and xanthate collector molecule, as a function of bacterial conditioning pH. In all cases the flotation was conducted at pH 9.0. Also included in the Figure 6 are the results of pyrite natural flotation unassisted by the biotreatment and collector addition, and the results of pyrite flotation with xanthate as a collector molecule without prior biotreatment with the bacteria. Although flotation in the last two cases was also performed at pH 9.0, the suspensions containing the pyrite were treated in the same manner, that is conditioning at the lower pH values for the same duration of time, as proceeded in the case of pyrite with a combined biotreatment and collector addition. This was done to subject the pyrite surface to same pH changes and conditioning so

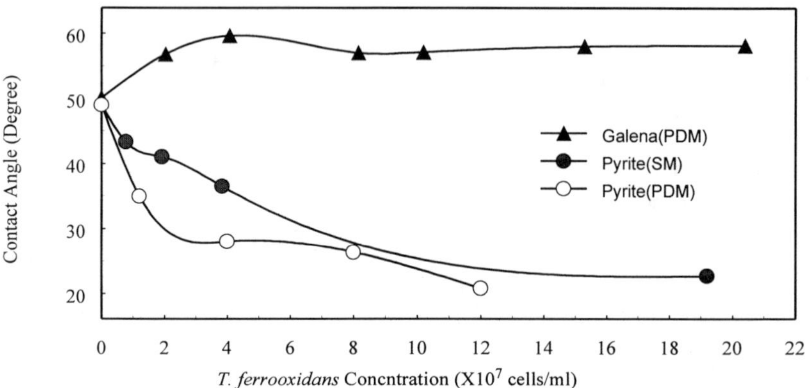

Figure 2. Contact angle as a function of *T. ferrooxidans* concentration

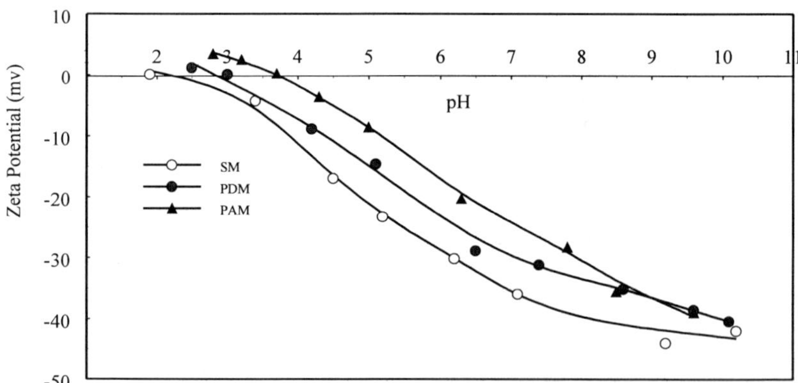

Figure 3. Zeta potential of *T. ferrooxidans* clutured in different media as a function of pH

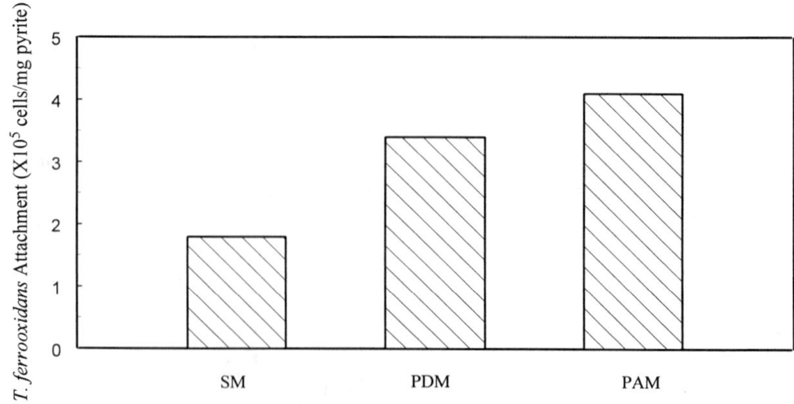

Figure 4. Attachment of *T. ferrooxidans* to pyrite

on the pyrite and galena at pH 2.0

that if any surface reactions should occur at low pH values comparisons can still be made for the cases presented in the Figure 6. It can be seen that in the case of pyrite natural flotation most of the pyrite was depressed and only 32% reported to the froth, which gradually decreased to <10% as the conditioning pH was increased. On the other hand, in the case of xanthate alone as a collector more than 93% pyrite reported to the froth and very little change occured as the pH was varied from 1.8 to 6.5. Finally, in the case of combined bio and collector treatment of pyrite the recovery was observed to be 43% at pH 1.2 and decreased slowly till pH 4.0 and decreased further steeply thereafter to 10%. Thus a large decrease in the pyrite recovery, in comparison to the case of xanthate alone treatment, may indicate that the pyrite surface was largely covered by the bacteria and access to the xanthate molecule is prevented.

Galena flotation was conducted with the bacteria obtained from PDM. Figure 7 shows the effect of *T. ferrooxidans* on galena flotation. For comparison, the results obtained with pyrite using the same bacteria were also included in Figure 7. It can be seen that *T. ferrooxidans* depresses galena less than pyrite in the flotation. A comparison indicates that at *T. ferrooxidans* concentration of 7.2×10^7 cells/ml, the galena flotation recovery was 84% while under similar conditions the pyrite recovery was observed to be 30%. Thus the biotreatment of a mixture containing pyrite and galena, using *T. ferrooxidans* grown in PDM, offers an effective method of separating these two minerals.

Mechanisms of Sulfide Minerals Flotation and Depression Mechanism of the xanthate assisted sulfide minerals flotation is well known. It involves the adsorption of xanthate, with a subsequent formation of dixanthogen, on the surface of the mineral which increases the hydrophobicity and reports to the froth phase in the flotation cell. However, the depression of pyrite with a prior biotreatment in the flotation is believed to be the result of two possible viz., direct and indirect mechanisms.

In the direct mechanism adhesion of *T. ferrooxidans* to pyrite surface renders the mineral hydrophilic which causes the depression (Srihari et al., 1991; Murthy et al., 1992). The adsorption (or adhesion) of bacteria onto the mineral and its pH dependency can be adequately explained with the help of DLVO theory (Misra et al., 1993).

The indirect mechanism requires no intimate contact or adhesion of the bacteria on the mineral surface. However, it operates through the chemical action of ferric sulphate, a product produced by the bacterial metabolism occurring on the surface of the mineral. Surface oxidation of the pyrite enables it to be hydrophilic resulting in the depression of the mineral. In a recent study Pesic and Kim (1991) have reported that surface oxidation at the pyrite mineral surface, caused by the *T. ferrooxidanss* over a period of five days, resulted in the formation of a layer of jarosite. The x-ray diffraction method was used to cofirm the formation of jarosite. The product jarosite is formed in an acid producing reaction as shown below:

$$3Fe^{3+} + X^{+} + 2HSO_4^{-} + 6H_2O \quad \Leftrightarrow \quad XFe_3(SO_4)_2(OH)_6 + 8H^{+} \qquad (2)$$

and has the general formula $XFe_3(SO_4)_2(OH)_6$, where X represents K^+, Na^+, NH_4^+, or H_3O^+. Precipitation of jarosite adversely affects bacterial leaching systems due to the formation of impenetrable barrier. However, in the biotreated flotation of pyrite, if jarosite is formed, it may help depress the mineral.

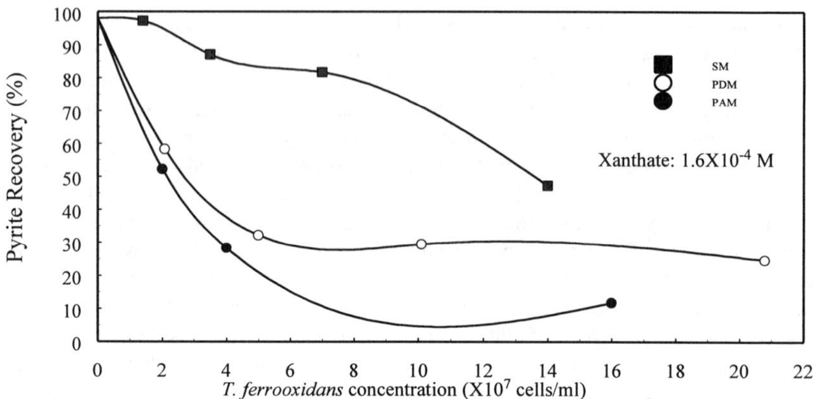

Figure 5. Pyrite flotation recovery as a function of *T. ferrooxidans* concentration

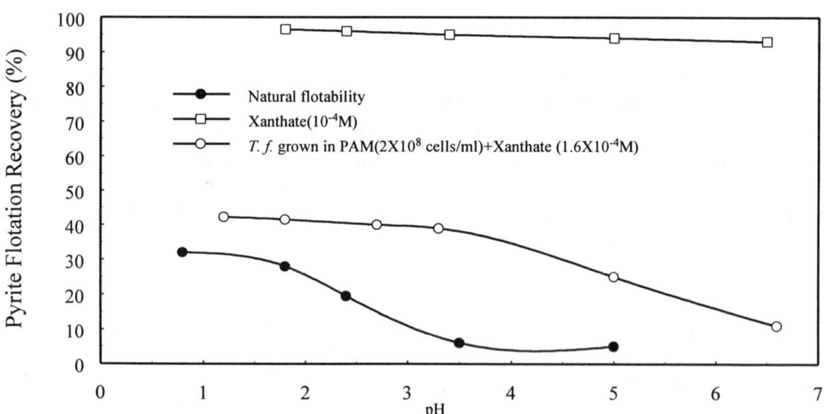

Figure 6. Flotation of Pyrite as a function of bioconditioning pH
-Flotation conducted at constant pH of 9

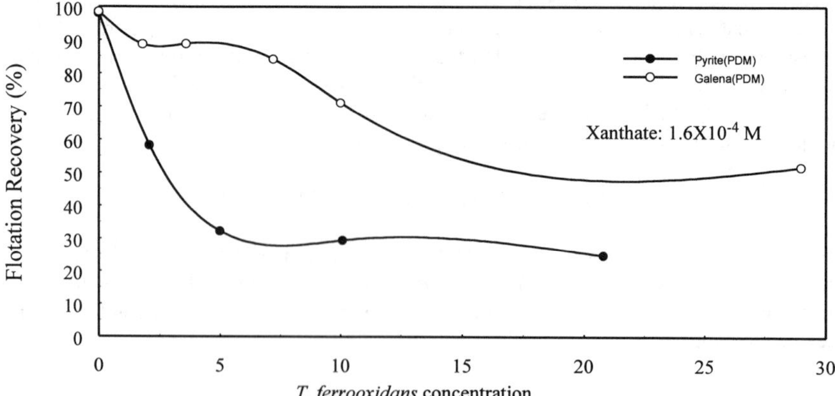

Figure 7. Flotation recovery of pyrite and galena as a function of *T. ferrooxidans* concentration (X10^7 cell/ml)

SUMMARY

Bioconditioning of pyrite with *Thiobacillus ferrooxidans* produces a significant surface modification of the mineral surface. The flotability of pyrite is significantly reduced depending on the bacterium concentration, conditioning pH, and the nutritional composition of the culture medium in which the bacterium is grown. Decrease in the pyrite flootability is attributed to the adhesion of *T. ferrooxidans* on its surface, hydrophobic sulfur oxidation, and the formation of jarosite on the surface of pyrite. From the results obtained in this study the following conclusions were made:

1. *T. ferrooxidans,* in conjunction with the xanthate collector, can significantly depress pyrite in the flotation. A similar treatment of galena has indicated no such alteration in its hydrophobic/hydrophilic balance. Thus using a combination of biotreatment involving *T. ferrooxidans* and xanthate collector it is possible to separate the pyrite and galena from a mixture containing both the minerals.
2. The surface properties of the bacterium can be altered by changing the conditions and nutritional composition of the culture medium. The three different media used for the growth of *T. ferrooxidans.* viz., 9K standard, phosphate-deficient, and pyrite added media have produced bacteria with notable differences in the surface properties, such as contact angles and zeta potentials. The largest change was observed in the bacteria PZC values.
3. The mechanism of pyrite depression in the flotation cell involves the adsorption of *T. ferrooxidans* on the mineral surface making it hydrophilic. The adsorption of *T. ferrooxidans* on pyrite can be explained by the DLVO theory. It is also believed that *T. ferrooxidans* oxidizes the elemental sulfur on the surface of pyrite leading to the formation of hydrophilic jarosite.

REFERENCES

Albertson, N. H., Jones, G. W. and Kjelleberg, S., 1987, “The Detection of Starvation-Specific Antigens in Two Marine Bacteria,” J. Gen. Microbiol., Vol. 133, pp. 225-2231.

Amaro, A. M., Seeger, M., Arredondo, A., Moreno, M. and Jerez, C. A., 1993, "The Growth Conditions Affect *Thiobacillus ferrooxidans* Attachment to Solids," Biohydrometallurgical Technologies, ed. by A. E. Torma, M. L. Apel and C. L. Brierley, The Minerals, Metals & Materials Society, pp. 577-585.

Andrew, G. F., 1988, “The Selective Adsorption of *Thiobacilli* to Dislocation Site on Pyrite Surfaces,” Biotech. Bioeng., Vol. 31, pp. 378-381.

Bennett, J. C., Tribusch, H., 1978, “Bacterial Leaching Patterns on Pyrite Crystal Surfaces,” J. Bacteriol., Vol. 134, pp. 310-317.

Devasia, D., Natarajan, K. A., Sathyanaryana, D. N. and Rao, G. R., 1993, “Surface Chemistry of *Thiobacillus ferrooxidans* Relevant to Adhesion on Mineral Surfaces,” Applied and Envir. Micro., Vol. 59, No. 12, pp. 4051-4055.

Dogan, M. Z., Ozbayoglu, G., Hicyilmaz, C., Sarikaya, M. and Ozcengiz, G., 1985, "Bacterial Leaching versur Bacterial Conditioning and Flotation in Desulphurization of Coal," XV International Mineral Processing Conference, Cannes, Jun 2-9, 1985, pp. 304-313.

Dogan, M. Z., Ozbayoglu, G., Hicyilmaz, C., Sarikaya, M. and Ozcengiz, G., 1986, “Bacterial Leaching versur Bacterial Conditioning and Flotation in Desulphurization of Three Different Coal,” Fundamental and Applied Biohydrometallurgy. (Eds) R. W. Lawrence, R. M. R. Branion and H. G. Ebner, New York, pp 165-170.

Doyle, R. J. and Rosenberg, M., 1990, “Microbial Cell Surface Hydrophobicity,” (American Society for Microbiology).

Harada, T. and Kuniyoshi, N., 1985, “Effects of Bacterial Oxidation on the Floatability of Pyrite,” J. Min. Metal. Inst. Japan, Vol. 101, pp. 719-724.

Jerez, C. A., Seeger, M. and Amaro, A. M., 1992, “Phosphate Starvation Affects the Synthesis of Outer Membrane Proteins in *Thiobacillus ferrooxidans*,” FEMS Microbiol. Lett., Vol. 98, pp. 29-34.

Marshall, K. C., 1976, "Solid-liquid and solid-gas interfaces", Interfaces in Microbial Ecoloizy, ed. K. C. Marshall, (Harvard University Press, London. 1976), pp. 27-52.

Misra, M., Chen, S., Smith, R. W. and Raichur A. M, 1993, “Adhesion of Hydrophobic Microorganism on Hematite and its Effect on Flotation," Minerals and Metallurgical Processing, Vol. 10, No. 4, (Nov. 1993), pp. 170-175.

Misra, M., Miller, J. D., 1984, “The Effect of SO_2 in the flotation of Sphalerite and Chalcopyrite in Flotation of Sulphide Minerals,” Advances in Sulfide Mineral Flotation, ed. Erie Foorsberg.

Murthy, K. S. and Natarajan, N., 1992, “The Role of Surface of Attachment of *Thiobacillus ferrooxidans* on the Biooxidation of Pyrite,” Miner. Metellurg. Processing, Vol. 9, pp. 20-24.

Ohmura, N. and Saiki, H., 1994, “Desulfurization of Coal by Microbial Column Flotation,” Biotechno. and Bioeng., Vol. 44, pp. 125-131.

Pesic, B. and Kim, I., “Electrochemistry of *Thiobacillus ferrooxidans* Interaction with Pyrite,” Mineral Bioprocessing, eds. R. W. Smith and M. Misra, TMS, pp.413-432.

Shrihari, R., Kumar, K. S. Gandhi, K. S. and Natarajan, K. A., 1991. The Role of Cell Attachment in Leaching of Chalcopyrite Mineral by *Thiobacillus ferrooxidans,”* Appl. Microbiol. Biotechnol., Vol 36, pp. 278-282.

Smith, G., Steaver, M. H., Lugtenberg, B. J. J. and Kijne, J. W., 1992, "Flocculence of *Saccharomyces cerevisiae* cells is induced by nutrient limitation, with cell surface hydrophobicity as a major determinant," Appl. Environ. Microbiol. vol. 55, pp. 3709-3714.

Sugio, T., Mizunashi, W., Inagaki, K., and Tano, T., 1987, “Purification and Some Properties of Sulfur: Ferric Ion Oxidoreductase from *Thiobacillus ferrooxidans*,” J. Bacteriol., Vol. 169, pp. 4916-4922.

Townsley, C. C., Atkins, A. S. and Davis, A. J., 1987, “Suppression of Pyrite Sulphur during Flotation by *Thiobscillus ferrooxidans,”* Biotech. and Bioengg., vol. 30: pp.1-8.

Wibawan, I. W. T., Lammler, C. and Pasaribu, F. H., 1992, “Role of Hydrophobic Surface Proteins in Mediating Adherence of Group B Streptococci to Epithelial Cells,” J. Gen. Microbiol., Vol. 138, pp. 1237-1242.

Zeky M. El, and Attia, Y. A., 1987, “Coal Slurries Desulfurization by Flotation using *Thiophilic* Bacteria for Pyrite Depression,” Coal Preparation, Vol. 5, pp. 15-37.

AUTHORS